NMR-MRI, µSR and Mössbauer Spectroscopies

in Molecular Magnets

Pietro Carretta   Alessandro Lascialfari

NMR-MRI, µSR and Mössbauer Spectroscopies in Molecular Magnets

Pietro Carretta
Alessandro Lascialfari
Dipartimento di Fisica "A. Volta", Università degli Studi di Pavia

Library of Congress Control Number: 2007923021

ISBN 978-88-470-0531-0 Springer Milan Berlin Heidelberg New York

Springer is a part of Springer Science+Business Media
springer.com

© Springer-Verlag Italia 2007

Cover design: Simona Colombo, Milano
Typeset by the authors using a Springer Macro package

Springer-Verlag Italia
Via Decembrio, 28
20137 Milano

Printed on acid-free paper

Preface

This book is a collection of scientific articles on the basic aspects of nuclear magnetic resonance (NMR), muon spin rotation (μSR) and Mössbauer spectroscopies, applied to the study of molecular magnets and to related systems, such as low-dimensional magnets and contrast agents for magnetic resonance imaging (MRI). These articles gather, to a certain extent, the lecture notes presented by the authors at the *Training School on NMR-MRI, μSR and Mössbauer Techniques*, held in Pavia from the 17th to the 30th of September 2006.

The motivation for the School and for the publication of this book originates from the growing interest of a broad scientific community to the field of molecular magnetism where the aforementioned techniques play a key role. Nowadays several groups of physicists, of chemists and some groups of biologists work in this field in order to unravel the fundamental physical aspects of molecular magnets which are of interest for their future applicability as storage units, in quantum computation, in hybrid superconducting-magnetic systems or as contrast agents, for example. Nevertheless, the use and the knowledge of NMR-MRI, μSR and Mössbauer spectroscopies is still limited to a small scientific community. The purpose of the School and of the book was to introduce these techniques to a broader scientific community working on molecular magnets and related systems. In particular, to give the basic principles of each technique, to show which information is derived, for example, from the spectra, from the muon depolarization curves or from the relaxation rates, and how it is complementary to the one obtained by other techniques. Moreover, it will be shown how from the experimental results derived by these techniques one can tailor the properties of molecular magnets and nanoparticles which, in turn, can be used as contrast agents for MRI.

The book is addressed to graduate and senior researchers working on molecular magnets or closely related areas, having a background on quantum mechanics. It is divided into three main parts, each one dedicated to a different spectroscopy. The first chapter of each part is a short tutorial introduction to each one of the spectroscopies. Since an exhaustive introduction is out of the aim of this book, reference is properly made to text books and to web sites which allow a deeper understanding of the basic principles underlaying these techniques. This introductory chapter is followed by more specialized ones giving an overview on recent results obtained by NMR, μSR and Mössbauer spectroscopies in molecular magnets and in strictly related areas. Two chapters of the first part are dedicated to MRI and to the investigation of contrast agents suitable for applications in MRI.

We would like to acknowledge MAGMANet network of excellence for supporting the School and the publication of this book. CNR-INFM and CNISM secretaries Stefania Riboni, Antonella Biondi, Elisa Bolognesi and Maria Grazia Angelini are also acknowledged for their precious help.

Pavia,
March 2007

Pietro Carretta
Alessandro Lascialfari

Contents

Part III Mössbauer Spectroscopy

List of Contributors

Stephen J. Blundell
Oxford University
Department of Physics
Clarendon Laboratory,
ParksRoad, Oxford OX1-3PU
(United Kingdom)
s.blundell@physics.ox.ac.uk

Ferdinando Borsa
Dipartimento di Fisica "A.Volta" -
University of Pavia
Via Bassi, 6 -27100 Pavia
(Italy)
borsa@fisicavolta.unipv.it

Pietro Carretta
Dipartimento di Fisica "A.Volta" -
University of Pavia
Via Bassi, 6 -27100 Pavia
(Italy)
carretta@fisicavolta.unipv.it

Luciano Cianchi
Istituto dei Sistemi Complessi
(Sezione di Firenze)-CNR,
Via Madonna del Piano, 10
50019 Sesto Fiorentino, Firenze
(Italy)
luciano.cianchi@isc.cnr.it

Maurizio Corti
Dipartimento di Fisica "A.Volta" -
University of Pavia
Via Bassi, 6 - 27100Pavia
(Italy)
corti@fisicavolta.unipv.it

Roberto De Renzi
Department of Physics and CNISM,
Viale G.P. Usberti, 7a
I-43100 Parma
(Italy)
Roberto.DeRenzi@unipr.it

George Filoti
National Institute for Materials
Physics
P.O. Box MG-07, 77125
Bucharest-Magurele
(Romania)
filoti@infim.ro

Amit Keren
Technion-Israel Institute of
Technology, Physics Department
Haifa 32000
(Israel)
keren@physics.technion.ac.il

Alessandro Lascialfari
Institute of General Physiology and
Biological Chemistry "G. Esposito"
University of Milano
Via Trentacoste 2, I-20134 Milano

(Italy)
lascialfari@fisicavolta.unipv.it

Sophie Laurent
Department of General, Organic and
Biomedical Chemistry
NMR and Molecular Imaging
Laboratory
University of Mons-Hainaut
B-7000 Mons
(Belgium)
Sophie.Laurent@umh.ac.be

Robert N. Muller
Department of General, Organic and
Biomedical Chemistry
NMR and Molecular
Imaging Laboratory
University of Mons-Hainaut
B-7000 Mons
(Belgium)
robert.muller@umh.ac.be

Attilio Rigamonti
Dipartimento di Fisica "A.Volta" -
University of Pavia
Via Bassi, 6 -27100 Pavia
(Italy)
rigamonti@fisicavolta.unipv.it

Alain Roch
Department of General, Organic and
Biomedical Chemistry
NMR and Molecular Imaging
Laboratory
University of Mons-Hainaut
B-7000 Mons
(Belgium)
Alain.Roche@umh.ac.be

Gabriele Spina
Dipartimento di Fisica
University of Firenze
Via G. Sansone,1
50019 SestoFiorentino, Firenze
(Italy)
gabriele.spina@unifi.it

Fabio Tedoldi
Bruker BioSpin S.r.l.
Via Pascoli 70/3, I-20133 Milano
(Italy)
fabio.tedoldi@bruker.it

Luce Vander Elst
Department of General, Organic and
Biomedical Chemistry
NMR and Molecular
Imaging Laboratory
University of Mons-Hainaut
B-7000 Mons
(Belgium)
Luce.Vanderelst@umh.ac.be

Part I

Nuclear Magnetic Resonance

A short introduction to Nuclear Magnetic Resonance

Pietro Carretta

Dipartimento di Fisica "A.Volta" and CNISM - University of Pavia - Via Bassi, 6 - 27100 Pavia (Italy)

The aim of this manuscript is to introduce some of the physical principles involved in nuclear magnetic resonance (NMR) spectroscopy, in order to allow a better understanding of the following parts of the book. The foundations of NMR have been settled decades ago [1] and its basic aspects are described in detail in several text books where also a complete formal derivation can be found. The reader can refer, for example, to the excellent monograph by C.P. Slichter [2] or to the book by A. Abragam [8] in order to unravel the physics underlying this spectroscopy. Although in some parts of this manuscript continuous wave (CW) NMR spectroscopy will be recalled we shall be dealing most of the time just with pulsed NMR technique.

1 Precessing moments and the rotating frame

A first issue to address is what is being measured by pulsed NMR technique. The answer to this question is not unique but still one can state on a rather general ground that by means of pulsed NMR one detects the time evolution of nuclear magnetization. Since the thermal energy $k_B T$ is in general much larger than the interaction among nuclear magnetic moments, in order to produce a non-zero magnetization one has to apply a magnetic field $\boldsymbol{H}_0$. Accordingly a nuclear magnetization $M_0 = \gamma^2 \hbar^2 I(I+1) H_0 / 3 k_B T$ arises, where I is the nuclear spin and γ its gyromagnetic ratio.

Let us first consider what is the effect of the magnetic field on the time evolution of a single magnetic moment $\boldsymbol{M}$. This simple problem can be solved within a classical approach and one finds that $\boldsymbol{M}$ precesses around $\boldsymbol{H}_0$ (see Fig. 1) at the Larmor frequency $\omega_0 = \gamma \boldsymbol{H}_0$. However, we are rather interested on the time-evolution of $\boldsymbol{M}$ due to the interactions with the other degrees of freedom of our system (e.g. the electron spins, the other nuclear spins, the ionic charge distribution, etc...), not on the precessional motion around H_0. Remember that actually $\boldsymbol{H}_0$ was applied just to generate a nuclear magnetization, as this is the quantity that will produce the NMR signal. Therefore it may be convenient to study the time-evolution of the nuclear magnetization in a frame of reference, different from the laboratory one, where one can get rid of the effect of $\boldsymbol{H}_0$ on the time evolution of $\boldsymbol{M}$.

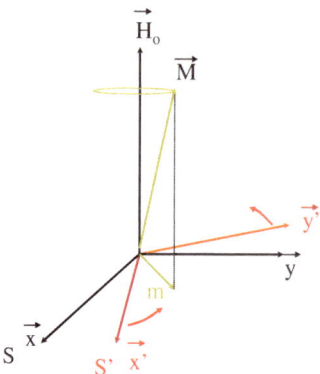

Fig. 1. Sketch of the precessional motion of a magnetic moment $\boldsymbol{M}$ around $\boldsymbol{H}_0$ in the laboratory frame of reference S. The frame of reference S' rotating around $\boldsymbol{H}_0$ at a frequency ω is also shown. m is the component of $\boldsymbol{M}$ in the xy plane.

This frame of reference is the one (S') rotating around $\boldsymbol{H}_0$ at a frequency $\omega \equiv \omega_0$ (see Fig. 1). In a frame of reference rotating at a frequency ω around $\hat{z}$, the magnetic moment is observed to precess around an effective field $\boldsymbol{H}_e = (H_0 - \omega/\gamma)\hat{z}'$. This effective field vanishes for $\omega = \omega_0$ so that $\boldsymbol{M}$ no longer precesses in S'. Then any time-dependent perturbation at a frequency $\omega \rightarrow \omega_0$ (time-independent in S') would produce significant effects on the time-evolution of $\boldsymbol{M}$. For instance, let us consider the effect of a small RF field $\boldsymbol{H}_1$ rotating at frequency ω perpendicular to $\boldsymbol{H}_0$, corresponding to a static field along $\hat{x}'$ in S'. Then in S' the magnetic moment will precess around (see Fig. 2)

$$\boldsymbol{H}_e = (H_0 - \frac{\omega}{\gamma})\hat{z}' + H_1\hat{x}'$$

and for $\omega = \omega_0$ $\boldsymbol{M}$ will precess just around $\boldsymbol{H}_1$, at a frequency $\omega_1 = \gamma H_1$. Then one realizes it is convenient to analyze the time evolution of nuclear magnetization in a frame of reference rotating at ω_0 and, as we shall see later on, the detected NMR signal is indeed the one in the rotating frame S'.

For $\omega = \omega_0$ the RF field will drive $\boldsymbol{M}$ away from the $\hat{z}$ axes by an angle $\alpha(t)$ given by

$$cos(\alpha(t)) = 1 - 2sin^2(\frac{\omega_1 t}{2}) \,. \tag{1}$$

Hence even a very small perturbation can drive the magnetization away from the $\boldsymbol{H}_0$ axes, provided that its frequency is very close ($\omega_1 \gg |\omega - \omega_0|$) to Larmor frequency. This is the classical description of the resonance process,

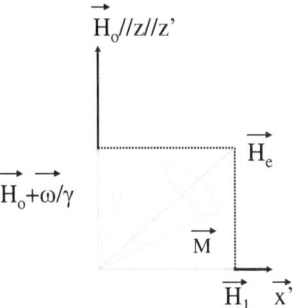

Fig. 2. Sketch of the time-evolution of M in S'. The magnetic moment precesses around an effective field H_e which for $\omega = \omega_0$ coincides with H_1. Notice that H_0 and ω have opposite orientation.

which is the analogue of the resonance absorption driven by the magnetic-dipole transition mechanism, namely the one induced by the perturbation $\mathcal{H}_1(t) = -\gamma\hbar H_1(t)$ of the Zeeman hamiltonian $\mathcal{H}_0 = -\gamma\hbar H_0$. If, for simplicity, we refer to a spin $I = 1/2$ the energy level diagram is the one reported below where the notation $|2m_I >$ for the eigenstates is used.

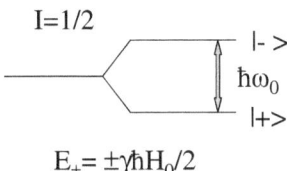

Fig. 3. Energy level diagram for a spin $I = 1/2$ in a magnetic field H_0 and illustration of the transition induced by the RF field at frequency ω_0.

One can estimate the expectation value

$$< I_z >= (1/2)(p_+ - p_-),$$

with p_+ and p_- the probability to find the spin in state $|+ >$ and $|- >$, respectively. Since $p_+ + p_- = 1$ one has

$$p_- = \frac{1 - 2 < I_z >}{2}$$

Then, if at $t = 0$ $p_+ = 1$ and $p_- = 0$ (i.e. $I \parallel H_0$), at time t (see Eq. 1 for comparison)

$$p_-(t) = \frac{1}{2}(1 - cos\alpha(t)) = sin^2\frac{\omega_1 t}{2}$$

for $\omega = \omega_0$. This equation corresponds to the well known Rabi equation giving the probability to pass from state $|+>$ to state $|->$.

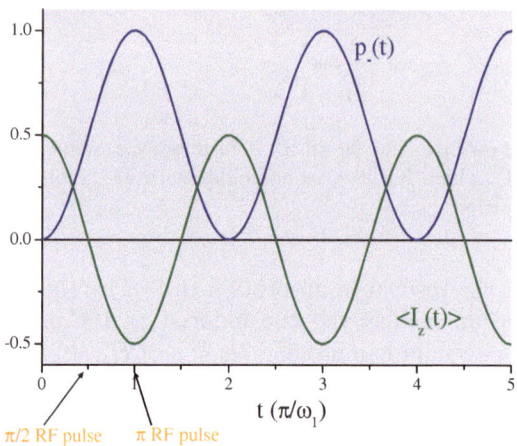

$\pi/2$ RF pulse π RF pulse

Fig. 4. Time evolution of $p_-(t)$ and of $< I_z >$. It is pointed out that if the RF field is switched off after a time $\pi/2\omega_1 < I_z >$ is zero, namely the spin expectation value has turned by $\pi/2$, from parallel to perpendicular to z. Then one says that a $\pi/2$ pulse was applied. In general, to turn by an angle $\theta < \boldsymbol{I} >$ one has to apply an RF pulse of duration θ/ω_1.

After a RF pulse (a $\pi/2$ pulse for example (see Fig. 4)) one can detect, with the same coil used to apply the RF, the voltage induced by the time evolution of nuclear magnetization. According to Faraday-Maxwell's law this voltage is proportional to the precessional frequency in the laboratory frame of reference S and to the nuclear in-plane magnetization $< M_{x,y}(t) >$. The signal detected by the coil in S can be transformed into the one in S' by mixing it with a reference signal oscillating at ω_0 (e.g. $cos(\omega_0 t)$). The low-frequency signal at the output of the mixer is the one detected in pulsed NMR experiments, the so-called FID (free induction decay) signal. The bloch scheme of a typical pulsed NMR spectrometer is equivalent to the one of an MRI spectrometer (see Fig. 10 in the chapter by Lascialfari and Corti in this book), except for the pulsed gradient unit.

2 Time-evolution of nuclear magnetization and the Bloch equations

So far we have just considered the time evolution of a single spin. However, in general experiments are performed on a sample containing a statistical ensemble of nuclear spins and hence one has to consider how the statistical average of the spin components, namely the components of the nuclear magnetization, evolve in time. Then one can write for the statistical average

$$< \overline{I_{x,y,z}} >= \sum_m < m|I_{x,y,z}|m > \frac{e^{-E_m/k_B T}}{Z}$$

with $|m>$ eigenstates of $\mathcal{H} = -\gamma \hbar I_z H_0$. For simplicity, if one considers an ensemble of nuclei with $I = 1/2$ in a magnetic field $\boldsymbol{H}_0$ one finds

$$< \overline{I_z} >= \frac{1}{2} \frac{e^{-\gamma \hbar H_0/2k_B T} - e^{\gamma \hbar H_0/2k_B T}}{e^{-\gamma \hbar H_0/2k_B T} + e^{\gamma \hbar H_0/2k_B T}}$$

and

$$< \bar{I}_{x,y} >= 0 \ .$$

Notice that for a single spin $< I_{x,y} >$ is non-zero, while its average value is zero owing to the random phase of the x, y components of the spins. In general one can calculate the average statistical value starting from the density matrix $\rho = exp(-\beta \mathcal{H})/Z$, with Z the partition function. Then

$$< \overline{I_z(t)} >= Tr\{\rho(t)I_z\}$$

where the time evolution of $\rho(t)$ is given by

$$\frac{d\rho}{dt} = \frac{i}{\hbar}[\rho, \mathcal{H}]$$

Consider now the experimental configuration introduced in the previous section, where $\boldsymbol{H}_0 \parallel \hat{z}$ and the RF field $\boldsymbol{H}_1$ was perpendicular to it. Then $\mathcal{H} = \mathcal{H}_0 + \mathcal{H}_1(t)$, where $\mathcal{H}_1 = -\gamma \hbar H_1 I_x$ can be treated as a perturbation of $\mathcal{H}_0 = -\gamma \hbar H_0 I_z$. Then one finds that

$$< \overline{M_y(t)} >= M_0 sin(\omega_1 t)$$

in the rotating frame. Again we observe that if the RF field is applied at a frequency ω_0 in S' $< \boldsymbol{M}(t) >$ precesses just around $\boldsymbol{H}_1$.

Although in certain cases it is rather simple to describe the time evolution of $< \overline{M_{x,y}(t)} >$ in general, when one has to consider all hyperfine interactions present in a sample, this problem can become rather cumbersome. Therefore, it is convenient to consider the phenomenological equations devised by Bloch

[4] to describe the time evolution of the components of the nuclear magnetization in the lattice [1]:

$$\frac{dM_z}{dt} = \gamma(\boldsymbol{M} \times \boldsymbol{H}_0)_z + \frac{M_0 - M_z}{T_1}$$

$$\frac{dM_{x,y}}{dt} = \gamma(\boldsymbol{M} \times \boldsymbol{H}_0)_{x,y} - \frac{M_{x,y}}{T_2}$$

Here two characteristic decay times have been introduced: T_1 the spin-lattice
relaxation time and T_2 the spin-spin relaxation time. T_1 describes the time
evolution of M_z and is therefore directly related to the modifications in the
population of the Zeeman levels, which can occur after the exchange of energy
with the lattice excitations. The decay of $M_{x,y}$ occurs in a time T_2, which
is not only affected by the processes involved in the spin-lattice relaxation,
but also by other processes which do not imply an exchange of energy with
lattice excitations. For instance, as we shall see later on, the nuclear dipole-
dipole interaction yields a spread in the resonance frequencies of the nuclei and
accordingly a dephasing of the in-plane components of each nuclear moment
takes place, yielding a decrease in $M_{x,y}$.

It is interesting to treat the resonance absorption starting from Bloch
equations. Let us consider the effect of an RF field $H_1 \ll H_0$. Then one has
to replace in the above equations $\boldsymbol{H}_0$ with

$$\boldsymbol{H} = H_1(\hat{x}cos(\omega t) - \hat{y}sin(\omega t)) + H_0\hat{z}$$

For a negligible RF field, namely $\gamma H_1 \ll 1/\sqrt{T_1 T_2}$, so that the magnetization

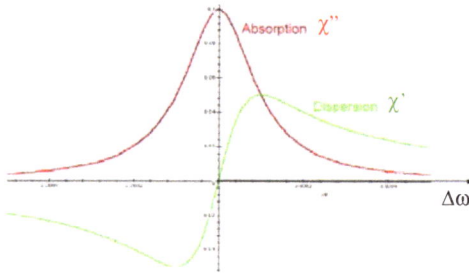

Fig. 5. Real ($\chi'(\omega)$) and imaginary ($\chi''(\omega)$) parts of the nuclear dynamical spin
susceptibility.

is only slightly tilted from the z axes (i.e. $M_z \simeq M_0$), by considering solutions

---

[1] From now on we shall indicate the components of nuclear magnetization with
$M_{x,y,z} = < \overline{M_{x,y,z}} >$.

of the form $M_x = m\cos(\omega t + \phi)$ and $M_y = -m\sin(\omega t + \phi)$ one finds that the in-plane component of the magnetization (the one which can be detected with a coil in the xy plane)

$$m \simeq \frac{\omega_1 M_0 T_2}{\sqrt{1 + (\omega - \omega_0)^2 T_2^2}} \ .$$

If one derives the in-plane components of the magnetization oscillating in phase and out of phase by $\pi/2$ with the RF field, one can estimate the real and imaginary part of the nuclear spin susceptibility. One finds:

$$\chi'(\omega) = \frac{M_0}{H_0} \frac{\omega_0(\omega_0 - \omega)T_2^2}{1 + (\omega_0 - \omega)^2 T_2^2} \ ,$$

$$\chi''(\omega) = \frac{M_0}{H_0} \frac{\omega_0 T_2}{1 + (\omega_0 - \omega)^2 T_2^2} \ ,$$

the dissipative part of the nuclear spin susceptibility is a Lorentzian of width $1/T_2$ centered around ω_0. In other words there is a peak in the absorption of the RF field at the Larmor frequency, which is spread over a certain width determined by the nuclear spin-spin relaxation time.

3 NMR spectra

When the sample under investigation is placed into the coil one observes a change in the inductance to

$$L = L_0[1 + 4\pi\chi(\omega)]$$

and a corresponding variation of the resistance by

$$\Delta R = L_0 \omega 4\pi \chi'' \ .$$

Then one can estimate that the average power adsorbed by the nuclei

$$P(\omega) = \frac{1}{2}\omega H_1^2 \chi''(\omega) V \ ,$$

where V is the sample volume. For $\omega = \omega_0$, by recalling the expression derived for χ'' in the weak field limit, one would find

$$P(\omega_0) = \frac{1}{2}\omega_0^2 H_1^2 \chi_0 V T_2$$

So, one can derive directly $\chi''(\omega)$ from the power adsorbed by the circuit. Indeed, in CW NMR one directly detects $\chi''(\omega)$.

Since the spacing between adjacent hyperfine levels is not the same for all nuclei it is convenient to introduce a distribution function $f(\omega)d\omega$ giving the

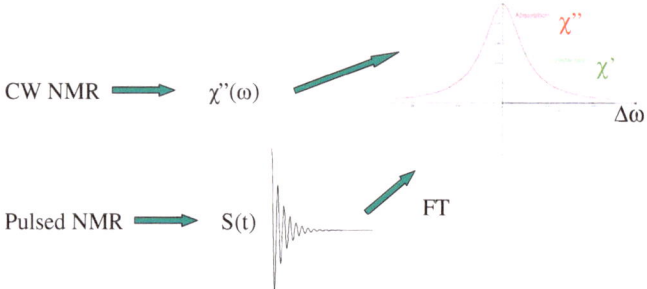

Fig. 6. Schematic illustration of the equivalence of the NMR spectra derived from CW NMR and from pulsed NMR spectroscopies after performing the Fourier transform (FT) of the FID signal.

fraction of nuclei with a resonance frequency between ω and $\omega + d\omega$. One can show that the energy adsorbed per unit time by the nuclei is

$$P(\omega) = \frac{\chi_0}{2}\omega\omega_0 H_1^2 f(\omega)2\pi$$

If we now compare the above equation with the expression previously derived for $P(\omega)$ one finds that

$$\chi''(\omega) = 2\pi f(\omega)\omega_0\chi_0$$

Then the signal detected in a CW NMR experiment, directly proportional to $P(\omega)$, is proportional to $\chi''(\omega)$ and gives the number of nuclei which are resonating with frequency between ω and $\omega + d\omega$, namely the NMR spectrum $f(\omega)$. On the other hand, by resorting to the fluctuation-dissipation theorem, one can show that

$$f(\omega) \propto \chi''(\omega) = \frac{\omega}{k_B T} \int_0^\infty e^{i\omega t} < \overline{M_x(t)M_x(0)} > dt$$

namely, the NMR spectrum is the Fourier transform at frequency ω of the correlation function for the transverse components of the nuclear magnetization, proportional to the signal measured in the pulsed NMR experiment (see Fig. 6).

4 Moment expansion of the NMR signal

In order to establish a priori the shape of the NMR spectra or of the FID signal one has to consider all the interactions acting on the nuclei. The nuclear hyperfine hamiltonian can be written in the form

$$\mathcal{H} = \mathcal{H}_Z + \mathcal{H}_{n-n} + \mathcal{H}_{n-e} + \mathcal{H}_{EFG}$$

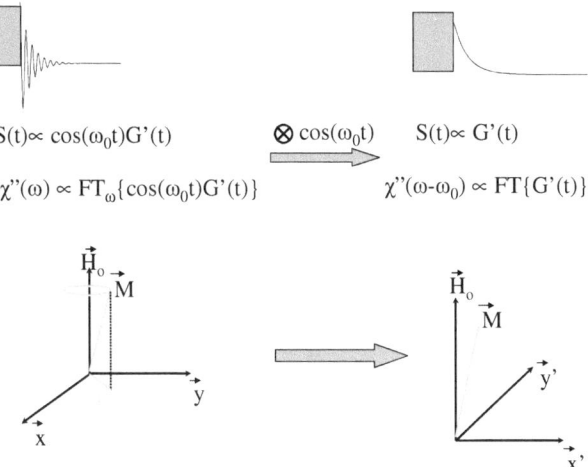

Fig. 7. Schematic view of the passage from the laboratory (left) to the frame of reference rotating at frequency ω_0. At the top the effect on the FID signal after a $\pi/2$ pulse is shown. The signal detected in the laboratory frame is transformed into the one in the rotating frame of reference by mixing it with a reference signal proportional to $cos(\omega_0 t)$. Then the difference in the expressions for the FID signal $s(t)$ and NMR spectra are shown. Finally, at the bottom, the vectorial illustration of the magnetization in the two reference frames is shown.

where the first term is the Zeeman-like interaction with $\boldsymbol{H}_0$

$$\mathcal{H}_Z = -\gamma\hbar \sum_i I_z^i H_0 \ .$$

$\mathcal{H}_{n-n}$ is nuclear dipole-dipole interaction, $\mathcal{H}_{n-e}$ describes the interaction between the nuclear and electron spins, while $\mathcal{H}_{EFG}$ is the quadrupole hamiltonian associated with the interaction of the nuclear quadrupole moment Q (non-zero for $I > 1/2$) with the electric field gradient (EFG) generated by the charge distribution around the nucleus.

Let us consider for simplicity just the first two terms of the hyperfine hamiltonian. Then

$$\mathcal{H} = -\gamma\hbar H_0 \sum_j I_z^j + \sum_{j<k} \frac{\hbar^2\gamma^2}{r_{jk}^3}\left(\boldsymbol{I}^j\boldsymbol{I}^k - 3\frac{(\boldsymbol{I}^j\boldsymbol{r}_{jk})(\boldsymbol{I}^k\boldsymbol{r}_{jk})}{r_{jk}^2}\right) \ .$$

It is convenient to write the nuclear dipole-dipole hamiltonian in the form

$$\mathcal{H}_{n-n} = \sum_{j<k} \frac{\hbar^2\gamma^2}{r^3}\left(A + B + C + D + E + F\right)_{jk}$$

with

$$A_{jk} = I_z^j I_z^k f(\theta, \phi) \quad \Delta m_T = 0$$

$$B_{jk} = -(I_+^j I_-^k + I_-^j I_+^k)\frac{f(\theta, \phi)}{4} \quad \Delta m_T = 0$$

$$C_{jk} = (I_z^j I_+^k + I_+^j I_z^k)g(\theta, \phi) \quad \Delta m_T = 1$$

$$D_{jk} = C_{jk}^* \quad \Delta m_T = -1$$

$$E_{jk} = (I_+^j I_+^k + I_+^j I_+^k)h(\theta, \phi) \quad \Delta m_T = 2$$

$$F_{jk} = E_{jk}^* \quad \Delta m_T = -2$$

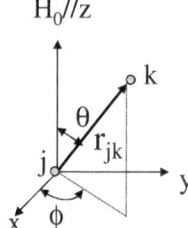

Since $\mathcal{H}_Z \gg \mathcal{H}_{n-n}$ only the terms of $\mathcal{H}_{n-n}$ that commute with $\mathcal{H}_Z$ will to the first order contribute to the time evolution of nuclear magnetization, namely the A and B terms. Then one has that the FID signal in the laboratory frame is given by

$$G(t) = cos(\omega_0 t) < M_x'(t)M_x(0) >$$

where

$$G'(t) = < M_x'(t)M_x(0) > = Tr\{e^{i\mathcal{H}'_{n-n}t/\hbar}M_x e^{-i\mathcal{H}'_{n-n}t/\hbar}M_x\}$$

with

$$\mathcal{H}'_{n-n} = \sum_{j<k} \frac{\hbar^2\gamma^2}{r_{jk}^3}\left(A + B\right)_{jk}$$

In the frame of reference rotating at frequency ω_0, the FID is directly given by $G'(t)$. Namely we have removed the term $cos(\omega_0 t)$ which was just describing the precessional motion around $\boldsymbol{H}_0$. Once more one realizes that if we work in the rotating frame we probe just the time-evolution of the nuclear magnetization due to the microscopic interactions within the sample.

For $\gamma^2\hbar t/r^3 < 1$ one can write

$$G'(t) = G'(0)[\mathcal{M}_0 + \mathcal{M}_2\frac{t^2}{2} + \mathcal{M}_4\frac{t^4}{4!} + ...] \ ,$$

where here $\mathcal{M}_n$ represents the nth moment of the NMR spectrum in the rotating frame ($\propto FT\{G'(t)\}$). The second moment, for example, is given by

$$\mathcal{M}_2 = -\frac{Tr\{[\mathcal{H}'_{n-n}, I_x]^2\}}{Tr\{I_x^2\}}$$

By writing $\mathcal{H}'_{n-n}$ in polar coordinates one finds

$$\mathcal{M}_2 = \gamma^4 \hbar^2 \frac{I(I+1)}{3} \sum_k \left(\frac{3}{2} \frac{1 - 3cos^2(\theta_k)}{r_k^3} \right)^2$$

Thus, one realizes that by deriving $\mathcal{M}_2$ from the FID signal one can estimate in a rather precise way the interatomic distances, for example.

If one considers the dipolar interaction with nuclei S of a different species one finds

$$\mathcal{M}_2 = \frac{4}{9} \gamma_I^2 \gamma_S^2 \hbar^2 \frac{S(S+1)}{3} \sum_{j<k} \left(\frac{3}{2} \frac{1 - 3cos^2(\theta_{jk})}{r_{jk}^3} \right)^2$$

It has be remarked that nuclear spins can interact also through and indirect coupling mediated by electron spins, usually much weaker than the classical one described above, of the form

$$\mathcal{H}_{n-n}^{ind} = J\boldsymbol{I}_i\boldsymbol{I}_j$$

with J a scalar. This interaction is of major relevance in high resolution NMR spectroscopy and is responsible for the appearance of multiplets formed by $2(n-1)I + 1$ lines, where n is the number of interacting nuclei of the same species. On the other hand, in certain cases, when the hyperfine coupling with electron spins is quite large, an indirect interaction (with J a tensor) larger than the direct one can be observed. Such a scenario occurs in the cuprates, for example [5].

5 Electron-nucleus hyperfine interaction

Let us now consider the interaction with the electron spins, which can often be written in the form

$$\mathcal{H}_{n-e} = -\gamma\hbar \sum_{i,k} \boldsymbol{I}_i \tilde{A}_{ik} \boldsymbol{S}_k$$

with $\tilde{A}_{ik}$ the hyperfine coupling tensor. Then, there will be an hyperfine field at the i-th nucleus given by $\boldsymbol{h}_i = \sum_k \tilde{A}_{ik} \boldsymbol{S}_k$. If the electron spins have a non-zero average polarization $< \boldsymbol{S} >$ then the local field probed by the nuclei will be $\boldsymbol{h}_i = \sum_k \tilde{A}_{ik} < \boldsymbol{S}_k >$ and one can directly estimate $< \boldsymbol{S} >$ from the

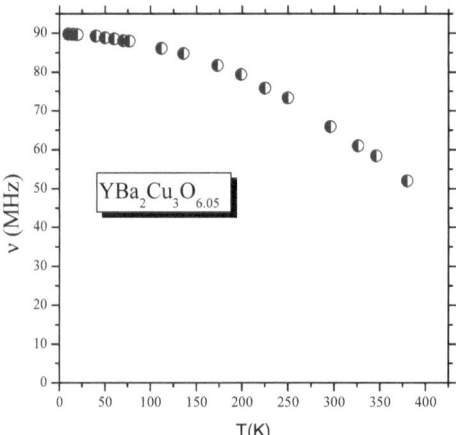

Fig. 8. Temperature dependence of ^{63}Cu $1/2 \rightarrow -1/2$ zero-field resonance frequency in the antiferromagnetic phase of YBa$_2$Cu$_3$O$_{6.05}$ vs. temperature. The resonance frequency is directly proportional to the order parameter, namely to the thermal average of Cu^{2+} spin $< \boldsymbol{S} >$ [6].

resonance frequency in the local field. When an external field is applied the nuclei will experience a magnetic field

$$\boldsymbol{H} = \boldsymbol{H}_0 + \sum_k \tilde{A}_{ik} < \boldsymbol{S}_k >$$

and the resonance frequency will be shifted to

$$\omega = \omega_0 (1 + \tilde{\Delta K})$$

with

$$\tilde{\Delta K} = \frac{\sum_k \tilde{A}_k < \boldsymbol{S}_k >}{H_0} .$$

Then one can write that $\tilde{\Delta K} = \sum_k \tilde{A}_k \chi(\boldsymbol{q} = 0, \omega = 0)$ and one notices that from the shift of the NMR resonance spectrum one can derive the static uniform susceptibility associated with those electron spins which interact with the nuclear ones. Another contribution to the shift, usually smaller than the one involving the electron spins, arises from the screening of the magnetic field by the electrons on the ligands and by the core electrons. This shift depends significantly on the chemical bonding around the nuclei and for this reason is called chemical shift.

The electric quadrupole hamiltonian describing the interaction between the nucleus quadrupole moment Q and the EFG tensor $\tilde{V}$ generated by the charge distribution around the nucleus is

$$\mathcal{H}_{EFG} = \sum_i \frac{e^2 Q V_{ZZ}}{4I(2I-1)} \left(3(I_z^i)^2 - I(I+1) + \frac{\eta}{2}[(I_+^i)^2 + (I_-^i)^2] \right)$$

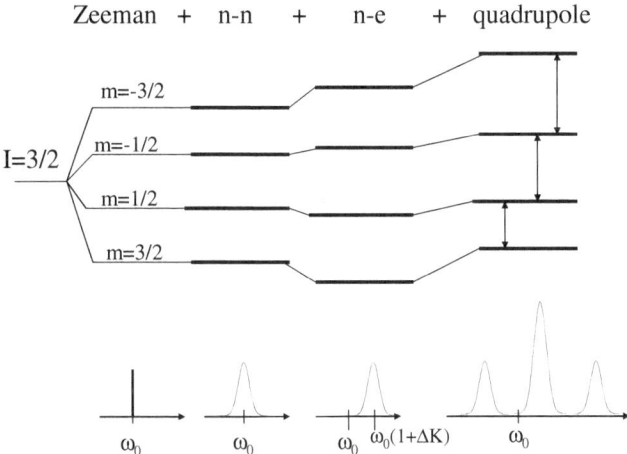

Fig. 9. Schematic illustration of the modifications in the hyperfine levels of $I = 3/2$ nuclei, due to the different terms of the nuclear hyperfine hamiltonian. The corresponding modifications in the NMR spectra are reported at the bottom of the figure.

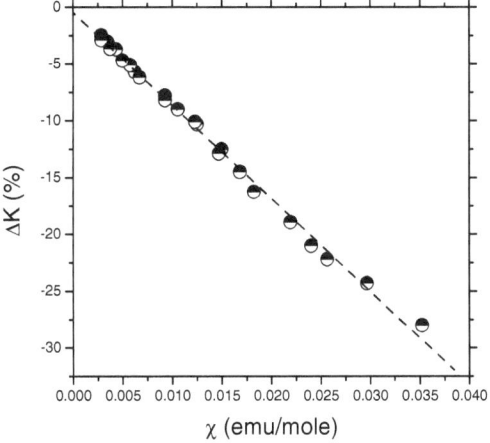

Fig. 10. The shift of ^{29}Si NMR frequency in MnSi is reported as a function of the macroscopic susceptibility measured with a SQUID magnetometer. The slope of the plot directly yields the hyperfine coupling, while the intercept gives the chemical shift (almost negligible here).

where V_{ZZ} is the main component of the EFG tensor written in the frame of reference of its principal axes (X, Y and Z)

$$\eta = \frac{V_{XX} - V_{YY}}{V_{ZZ}} \text{ with } |V_{ZZ}| \geq |V_{YY}| \geq |V_{XX}|$$

is the EFG tensor asymmetry parameter. If this is the only relevant hyperfine interaction the degeneracy between the spin components is partially resolved (Fig. 11) and one can induce transitions between the hyperfine levels by irradiating the nuclei at a frequency

$$\nu_Q = \frac{3eV_{ZZ}Q}{h2I(2I-1)}(1 + \frac{\eta^2}{3})^{1/2} \, .$$

This resonance frequency is extremely sensitive to the symmetry of the charge distribution around the nuclei and hence to the presence of lattice distortions, as shown in Fig. 11.

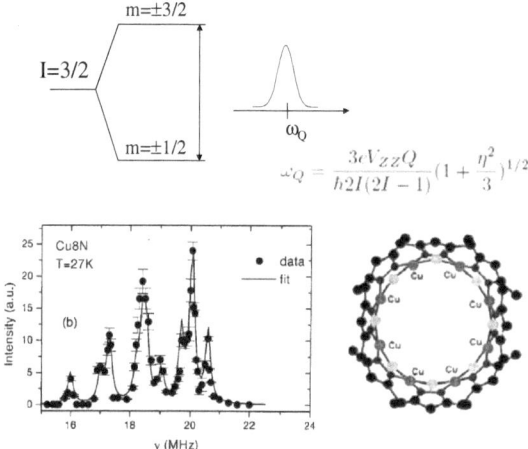

Fig. 11. At the top a schematic illustration of the transitions involved in an NQR experiment on $I = 3/2$ nuclei is presented. In the lower part of the figure ^{63,65}Cu ($I = 3/2$) NQR spectrum in Cu8 molecular ring is shown. It is evidenced that although in principle all Cu sites should be equivalent, a difference is present at the microscopic level, possibly associated with a lattice distortion [7].

When both Zeeman and quadrupole interactions are present the eigenvalues will depend on the orientation of the magnetic field with respect to the EFG principal axis. Then, in a powder sample there will be a distribution of possible orientations and of resonance frequencies. Therefore, the NMR powder spectra can become very broad if the quadrupole interaction is sizeable. The powder spectrum $f(\nu)$ can be estimated from the number of nuclei resonating at frequency ν which, for instance if $\eta = 0$, is simply given by

$f(\nu) \propto sin(\theta(\nu))/(d\nu/d\theta)$ (θ is the polar angle defining the orientation of $\boldsymbol{H}_0$ with respect to Z in the XYZ frame of reference).

6 Effect of the motions on the NMR spectra

Due to the time dependence of the hyperfine hamiltonian one can observe a modification in the NMR spectra. Let us consider a rather standard situation where the rigid lattice NMR spectrum is a Gaussian and the local field at the nuclei is fluctuating. Then the FID signal is given by

$$G'(t) \propto Tr\{e^{i\mathcal{H}'_P t/\hbar} I_x e^{-i\mathcal{H}'_P t/\hbar} I_x\}$$

where $\mathcal{H}'_P(t)$ is the time-dependent hamiltonian which can be considered as a perturbation of the Zeeman hamiltonian. Suppose that the time-dependence is induced by fluctuations of the local field at the nucleus (for example due to molecular motions, spin fluctuations, ionic diffusion, flux lines lattice motion in a superconductor, etc...). Then one can write

$$\mathcal{H}'_P(t) = -\hbar \sum_i I^i_z \Delta\omega_i(t) = -\hbar\gamma \sum_i I^i_z h^i_z(t)$$

where $\Delta\omega_i(t)$ describes the fluctuations in the resonance frequency of the i-th nucleus. If we consider a stationary gaussian distribution for the fluctuations with a mean-square amplitude $< \Delta\omega^2 >$, then one finds that

$$G'(t) = G'(0)exp\left(- < \Delta\omega^2 > \int_0^t (t - \tau)g(\tau)d\tau\right) \qquad (2)$$

where $g(\tau) =< \Delta\omega(t + \tau)\Delta\omega(t) > / < \Delta\omega^2 >$ is the normalized correlation function describing the fluctuations of the resonance frequency.

Now one can introduce the corresponding correlation time

$$\tau_c = \int_0^\infty g(\tau)d\tau$$

which describes the characteristic decay time for $g(\tau)$. Without making any assumption on the analytical form of $g(\tau)$ one can distinguish two limiting cases:

a) Slow motions regime. The correlation time is extremely long and the FID is recorded over a time $t \ll \tau_c$. Then in Eq. 2 one can set $g(\tau) \simeq g(0) \simeq 1$, so that

$$G'(t) \simeq G'(0)e^{-<\Delta\omega^2>t^2/2}$$

namely a Gaussian decay, as it has to be expected since we have assumed a rigid lattice Gaussian spectrum with a second moment $< \Delta\omega^2 >$.

b) Fast motions regime. The FID signal is recorded over a time $t \gg \tau_c$. Then one can set the upper limit of the integral in Eq. 2 to ∞ and neglect τ with respect to t, since $g(\tau)$ has already vanished over the time t. Then

$$G'(t) \simeq G'(0)e^{-<\Delta\omega^2>t\tau_c} = G(0)e^{-t/T_2'}$$

where

$$\frac{1}{T_2'} = <\Delta\omega^2> \tau_c = \gamma^2 \int_0^\infty <h_z(t)h_z(0)> dt$$

is the relaxation rate of the FID signal, namely of the transverse magnetization. One observes that now the FID decay is exponential and thus the corresponding NMR spectrum is a lorentzian with full width at half maximum equal to $1/T_2'$. Upon decreasing τ_c the linewidth decreases and one observes the motional narrowing of the NMR line.

If $g(\tau) = exp(-\tau/\tau_c)$ then one can write

$$G'(t) = G'(0)e^{-<\Delta\omega^2>\tau_c^2[exp(-t/\tau_c)-1+(t/\tau_c)]}$$

which nicely interpolates between fast and slow motions regime.

7 The spin echo

In general one observes that the decay of the FID signal is affected not only by intrinsic effects but also by extrinsic effects as magnetic field inhomogeneities associated, for example, with a distribution of paramagnetic shifts or simply to the inhomogeneity of the magnetic field generated by the magnet over the sample volume. Then one has an additional contribution to the decay of the FID, namely

$$G(t) = G(0)exp(-t/T_2')exp(-\gamma\Delta Ht) = G(0)exp(-t/T_2^*)$$

with ΔH the magnetic field distribution. Sometimes T_2^* can be so short that it is not even possible to record the FID signal. To avoid this problem one can resort to the **spin-echo** technique. Let us suppose that the magnetic field distribution is static and we apply the pulse sequence in Fig. 12.

The effect of the π RF pulse is to reverse the time evolution of the in-plane components of the nuclear spins. So, one observes that the dephasing of the spins during the first half of the sequence (between 0 and t) is recovered in the second half, after the π pulse, and the in-plane magnetization recovers at $2t$ and produces an echo signal. It is evident that the refocussing of the nuclear spins can occur as far as the resonance frequency of each nucleus during the first half and the second half of the pulse sequence is the same.

As an illustrative example let us consider what happens if the magnetic field is changed at time t, when the π RF pulse is applied. This is done, for example, in a controlled way in a spin-echo double resonance (SEDOR)

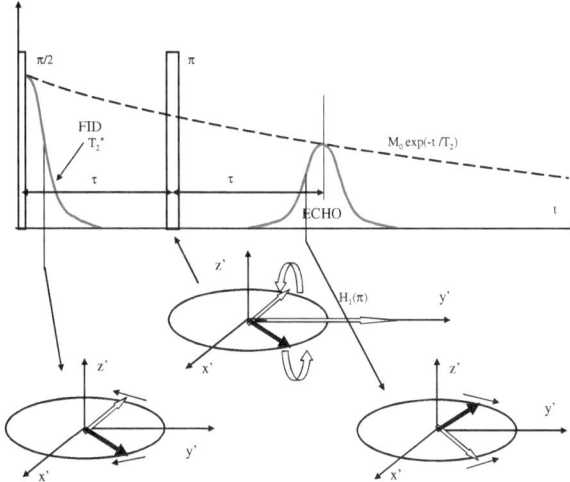

Fig. 12. Schematic representation of the motions of the nuclear spins generating the echo signal upon application of a sequence of $\pi/2$ and π RF pulses. It is shown how the FID signal after the $\pi/2$ pulse decays in a rather short time T_2^* owing to the dephasing of the nuclear spins. An echo is formed at time $2t$ thanks to the refocussing of the spins after the π RF pulse. The decay of the echo occurs over a time $T_2 \geq T_2^*$. It should be remarked that with pulse techniques, by switching the phase of the RF field it is possible to apply the second pulse (at time t) along a direction different from the one of the first pulse at $t = 0$ (e.g. from x to y in the rotating frame).

experiment, where at time t one applies a π pulse also on the nuclei S which are coupled to I nuclei through a $\mathcal{H}_{IS} = a\mathbf{I}.\mathbf{S}$. The π pulse on S yields a change in the resonance frequency of the I nuclei during the second half. Accordingly a decrease in the echo signal of the I nuclei is detected and the coupling between the two nuclear species is estimated.

Let us now consider the more general situation where the magnetic field fluctuations are described by a gaussian stationary distribution function. If now one calculates the dephasing of the nuclear spins between the $\pi/2$ and π $(0 - t)$ pulse and then between the π pulse and the echo $(t - 2t)$ one derives that the echo amplitude at time $2t$ is given by

$$E(2t) = E(0)exp\left(- <\Delta\omega^2 > \left[2\int_0^\infty (t-\tau)g(\tau)d\tau - \int_0^t \tau g(\tau)d\tau - \right.\right.$$
$$\left.\left. - \int_t^{2t} (2t-\tau)g(\tau)d\tau\right]\right) \ .$$

Without making any assumption on $g(\tau)$, in the fast motions regime $(\sqrt{<\Delta\omega^2 >}\tau_c \ll 1)$ one finds that

$$E(2t) = E(0)e^{-<\Delta\omega^2>\tau_c 2t} = E(0)e^{-2t/T_2'}$$

namely the echo decays with the same characteristic time of the FID. In the very slow motions regime, i.e. $\sqrt{<\Delta\omega^2>}\tau_c \gg 1$ the echo sequence allows to rephase completely the nuclear magnetization and then

$$E(2t) \to E(0) \ .$$

On the other hand, if $\sqrt{<\Delta\omega^2>}\tau_c \simeq 1$ one finds

$$E(2t) = E(0)e^{-\frac{<\Delta\omega^2>(2t)^3}{3\tau_c}}$$

If $g(\tau) = exp(-\tau/\tau_c)$ one can derive an expression valid in any limit [8]

$$E(2t) = E(0)exp\left(-<\Delta\omega^2>\tau_c[2t - \tau_c(1 - exp(-t/\tau_c))(3 - exp(-\tau/\tau_c))]\right)$$

One notices that if $\sqrt{<\Delta\omega^2>}\tau_c \simeq 1$ the decay of the echo amplitude can be quite fast. After a time $2t$ the echo would have decayed by a factor $exp(-<\Delta\omega^2>(2t)^3/3\tau_c)$. What happens if during the time $2t$ we apply n π pulses separated by a shorter delay $2t'$, so that $2t \simeq 2nt'$ (see figure below)? Then, after every π pulse there will be a decay of the echo by a factor $exp(-<\Delta\omega^2>(2t')^3/3\tau_c)$ and after n π pulses by

$$exp(-\frac{<\Delta\omega^2>n(2t')^3}{3\tau_c}) \gg exp(-\frac{<\Delta\omega^2>(2t)^3}{3\tau_c})$$

Hence the effect of the motions on the decay of the echo amplitude is significantly reduced. This particular echo sequence was first introduced by Carr and Purcell [9].

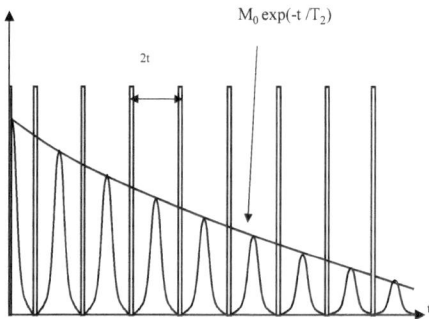

$M_0 exp(-t/T_2)$

Fig. 13. Carr-Purcell sequence with Meiboom-Gill phase alternation. The π pulses are generated by an RF field which in S' is alternatively along $+x'$ and $-x'$.

8 Nuclear spin-lattice relaxation rate

As we have seen in Bloch equations the longitudinal component of nuclear magnetization relaxes back to its equilibrium value, determined by the temperature of the lattice, with a characteristic relaxation time T_1. A simple RF pulse sequence which allows one to determine T_1 is shown in Fig. 14. After flipping the magnetization along x' with a $\pi/2$ pulse one waits for a delay τ and then applies a second $\pi/2$ pulse. The second $\pi/2$ will flip back along x' the fraction of magnetization which during the time τ has relaxed back to equilibrium. One can repeat the same experiment for different τ values and then derive T_1.

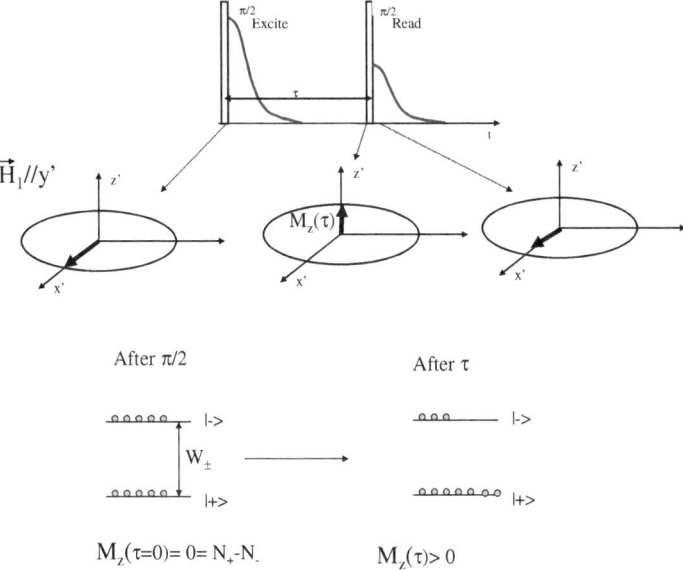

Fig. 14. (Top) Example of a simple RF pulse sequence used to measure the recovery of $M_z(\tau)$. The first $\pi/2$ turns the magnetization into the xy plane and the second pulse reads the amount of magnetization $M_z(\tau)$ which has relaxed back to equilibrium during the delay τ. The experiment is repeated for several values of τ to derive the recovery law of the magnetization. (Bottom) The effect of the first $\pi/2$ pulse on the statistical population on the hyperfine levels and its modification after τ are reported.

The recovery of nuclear magnetization towards equilibrium is determined by the transition probability among the hyperfine levels associated with the time-dependent part of the hamiltonian, namely by the lattice excitations.

The effect of the previous pulse sequence on the longitudinal magnetization can also be understood from the analysis of the statistical populations on the hyperfine levels (see Fig. 14). In fact, if we consider for simplicity nuclei with $I = 1/2$, then $M_z(\tau) \propto N_+ - N_-$, the difference of population between the two levels.

In general for nuclei with spin I one has $2I + 1$ states and one has to solve a system of $2I + 1$ differential equations

$$\frac{dN_m}{dt} = \sum_{n \neq m} (N_n W_{nm} - N_m W_{mn})$$

to derive the time evolution of the population difference between the levels which are being irradiated. For the simple case of $I = 1/2$ one finds that

$$M_z(\tau) = M_z(\tau \to \infty)(1 - e^{-\tau/T_1})$$

or equivalently

$$y(\tau) = \frac{M_z(\infty) - M_z(\tau)}{M_z(\infty)} = e^{-\tau/T_1}$$

with

$$\frac{1}{T_1} \equiv 2W_\pm$$

It should be noticed that in order to have spin-lattice relaxation $W_\mp > W_\pm$, but since usually $\hbar\omega_0 \ll k_B T \; W_\mp \simeq W_\pm$ and one can express T_1 just in terms of one of the two transition probabilities. For $I > 1/2$ one has to consider all possible transitions which are driven by the time-dependent part of the hamiltonian. If the fluctuations are associated with an effective fluctuating magnetic field (e.g electron spin fluctuations) then just transitions with $\Delta m = \pm 1$ have to be considered in solving the system of differential equations. If the fluctuations are the ones of the electric field gradient, since $\mathcal{H}_{EFG}$ is quadratic in the spin components, one has to consider also $\Delta m = \pm 2$ transitions. In general one finds a recovery law for nuclear magnetization

$$y(\tau) = \sum_j c_j e^{-\alpha_j \tau/T_1}$$

still with $1/T_1 = 2W_\pm^{I=1/2}$.

Let us suppose that the time-dependent part of the hamiltonian is associated with an effective magnetic field fluctuation

$$\mathcal{H}_P(t) = -\gamma\hbar \boldsymbol{I} \boldsymbol{h}(t)$$

which can be considered as a perturbation of the Zeeman hamiltonian. Then, starting from time-dependent perturbation theory one can express $W_\pm$ in terms of the correlation function

$$g(\tau) =< \overline{\mathcal{H}_P(\tau)\mathcal{H}_P(0)} >$$

describing the fluctuations of $\mathcal{H}_P(t)$. If $g(\tau)$ decays with a correlation time $\tau_c \ll T_1$ and $T_1\omega_0 \gg 1$, conditions which are usually satisfied, then one can write

$$\frac{1}{T_1} = \frac{\gamma^2}{2} \int_{-\infty}^{+\infty} e^{i\omega_0 t} < h_+(t)h_-(0) > dt \tag{3}$$

This fundamental expression shows that $1/T_1$ is driven by the transverse components of the fluctuating field at the nucleus, to comply with magnetic-dipole selection rules, and that $1/T_1$ is proportional to the Fourier transform of the correlation function at the resonance frequency, to comply with energy conservation. In other terms $1/T_1$ probes the spectral density at ω_0.

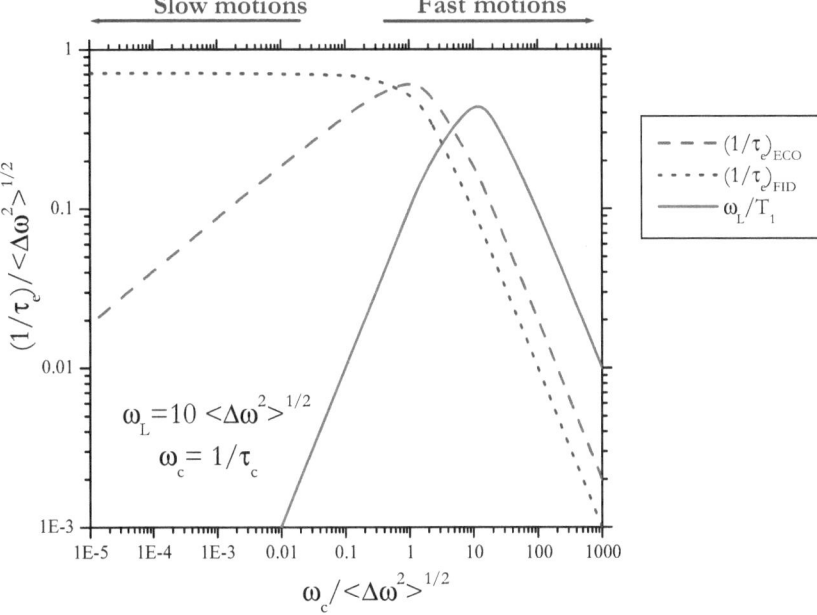

Fig. 15. Effect of the frequency of the fluctuations on the decay of the FID, of the echo and on $1/T_1$. $< \Delta\omega^2 >$ is the second moment for the amplitude of the frequency fluctuations. τ_e is the characteristic time at which the FID or the echo decreases by a factor e. It has been assumed at the sake of illustration that the Larmor frequency is 10 times $\sqrt{< \Delta\omega^2 >}$.

Suppose that $< h_+(t)h_-(0) >=< \Delta h_\perp^2 > exp(-t/\tau_c)$. Then from Eq. 3 one derives that

$$\frac{1}{T_1} = \frac{\gamma^2}{2} < \Delta h_\perp^2 > \frac{2\tau_c}{1 + \omega_0^2\tau_c^2} \tag{4}$$

One can distinguish three regimes:

a) Fast motions, $\omega_0 \tau_c \ll 1$, then

$$\frac{1}{T_1} = \gamma^2 < \Delta h_\perp^2 > \tau_c$$

Then one notices that if the fluctuations are isotropic so that $< \Delta h_\perp^2 >= 2 < \Delta h_z^2 >$ then if $\omega_0 \tau_c \ll 1$ and $\sqrt{< \Delta \omega^2 >} \tau_c \ll 1$ one finds $1/T_1 = 2/T_2'$.

b) Slow motions, $\omega_0 \tau_c \gg 1$, then

$$\frac{1}{T_1} = \gamma^2 < \Delta h_\perp^2 > \frac{1}{\omega_0^2 \tau_c}$$

c) $\omega_0 \tau_c = 1$ then

$$\frac{1}{T_1} = \frac{\gamma^2}{2} < \Delta h_\perp^2 > \frac{1}{\omega_0}$$

and one has a maximum in $1/T_1$, as it has to be expected since the highest transition probability would take place when the characteristic frequency for the fluctuations corresponds to the resonance frequency ω_0.

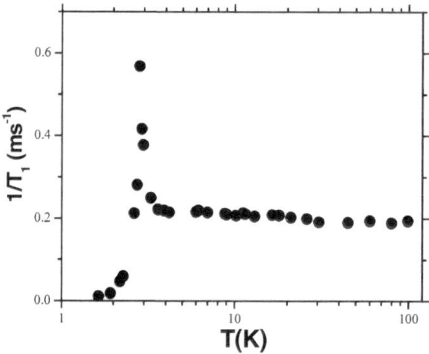

Fig. 16. Temperature dependence of ^{7}Li $1/T_1$ in $Li_2 VOSiO_4$. It is evident that at temperatures larger than the exchange coupling ($\simeq 9$ K) the spin-lattice relaxation rate is temperature independent.

Let us now derive some typical expressions for nuclear spin-lattice relaxation rate which apply to some model systems. In a paramagnetic insulator the fluctuations are associated with electron spin fluctuations $\boldsymbol{h}(t) = \sum_i \tilde{A}_i \boldsymbol{S}_i(t)$. At high temperature, namely $k_B T \gg J$ (J the exchange coupling among the spins), the correlation function for the spin components is [10]

$$< S_{x,y,z}^i(t) S_{x,y,z}^i(0) >= |S_{x,y,z}|^2 e^{-\omega_e^2 t^2 / 2}$$

with $\omega_e = (J/\hbar)\sqrt{2zS(S+1)/3}$ the Heisenberg exchange frequency. Then

$$\frac{1}{T_1} = \frac{\gamma^2}{2} \sum_i \left([A_{xx}^i]^2 + \ldots \right) \frac{S(S+1)}{3} \frac{\sqrt{2\pi}}{\omega_e} \ ,$$

temperature independent. One notices that as $\omega_e \gg \omega_0$ (fast motions) $1/T_1 \propto 1/\omega_e$ and does not depend on ω_0.

In general, when collective spin excitations are present one can write

$$\boldsymbol{h}(t) = \frac{1}{\sqrt{N}} \sum_{\boldsymbol{q}} \sum_i e^{i\boldsymbol{q}\boldsymbol{r}_i} \tilde{A}_i \boldsymbol{S_q}(t)$$

and by writing the transverse components of $\boldsymbol{h}(t)$ in $1/T_1$ expression one finds that

$$\frac{1}{T_1} = \frac{\gamma^2}{2} \frac{1}{N} \sum_{\boldsymbol{q}} \left(|A_{\boldsymbol{q}}|^2 S_{\alpha\alpha}(\boldsymbol{q}, \omega_0) \right)_\perp \ .$$

Here $|A_{\boldsymbol{q}}|^2$ is the form factor giving the hyperfine coupling of the nuclei with the spin excitations at wave-vector $\boldsymbol{q}$. $S_{\alpha\alpha}(\boldsymbol{q}, \omega_0)$ is the component of the dynamical structure factor at the resonance frequency. The term $\perp$ indicates that one has to consider the products $|A_{\boldsymbol{q}}|^2 S_{\alpha\alpha}(\boldsymbol{q}, \omega_0)$ associated with the perpendicular components of the hyperfine field at the nucleus. From the fluctuation-dissipation theorem, by recalling that usually $k_B T \gg \hbar\omega_0$ one can also write

$$\frac{1}{T_1} = \frac{\gamma^2}{2} \frac{k_B T}{\hbar} \frac{1}{N} \sum_{\boldsymbol{q}} \left(|A_{\boldsymbol{q}}|^2 \frac{\chi''_{\alpha\alpha}(\boldsymbol{q}, \omega_0)}{\omega_0} \right)_\perp$$

One can now introduce the expressions for the dynamical susceptibility of a certain system and derive the corresponding expression of $1/T_1$. In a metal one finds

$$\frac{1}{T_1} = \left(\frac{16}{3}\right)^2 \pi^3 \hbar^2 \gamma^2 \mu_B^2 (|\psi(0)|^2)_{FS} D^2(E_F) k_B T \ .$$

where $D(E_F)$ is the density of states at Fermi level (E_F) and $|\psi(0)|^2$ is the probability that the electron is at the nuclear site. One observes that $1/T_1$ is linear in T, at variance with what was derived for a paramagnetic insulator. This is a simple evidence of Fermi-Dirac statistics, which tell us that, owing to Pauli principle, only a fraction $\simeq k_B T/E_F$ of all electron spins can flip and cause the relaxation.

In certain cases T_1 can be so short to prevent its direct measurement. However, under certain circumstances it still be possible to estimate it. Let us call S the spin of the fast relaxing nuclei and I the spin of another species of nuclei which one can suitably investigate. If S and I spins are coupled through nuclear dipole-dipole interaction, then the flip of spin $\boldsymbol{S}$ due to T_1^S processes is detected as a fluctuation of the local field at the $\boldsymbol{I}$ nuclear spin. Namely, one detects a dynamic with a correlation time $\tau_c = T_1^S$. This dynamic can manifest itself on the decay of the echo amplitude, for instance. If $\sqrt{<\Delta\omega_I^2>}T_1^S \ll 1$, where $< \Delta\omega_I^2 >$ is the second moment of the frequency distribution associated

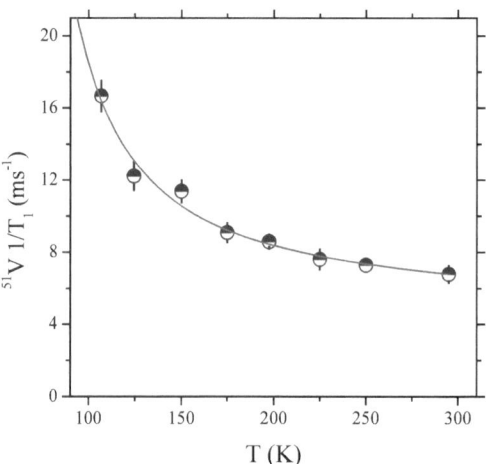

Fig. 17. Temperature dependence of ^{51}V $1/T_1$ in MoVO$_5$ estimated by measuring the decay rate of ^{95}Mo echo signal [11].

with the nuclear dipole interaction between S and I spins, then the decay of the echo amplitude of I spins is given by

$$E(2t) = E(0)e^{-<\Delta\omega_I^2>T_1^S 2t}$$

and then one can estimate T_1^S.

In pulsed NMR spectroscopy often one looks for the effect of the nuclear spins of a species S on the spins of another species I. This is the basis of the double resonance techniques, as the SEDOR one previously mentioned, of the Overhauser and cross-polarization techniques which allow to transfer polarization between different spin ensembles and of multi-dimensional techniques which are just a part of the marvellous realm of the nuclear magnetic resonance spectroscopy, which nowadays is employed in many different research areas [12].

References

1. E.M. Purcell, H.C. Torrey and R.V. Pound, Phys. Rev. 69, 37 (1946); F. Bloch, W.W. Hansen and M. Packard, Phys. Rev. 69, 127 (1946)
2. C. P. Slichter in *Principles of Magnetic Resonance* (Springer, Berlin, 1990) 3rd Ed.
3. A. Abragam in *Principles of Nuclear Magnetism* (Oxford University Press, New York, 1961)
4. F. Bloch, Phys. Rev. 70, 460 (1946)

5. C. H. Pennington, D. J. Durand, C. P. Slichter, J. P. Rice, E. D. Bukowski and D. M. Ginsberg, Phys. Rev. B 39, 274 (1989)
6. C. Bucci, P. Carretta, R. De Renzi, G. Guidi, S. Jang, E. Rastelli, A. Tassi, and M. Varotto Phys. Rev. B 48, 16769-16774 (1993)
7. A. Lascialfari, Z. H. Jang, F. Borsa, D. Gatteschi, A. Cornia, D. Rovai, A. Caneschi, and P. Carretta Phys. Rev. B 61, 6839-6847 (2000)
8. M.Takigawa and G. Saito, J.Phys.Soc. 55, 1233(1986)
9. H.Y. Carr and E.M. Purcell, Phys. Rev. 94, 630 (1954)
10. T. Moriya, Prog. Theor. Phys. 16, 23 (1956)
11. P. Carretta, N. Papinutto, C. B. Azzoni, M. C. Mozzati, E. Pavarini, S. Gonthier, and P. Millet Phys. Rev. B 66, 094420 (2002)
12. see for instance H. Friebolin in *Basic One- and Two-dimensional NMR Spectroscopy* (VCH, Weinheim, 1993) and http://www.cis.rit.edu/htbooks/nmr/.

NMR in magnetic single molecule magnets

Ferdinando Borsa

Dipartimento di Fisica "A.Volta" - University of Pavia - Via Bassi, 6 - 27100 Pavia (Italy) borsa@fisicavolta.unipv.it

Contents

Introduction

In the last years there has been a great interest in magnetic systems formed by a cluster of transition metal ions covalently bonded via superexchange bridges, embedded in a large organic molecule. Following the synthesis and the structural and magnetic characterization of these magnetic molecules by chemists, the physicists realized the great interest of these systems as a practical realization of zero-dimensional model magnetic systems. In fact the magnetic molecules can be synthesized in crystalline form whereby each molecule is magnetically independent since the intramolecular exchange interaction among the transition metal ions is dominant over the weak intermolecular, usually dipolar, magnetic interaction.

Magnetic molecules (see Fig.1 for some example systems) can be prepared now a day with an unmatched variety of parameters: (i) the size of the magnetic spin can be varied spanning from high "classical" spins to low "quantum" spins by using different transition metal ions i.e. Fe^{3+}, Mn^{2+} (s=5/2), Mn^{3+} (s=2), Cr^{3+}, Mn^{4+} (s=3/2), Cu^{2+}, V^{4+}(s=1/2); (ii) the exchange interaction can go from antiferromagnetic (AFM) to ferromagnetic (FM) with values of the exchange constant J ranging from a few Kelvins to more than 1000K; (iii) the geometrical arrangement of the magnetic core of the molecule can be as simple as a coplanar regular ring of magnetic ions as found in many Fe and Cr rings to totally asymmetric three dimensional clusters such as $[Fe_8(N_3C_6H_{15})_6O_2(OH)_{12}]^{8+} \cdot [Br_8 \cdot 9H_2O]^{8-}$ (in short Fe8); (iv) the symmetry of the magnetic Hamiltonian can go from isotropic Heisenberg type as in most cases to easy axis or easy plane. Choosing from this variety of model systems one can investigate fundamental problems in magnetism taking advantage of the fascinating simplicity of zero-dimensional systems. Examples of issues of interest are the transition from classical to quantum behavior, the effect of geometrical frustration and of symmetry breaking by substitution in heterometallic rings, the form of the spectral density of the magnetic fluctuations, the spectrum of the low-lying excitations with the connected problem of quantum spin dynamics and tunneling and the problem of long range magnetic order due to the weak (dipolar) intermolecular interactions.

NMR has proved to be a powerful tool to investigate both static and dynamic properties of magnetic systems. In particular it has been very successful in addressing some special features in low dimensional magnetic systems. In molecular nanomagnets there are additional features specific to the finite size of the system which require the development of novel models and theories to interpret the data.

In the first lecture we will mention some of the problems encountered in doing NMR in molecular nanomagnets. These are related for example to the presence of many non equivalent nuclei, to very broad and structured resonance lines, to very short relaxation rates, to vanishingly weak signals. Following this we will discuss the static magnetic properties which can be extracted from the measurements of the NMR spectra. The hyperfine interaction of nuclei with the magnetic moments in the cluster offers an unique tool to determine the local spin moment configuration and distribution in the nanomagnet. The subsequent four lectures will discuss the information that one can obtain from the nuclear relaxation measurements. These lectures will be organized somewhat arbitrarily according to the temperature range. In fact the physical issues regarding the magnetic properties and the spin dynamics of the molecular nanomagnets depend on the relative ratio of the thermal energy $k_B T$ and the magnetic exchange energy J. At high T the individual magnetic ions in the molecule behave as weakly correlated paramagnetic ions; at very low T the individual spins are locked into a collective quantum state of total spin S; at intermediate T the interacting spins develop strong correlations in a way similar to what happens in magnetic phase transitions in three dimensional systems. In the illustration of the physical issues encountered in the different temperature ranges we utilize the most representative results for different kind of molecules.

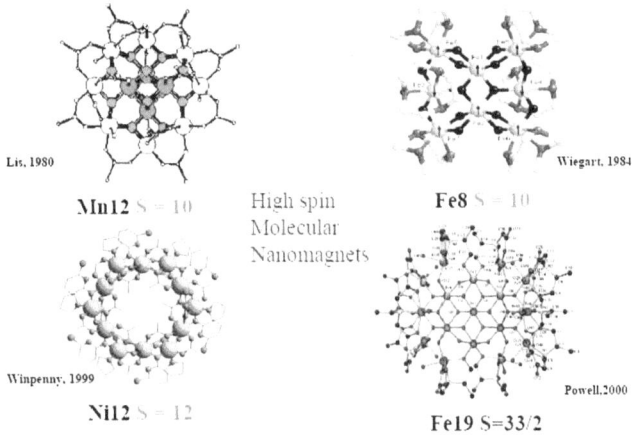

Fig. 1.

1 General features of NMR in molecular nanomagnets. NMR spectra and hyperfine interactions

1.1 General features of NMR in molecular nanomagnets

Molecular nanomagnets offer a wide variety of nuclei which can be used to probe the magnetic properties and the spin dynamics. Most of the measurements were done on proton NMR. In this case the signal is very strong but the width of the spectrum and the presence of many inequivalent protons in the molecule require some special attention in the analysis of the results. Due to the above reasons the recovery of the nuclear magnetization was found in many cases to be strongly non exponential. There are two sources for non exponential recovery. The first is due to incomplete saturation of the broad NMR spectrum which leads to an initial fast recovery of the nuclear magnetization due to spectral diffusion. If the spectrum is not too wide (at most twice the spectral width of the rf pulse) one can still saturate the whole line by using a sequence of rf pulses provided that the condition of T_1 much longer and T_2 much shorter than the pulse spacing can be met. In nanomagnets with magnetic ground state like $[Mn_{12}O_{12}(CH_3COO)_{16}(H_2O)_4]$ (in short Mn12) and in Fe8 the proton spectrum can be as wide as 4MHz and structured at low temperature. In this case the spectrum has to be acquired by sweeping the field and/or the frequency and plotting the amplitude of the echo signal after proper correction for changes of irradiation conditions. In this case the relaxation can only be measured at given points of the spectrum and the spectral diffusion effect cannot be avoided. One can try to establish the percentage of the fast initial recovery which is affected by spectral diffusion and exclude that from the measurement but a large systematic error can still be unavoidable. The second source of non exponential recovery is due to the presence in the molecule of protons having a different environment of magnetic ions and thus having a different relaxation rate. If T_2 is fast compared to T_1 then a common spin temperature is achieved during the relaxation process and the recovery is exponential with a single T_1^{-1} which is a weighted average of the rates of the inequivalent protons in the molecule. In the opposite limit encountered when T_1 is comparable to T_2 each nucleus or group of nuclei relax independently with its own spin temperature and the recovery of the nuclear magnetization results in the sum of exponentials:

$$n(t) = [M(\infty) - M(t)]/M(\infty) = \Sigma_i p_i exp(-t/T_{1i}) \qquad (1)$$

If there is a continuous distribution of T_1's the recovery follows a stretched exponential function $exp(-(t/T_1^*)^\beta)$ where $\beta < 1$ is the smaller the wider is the distribution and T_1^* is a relaxation parameter related to the distribution of T_1's in a non trivial manner. When the recovery is non exponential it is best to measure the T_1 parameter from the recovery of the nuclear magnetization at short times. In fact the slope at t→0 of the semilog plot of n(t) vs t yields an average relaxation rate $T_1^{-1} = \Sigma_i p_i \exp(-t/T_{1i})$. Unfortunately, in most cases the situation is intermediate between the two above limiting cases. In this circumstance there is no simple way that one can define a spin-lattice relaxation parameter. Since in many instances one is interested in the relative changes vs T and H one can simply define an effective relaxation parameter R by taking the time at which the recovery curve n(t) reduces to 1/e of the initial value.

Other isotopes which have been utilized for NMR studies of magnetic molecules include ^{2}H, ^{13}C, ^{7}Li, ^{23}Na, ^{63,65}Cu. The disadvantage of a weaker signal in ^{13}C is in part compensated by the advantage of having a nucleus with strong hyperfine coupling to the magnetic ions and with less number of inequivalent sites with respect to protons. For the remaining quadrupole nuclei there is the additional information obtained by the quadrupole coupling with the electric field gradient. When the quadrupole interaction is sufficiently strong to remove the satellite transition from the central line the non exponential decay of the nuclear magnetization becomes very difficult to analyze because besides the non equivalent sites one has to take into account the intrinsic non exponential decay due to unequal separation of the Zeeman levels. The ^{63,65}Cu case is the only one where, to our knowledge, a pure NQR experiment has been performed in molecular clusters (i.e. $[Cu_8(dmpz)_8(OH)_8] \cdot 2C_6H_5NO_2$, in short Cu8). The NQR spectrum was found to contain several lines in the frequency range 16-21MHz.

Very useful information were obtained from the ^{55}Mn and ^{57}Fe NMR in Mn12 and Fe8 clusters respectively. The NMR of the above nuclei can be observed only at low temperature (T< 4K) since with increasing temperature the relaxation times T_1 and T_2 become too short. The ^{55}Mn and ^{57}Fe (in isotopically enriched sample) NMR were detected both in zero field and in an externally applied field. The zero field ^{55}Mn NMR spectrum in Mn12 consists of three quadrupole broadened lines (i.e. each several MHz wide) in the frequency range 230-370 MHz while the ^{57}Fe NMR spectrum in Fe8 is made of eight different lines rather narrow (i.e. 100 kHz) in the frequency range 63-73 MHz. Both Mn12 and Fe8 are ferrimagnetic molecules at low temperature. However, since there are no domain walls and the anisotropy is very high no signal enhancement due to the rf enhancement in domain walls and/or domains is present contrary to normal ferro or ferri magnetic long range ordered systems. As a consequence the NMR signal intensity in zero external field is small (particularly in Fe8) even at low temperature since the frequency range of the overall spectrum is quite broad.

1.2 NMR spectra and hyperfine interactions

The local magnetic properties can be obtained from the NMR spectra since the nuclei are coupled to the electronic moments via the hyperfine interactions. The local field at the nuclear site is given by the sum of dipolar fields and contact fields: $H_{int} = H_{dip} + H_{contact} = \mathbf{I} \cdot A \cdot \boldsymbol{S} + B \, \mathbf{I} \cdot \mathbf{S}$. The NMR

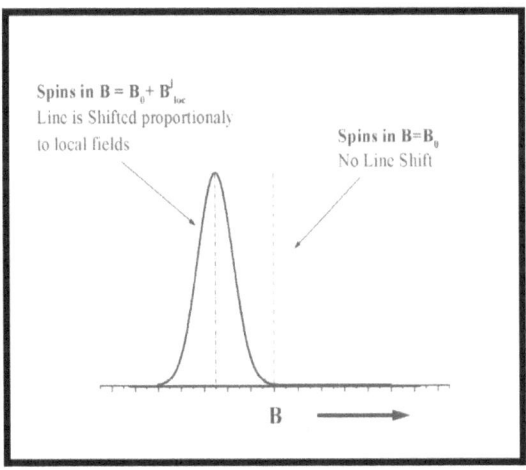

Fig. 2.

resonance frequency is given by $\nu_{res} = \gamma \, |\mathbf{H} + \mathbf{H}_{int}|$. If the molecular nanomagnet is at sufficiently high temperature, namely $k_B T \gg J$, then the magnetic system behaves as a normal paramagnet and one has that the local field is proportional to the local spin polarization i.e. the product of the magnetic susceptibility times the external field: $H_{int}^{j} \propto \Sigma_k A_{jk} \langle S_k \rangle \propto \langle S \rangle \propto \chi H$. If the dominant hyperfine interaction is dipolar as is the case of protons in molecular nanomagnets then the averaging over all proton sites and over all particle orientations results in a broadening of the line proportional to both the external field and to the magnetic susceptibility. This is shown in Fig.3 for the case of proton NMR in the molecular cluster Fe30 which has a very low J value (of order of 1K). The slope of the plot of the full width at half maximum (FWHM) vs H yields the product of the magnetic susceptibility times the average hyperfine interaction constant. In the graph below the slope is much larger at low T where the susceptibility is larger. If the susceptibility is known one can obtain information about the size of the dipolar hyperfine interaction. At low temperature the fluctuation time of the local spin becomes

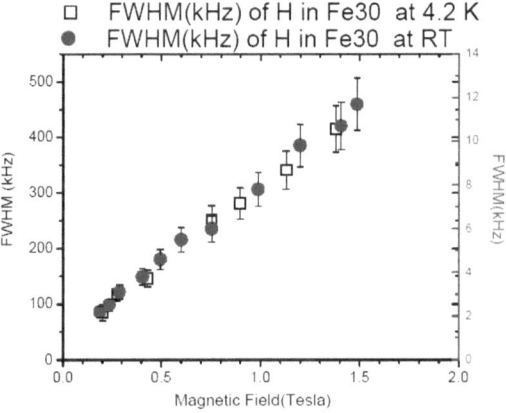

Fig. 3.

progressively longer as a combination of the onset of correlation among spins due to the magnetic interaction and of the 'freezing'in the direction of the crystal field anisotropy and/or of the external magnetic field. The local field $H_{int}= g \langle s \rangle A$ is seen as static by the resonant nucleus if the fluctuation time is longer than the inverse interaction energy (in frequency units) $\tau \gg A^{-1}$. In this case the NMR of the nucleus can be observed, in principle, also in absence of an applied magnetic field and from the relation $\nu_{res} = \gamma(H + g \langle s \rangle A)$ one can obtain a direct information on the local spin density $\langle s \rangle$ if the hyperfine constant A can be either calculated or measured independently. For example, at very low T ($\leq$ 4K) both the NMR of [55]Mn in Mn12 and the NMR of [57]Fe in isotopically enriched Fe8 can be detected in zero external field. In the case of Fe8 eight different lines where observed in zero field spanning over a wide frequency range (63-73 MHz)(see Fig.4). The different resonant frequency of the eight lines was interpreted as due primarily to a different local spin density of the eight Fe^{3+} ions in the molecular nanomagnet. The hyperfine constant A is dominated in this case by core polarization effects which create at the nuclear site a local field directed opposite to the polarization of the transition metal d electrons. From the formula $H_{int}=A\langle s\rangle$ and from the estimated theoretical value A= -126 kOe/μ_B one can derive the local spin density at each Fe site. It is noted that this information can be obtained only by NMR and by polarized neutron scattering.

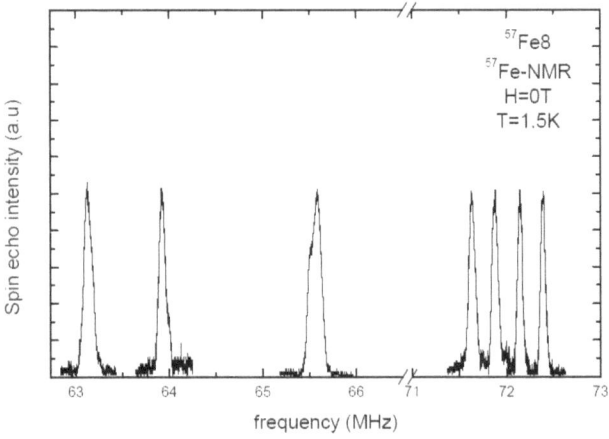

Fig. 4.

1.3 Determination of the local spin density and configuration

One important issue in molecular nanomagnets is the knowledge of the local spin configuration corresponding to the collective quantum state described by the total spin i.e S=10 for both Mn12 and Fe8. This issue has been addressed very successfully by NMR and we will show these two examples in the following.

a) Fe8

By applying an external field along the easy axis one can deduce the local spin configuration as follows. The field dependence of the eight ^{57}Fe resonance frequencies shown in Fig.4 as a function of a magnetic field applied along the main easy axis is linear as shown in Fig.5. As seen in the plot, two of the eight ^{57}Fe lines shift at higher frequency with increasing magnetic field while the other six shift to lower frequency. Since the resonance frequency is

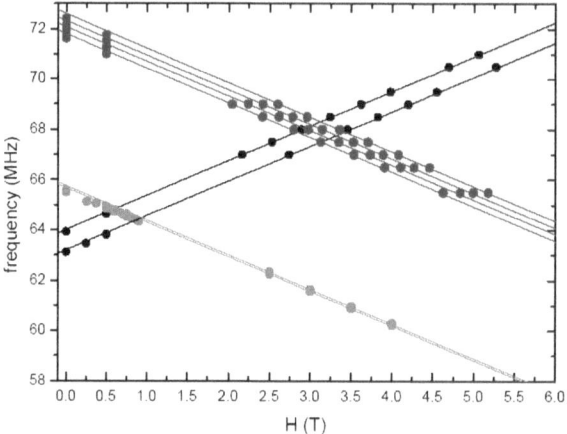

Fig. 5.

proportional to the vector sum of the internal field (H_{int}) and the external field (H_{ext}) i.e. $\nu_R = \gamma_N (\mathbf{H}_{int} + \mathbf{H}_{ext})$, this result indicates that the direction of the internal field at the site of two Fe^{3+} ions is opposite to the one at the other six sites. Since H_{int} originates mainly from the core-polarization, H_{int} is negative and the direction of the internal fields at nuclear sites is opposite to that of the Fe spin moment. Thus one can conclude that the spin direction of the two Fe moments is antiparallel to the external field, while that of the remaining six moments is parallel to the external field, corresponding to the standard spin structure of magnetic core of Fe8 cluster as postulated from results of magnetization. When the magnetic field is applied perpendicular to the main easy axis and parallel to the medium axis in the xy hard plane the field dependence of the resonance frequencies is non linear, as shown in Fig.6, since the external field and the local internal field are now perpendicular to each other. As described above, the resonance frequency is proportional to the effective internal field at the nuclear site, which is the vector sum of H_{int} due to spin moments and H_{ext} due to the external field i.e. $|\mathbf{H}_{eff}| = |\mathbf{H}_{int} + \mathbf{H}_{ext}|$. Thus the opposite field dependence of $|\mathbf{H}_{eff}|$ for the two lines and the other six lines indicates that the direction of two Fe spin moments remains antiparallel to that of the other six Fe spin moments. This leads to the conclusion that the individual spin moments do not cant independently along the direction of the transverse field but rather rotate rigidly maintaining the same relative spin configuration. From the quantitative analysis of the field dependence one can obtain information about the canting of the total magnetization.

b) Mn12

A similar study has been performed in Mn12 at 1.5K where the nanomagnet is in its ground state. Three ^{55}Mn NMR signals can be detected in Mn12 at

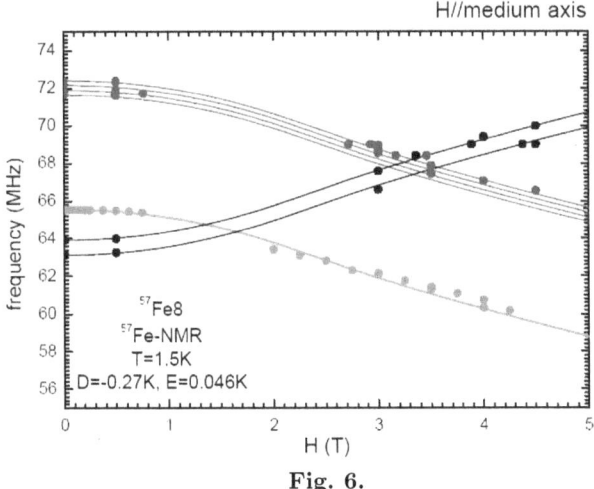

Fig. 6.

low T and in zero applied field. These correspond to three inequivalent groups of Mn ions. Two of the signals are much broader due to quadrupole effects of the ^{55}Mn nucleus. The frequency position of the three lines correspond to different local spin density of the corresponding Mn magnetic ion. With increasing parallel field, P1 shifts to higher frequency while the other two peaks P2 and P3 (pertaining to Mn^{3+} ions) shift to lower frequency. Since the resonance frequency is proportional to the vector sum of the internal field (H_{int}) and the external field (H_{ext}) i.e. $\nu_R = \gamma_N (|\mathbf{H}_{int} + \mathbf{H}_{ext}|)$, this result indicates that the direction of the internal field at the Mn sites for Mn^{3+} ions is opposite to that for Mn^{4+} ions. Since H_{int} originates mainly from the core-polarization, H_{int} is negative and the direction of the internal fields at nuclear sites is opposite to that of the Mn spin moment. Thus one can conclude that spin direction of Mn^{4+} ions is antiparallel to the external field, while that of Mn^{3+} ions is parallel to the external field, corresponding to the standard spin structure of magnetic core of Mn12 cluster as postulated from results of magnetization. The study of the field dependence when the field is applied perpendicular to the easy axis yields results similar to the ones illustrated above for Fe8.

c) Cr7Cd

Another example of determination by NMR of the local spin density is a recent study of the heterometallic antiferromagnetic (AF) ring Cr7Cd. The homometallic AF ring Cr8 is constituted of eight equivalent Cr^{3+} ions each carrying a spin s=3/2. Due to a strong antiferromagnetic exchange interaction J the ground state of Cr8 has a total spin S=0 (singlet non magnetic state). The local spin moment in the ground state is also zero i.e. $\langle s \rangle = 0$ thus yielding no hyperfine field at the ^{53}Cr site. When a diamagnetic Cd ion

is replaced for the magnetic Cr^{3+} ion,the symmetry is broken and the ground
state acquires a magnetic total spin value of S=3/2. By looking at the ^{53}Cr
NMR one can determine the staggered local spin moment around the heterometallic ring Cr7Cd. Since the ^{53}Cr nucleus has very low sensitivity and
low isotopic abundance the NMR signal could be observed only at very low
temperature and high magnetic field. Four different lines were detected two
of which overlapping into a broad unresolved line. The field dependence of
the three signals is shown in Fig.7. From the formula $\nu = \gamma[H + g \langle s \rangle A]$ one

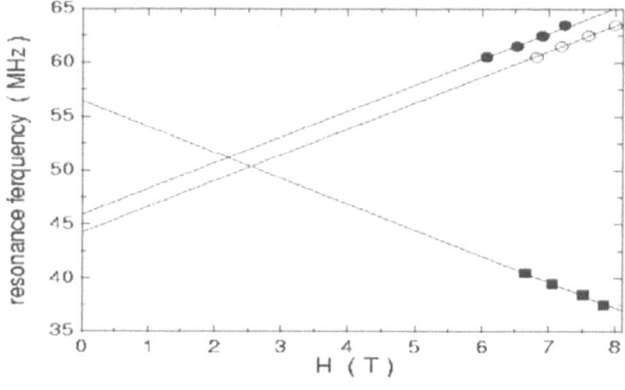

Fig. 7.

can derive the local spin moment $\langle s \rangle_i = \nu_i / g A$. In this particular case the
hyperfine constant A, which can be assumed to be the same for all sites, can
be determined independently from the condition that the sum of all local moments must equal the total spin S=3/2.

The sketch of the staggered local moment determined by NMR in Cr7Cd is
shown in Fig.8. The experimental values are found in excellent agreement with
the local moment distribution calculated from a model magnetic Hamiltonian
of the ground state which includes crystal field anisotropic terms. It should
be pointed out that what is measured here is the expectation value of the
spin component along the external field which is applied along the anisotropy
axis z. The applied field is sufficiently strong to align the total moment S=3/2
of the molecule along the field but is still smaller than the critical field for
level crossing to the S=5/2 excited state. It should be noted that the local
spin moment is strongly reduced from the single Cr^{3+} spin s=1.5 and is not
uniform along the ring. NMR is a unique tool for this kind of determination.

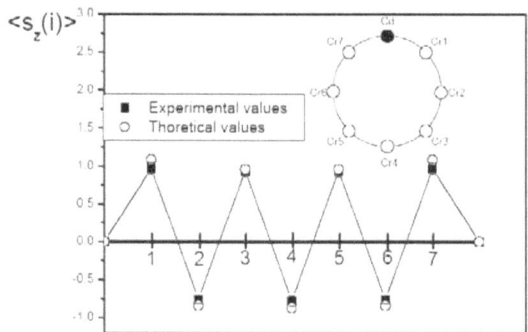

Fig. 8.

2 NMR and relaxation at high temperature (kT≫J). The spin dynamics in "zero"dimensions

Most of the magnetic molecular clusters investigated are characterized by exchange constants J which are well below the room temperature energy value $k_B T$. Exceptions to this is the Cu8 ring and to a certain extent also Mn12 and Fe8 clusters. If $k_B T \gg J$, the magnetic moments in the cluster are weakly correlated and the system behaves like a paramagnet at high temperature. Since the intermolecular magnetic interaction is negligibly small each molecular nanomagnet behaves as a paramagnet with a finite and small number of spins. Thus the spin dynamics is restricted to energy exchanges among a small number of spins and the situation can be defined as spin dynamics in "zero"dimensions. It is well known that the spin dynamics in paramagnets with limited dimensionality presents peculiar features absent in three dimensions. In fact as a consequence of the conservation of the total spin S^2 and Sz^2 for an isotropic Heisenberg hamiltonian J $\mathbf{S}_i \cdot \mathbf{S}_j$ the spin-spin pair and autocorrelation functions (CF) decay slowly at long times. For example in the one dimensional case an hydrodynamic spin diffusion approximation can be used at long times leading to a square root time decay of the CF. In general an approximate expression for the correlation function over all time intervals can obtained for an infinite Heisenberg classical chain at high temperature by matching the short time expansion to the long time diffusive behavior due to the conservation of the total spin and of its component in the direction of the applied field. For temperatures T $\gg$ J/k_B the conservation property can be incorporated for spins in a finite size nanomagnet by means of a discretized diffusion equation to which cyclic boundary conditions are applied. For this model it is found that the auto-correlation function (CF) decays rapidly at short times until it reaches a constant value which depends on the number of spins in the cluster. The plateau in the CF is reached after a time of the order of $10 \omega_D{}^{-1}$ where ω_D is the exchange frequency given at the simplest level of approximation by:

$$\omega_D = (2\pi J/h)[S(S+1)]^{1/2} \tag{2}$$

with J the exchange constant between nearest neighbor spins S. The same result is obtained for the CF by using a one dimensional hopping model on a closed loop or by calculating the spin correlation function with a mode-coupling approach. The levelling off of the time dependence of the CF at a value approximately given by 1/N with N the number of spins in the cluster is the result of the conservation of the total spin component for an isotropic spin-spin interaction. The time dependence of the CF in low dimensions can be contrasted with the one in three dimensional paramagnets where a short time expansion leads to a Gaussian decay of the CF. The situation is summarized as follows:

$\langle s_\pm(t)\, s_\pm(0)\rangle = 2\langle s_z(t)s_z(0)\rangle = 2/3s(s+1)\exp(-\omega_D{}^2 t^2)$ for 3D paramagnets
$\langle s_\pm(t)s_\pm(0)\rangle = 2\langle s_z (t)s_z(0)\rangle \propto t^{-1/2}$ for 1D paramagnetic chains
$\langle s_\pm(t)s_\pm(0)\rangle = 2\langle s_z(t)s_z(0)\rangle \propto 1/N$ for N spins nanomagnet

whereby the cases of restricted dimensionality hold for both auto and pair CF's in the long time limit.

NMR spin-lattice relaxation measurements are an ideal tool to probe the long time decay of the electronic spin correlation function since $1/T_1$ probes the spectral density of the fluctuations at low frequency i.e. the long time behavior of the CF. The relationship between the relaxation rate and the spectral density of the spin fluctuations can be obtained by a time dependent perturbation approach within the so called "weak collision" approximation. This approximation requires that the time dependent perturbation be small with respect to the Zeeman Hamiltonian so that many fluctuation events are needed to generate a nuclear spin relaxation transition. Alternatively one can say that the fluctuation time must be much shorter than the nuclear relaxation time: $\tau \ll T_1$.

One can start from the general formula for the nuclear relaxation transition probability:

$$1/T_1 = 2W = \gamma_N^2/2 \int \langle \delta H_\pm(t) \delta H_\pm(0)\rangle exp(-i\omega t) dt \tag{3}$$

The correlation function (CF) refers to the fluctuations of the local hyperfine field transverse to the quantization axis and calculated at the nuclear site.

In molecular nanomagnets the hyperfine interaction is the nuclear-electron dipolar interaction and/or the superexchange transferred contact interaction. If the hyperfine field in Eq.3 is expressed in terms of the electronic spin components and of the hyperfine coupling constants one obtains:

$$T_1^{-1} = 2(\gamma_N \gamma_e \hbar)^2 s(s+1) \sum_{ij} \{\alpha_{ij} \int_{-\infty}^{+\infty} \langle s_i^\pm(t)s_i^\pm(0)\rangle e^{i(\omega_N \pm \omega_e)t} dt +$$

$$\beta_{ij} \int_{-\infty}^{+\infty} \langle s_i^z(t) s_i^z(0)\rangle e^{i\omega_N t} dt \} \qquad (4)$$

and by introducing the spectral density functions:

$$T_1^{-1} = 2(\gamma_N \gamma_e \hbar)^2 s(s+1)(\Sigma_{ij}\alpha_{ij}J_{\pm}^{ij}(\omega_e) + \Sigma_{ij}\beta_{ij}J_z^{ij}(\omega_L)) \qquad (5)$$

where i,j number the electronic spins, ω_e and ω_N are the Larmor frequencies of the electron and of the nucleus respectively, α_{ij} and β_{ij} are geometrical factors and $J^{ij}{}_{\pm,z}$ are the transverse and longitudinal spectral densities of the spin fluctuations. It is noted that in a paramagnetic system at high T, $s_i(t)$ is the fluctuating electronic spin which coincides with the deviation $\delta s_i(t)$ from thermal equilibrium. Also we have factored out the Larmor precession of the electronic spin i.e. $\langle s^{\pm\prime}(t)\ s^{\pm\prime}(0)\rangle = \langle s^{\pm}(t)\ s^{\pm}(0)\ \rangle \exp(-i\omega_e t)$, an approximation valid only for an Heisenberg paramagnet at high T.

In a 3D paramagnet the spectral density is given by a Gaussian (or Lorenzian) and one expects no field dependence of the relaxation rate for $\omega_N, \omega_e \ll \omega_D$. This is shown in Fig.9 where $\omega_D = \omega_{ex}$. On the other hand

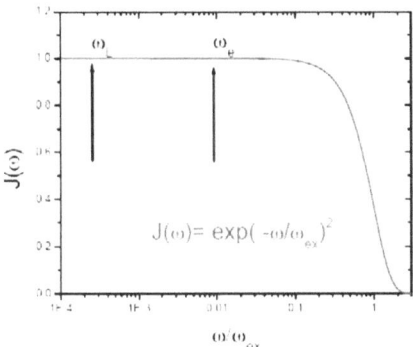

Fig. 9.

for 1D systems and for molecular nanomagnets the spectral density obtained from the Fourier transform of the CF in Eq.4 has a divergent behavior at low frequency and hence one expects a strong frequency (field) dependence of T_1^{-1}. In real low dimensional systems the CF at long times will in fact have a cut-off due to anisotropic terms in the Hamiltonian which do not commute with the Heisenberg Hamiltonian and thus allow for exchange of energy with the "lattice".

A sketch of the time decay of the CF and of the corresponding spectral density expected in finite size molecular clusters is shown in Fig.10. The initial fast decay is characterized by the constant $\Gamma_D^{-1} \approx \omega_D \equiv \omega_{ex}$ which depends on the intramolecular exchange interaction J (see Eq.1). The decay at long time of the CF has a cut-off at a time Γ_A^{-1} due to the anisotropic terms in the spin hamiltonian. As seen in the sketch of the spectral density of the spin fluctuations, if $\omega_e = \gamma_e H$ is of the order of the cut-off frequency Γ_A one expects a field dependence of the nuclear relaxation rate from the first term in Eq.5. In the following we will discus the magnetic field dependence of the nuclear relaxation rate at room temperature in terms of a simplified model which incorporates the theoretical understanding of the spin dynamics in clusters as described above. On the basis of the time dependence of the CF discussed

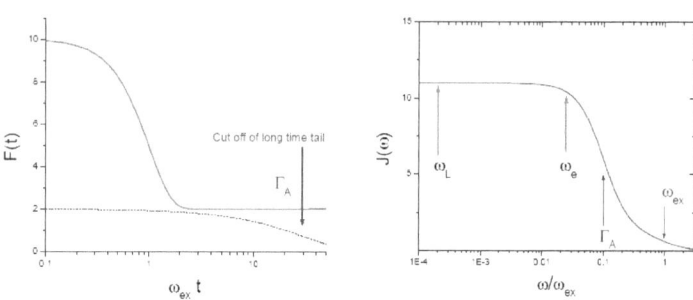

Fig. 10.

and sketched above, we model the spectral function in Eq.5 as the sum of two components:

$$\Phi^{\pm}(\omega) = \Phi^z(\omega) = \Phi'(\omega) + \Gamma_A/(\omega^2 + \Gamma_A^2) = \Gamma_D/(\omega^2 + \Gamma_D^2) + \Gamma_A/(\omega^2 + \Gamma_A^2) \quad (6)$$

where we assume the same CF for the decay of the transverse ($\pm$) and longitudinal (z) components of the spins. The first term in Eq.6 represents the Fourier transform (FT) of the initial fast decay of the CF while the second term represents the FT of the decay at long time of the CF due to anisotropic terms in the spin Hamiltonian and we model this second part with a Lorenzian function of width Γ_A.

From Eq.2 one can estimate that the exchange frequency ω_D is of the order of 10^{13} Hz for typical values of J/k_B (10-20K) and spin values S (1/2-5/2). The spectral function $\Phi'(\omega)$ in Eq.6 reaches a plateau and becomes almost frequency independent for $\omega < \omega_D/10$. For the magnetic field strength used in the experiment both ω_N and ω_e are smaller than $\omega_D/10$. Thus we will assume $\Phi'(\omega_n) = \Phi'(\omega_e) = \Gamma_D^{-1}$ in Eq.6 where the characteristic frequency Γ_D is of the same order of magnitude as $\omega_D/10$. Finally, by assuming $\omega_N \ll \Gamma_A$ in Eq.6, the relaxation rate can be rewritten as:

$$1/T_1 = K[1/2A^{\pm}(\Gamma_A/(\omega_e^2 + \Gamma_A^2) + 1/2A^{\pm}/\Gamma_D + A^z(1/\Gamma_D + 1/\Gamma_A)] \quad (7)$$

where the constants $A^{\pm}$ and A^z are averages over all protons in the molecule of the products of the hyperfine dipolar tensor components α_{ij} and β_{ij} respectively (see Eq.4). The constant K which has been factored out from the dipolar tensor coefficients is given by $K=(h\gamma_n\gamma_e)^2/(4\pi) = 1.94\cdot10^{-32}$ (sec^{-2}cm^6). The width Γ_A of the narrow component in the spectral function represents the frequency which characterizes the exponential time decay of the spin CF in the cluster due to anisotropic terms in the spin Hamiltonian.

The field dependence of $1/T_1$ was measured at room temperature in a large number of molecular nanomagnets. The results are shown in Fig.11. The experimental data were fitted by using an expression of the form of Eq.7,

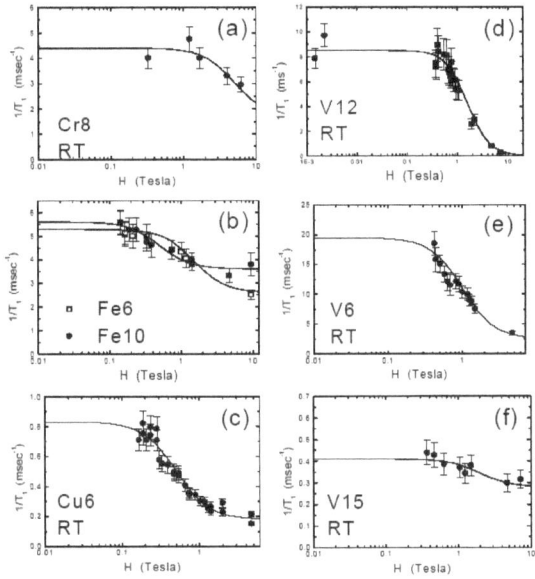

Fig. 11.

i.e.:

$$1/T_1 = A/(1 + (H/B)^2) + C(msec^{-1}) \quad (8)$$

where the magnetic field H is expressed in Tesla and $B = \Gamma_A / \gamma_e$ (Tesla). The fitting parameters for the different rings and clusters are summarized in Table 1.

The most significant parameter in the Table is B which measures the cut off frequency Γ_A of the electronic spin-spin correlation function. Except for

Table 1.

single molecule magnet	A $(msec^{-1})$	B (Tesla)	C $(msec^{-1})$
Cr8 (AFM ring- s=3/2)	2.7	5	1.7
Fe6 (AFM ring-s=5/2)	2.7	1.5	2.6
Fe10 (AFM ring-s=5/2)	2	0.5	3.6
Cu6 (FM ring-s=1/2)	0.65	0.5	0.18
V12 (AFM square-s=1/2)	8.5	1.6	≈ 0
V6 (AFM triangle-s=1/2)	17	1	2.5
V15 (AFM ring-s=1/2)	0.13	2	0.28

the Cr8 case (complete formula: $[Cr_8F_8Piv_{16}]$, Hpiv=pivalic acid) B is around 1 Tesla corresponding to $\Gamma_A \approx 10^{11}$ rad sec^{-1} or $h\Gamma_A/k_B \approx 1$ K.The cut-off effect is provided, in principle, by any magnetic interaction which does not conserve the total spin components. In practice, such small terms stem from a variety of mechanisms including intracluster dipolar and anisotropic exchange interaction, single ion anisotropies, inter-ring dipolar or exchange interactions etc... A detailed calculation for Fe6 based on intra-ring dipolar interaction yielded $\Gamma_{A=}$ 1.5 10^{11} sec^{-1}. A similar estimate for Cu6 based on known anisotropic nearest neighbor (exchange and dipolar) contributions to nearest neighbor interactions yielded 1.4×10^{11} sec^{-1}. Both these results can account very well for the experimental findings in the Table. From the comparison of Eqs.7 and 8 one has A= K $A^{\pm}$ $/2\Gamma_A$ and C$\approx$K A^z $/\Gamma_A$ (since $\Gamma_D \gg \Gamma_A$). Thus the order of magnitude of the hyperfine constants is $A^{\pm} \approx A^z \approx 1\div10$ 10^{46} cm^{-6}. Since $A^{\pm}$, A^z are the product of two dipolar interaction tensor components they are of order of r^{-6} where r is the distance between a ^{1}H nucleus and a transition metal local moment. For most of the rings the value of the hyperfine constants is consistent with a purely nuclear-electron dipolar interaction.

For V12 (complete formula: $(NHEt)_3[V^{IV}_8V^V_4As_8O_{40}(H_2O)]\cdot H_2O$) and V6 (complete formula of one variant: $Na_6[H_4(V_3L)_2P_4O_4]\cdot18H_2O$), the $A^{\pm}(\gg A^z)$ hyperfine constant is one order of magnitude higher indicating the presence of an additional contribution probably due to a contact interaction due to the admixing of the hydrogen s wave function with the d wave function of the Vanadium ions.

An alternative way to explain the anomalous values (A$\gg$C) for V6 and particularly for V12 in the Table is to go back to Eq.8 and assume that B = Γ_A/γ_N instead of B = Γ_A/γ_e. This implies that the cut-off frequency Γ_A is much less than in other clusters namely of order of the nuclear Larmor frequency in Eq.7. In this case the value of the constant C in Eq.8 is close to zero in agreement with the experiments as can be seen easily by modifying in the appropriate way the approximate Eq.7. It is, however, difficult to justify such small value for the cut-off frequency in V12.

By concluding, the experimental results seem to confirm very well the theoretical prediction for the dependence at long times of the CF in molecular nanomagnets which are a good experimental realization of spin dynamics in "zero"dimensions.

3 NMR and relaxation at intermediate temperatures (kT $\simeq$ J). The spin dynamics in a collective total spin S quantum state

As the temperature is lowered and it becomes comparable to the magnetic exchange interaction J strong correlations in the fluctuations of the magnetic moments of the molecule start building up. The situation is analogous to macroscopic three dimensional magnetic systems when the temperature approaches the critical temperature for the transition to long range magnetic order. In molecular magnets, as a result of the finite size of the system the low lying magnetic states are well separated among themselves. Therefore the correlation of the magnetic moments at low temperature has to be viewed as the result of the progressive population by the magnetic molecule of the collective low lying quantum total spin states without any phase transition. In this intermediate temperature range the nuclear spin lattice relaxation rate has to be described with a formalism different from the one utilized at high temperature when the system is totally uncorrelated. The starting expression is still given by Eq.4. In presence of strong spin-spin correlation $(k_B T \approx J)$ one has $\langle S_i S_{i+n} \rangle \neq 0$ which implies the presence of static short range order (SRO) in the magnetic system and $\lim_{t \longrightarrow \infty} \langle S_i(t) S_{i+n}(0) \rangle \neq \langle S_i(t) S_i(0) \rangle$ which implies dynamic spin correlation. Thus the semiclassical approach in terms of the time dependence of the autocorrelation function of the local spin operator is no longer practical and one should resort either to a quantum description of the CF or to a description in terms of collective spin variables. If the collective quantum states are well defined and underdamped the CF can be expressed in terms of the matrix elements of the local spin quantum operators:

$\langle S_i(t) S_j(0) \rangle = Tr \left\{ exp(-\beta H) exp(iHt/h) S_j exp(-iHt/h) S_i \right\} / Tr\{exp(-\beta t)\}$

leading to

$$\frac{1}{T_1} = \frac{\sum_{i,j} \sum_{n,m} exp(-\beta E_n)[A\langle n \mid j \mid m \rangle \langle m \mid S_i \mid n \rangle \delta(E_n - E_m - \omega_N)]}{\sum_n exp(-\beta E_n)} \quad (9)$$

The energy conservation condition requires that the magnetic states n,m are very close in energy for a direct relaxation transition probability to be different from zero. This situation is found e.g. in ordered 3D magnetic systems where the excited states are magnons. As we will argue below, in molecular nanomagnets the energy levels are well separated in energy so that no direct process is possible unless a finite broadening of the levels is introduced.

An alternative approach in presence of SRO and strong correlation among interacting electron spins, is to describe the spin dynamics of magnetic systems in terms of collective variables in q-space obtained by Fourier transforming the local spin variables in real space. One obtains:

$$\frac{1}{T_1} = \frac{(\hbar\gamma_n\gamma_e)^2}{4\pi} \int dt cos(\omega_N t) \int dq(\frac{1}{4}A^{\pm}(q)\langle S_q^{\pm}(t)S_{-q}^{\pm}(t)\rangle +$$
$$A_z(q)\langle S_q^z(t)S_{-q}^z(t)\rangle) \qquad (10)$$

Or by using the fluctuation dissipation theorem: $\int\langle S_q(t) S_{-q}(t)\rangle cos(\omega t)$ dt = 2 kT/ Im χ (q, ω)

$$1/T_1 = (\hbar\gamma_n\gamma_e)^2/(4\pi g^2\mu_B^2)k_B T[1/4\sum_q A^{\pm}(q)\chi^{\pm}(q)f_q^{\pm}(\omega_e) +$$
$$\sum_q A^z(q)\chi^z(q)f_q^z(\omega_n)] \qquad (11)$$

where γ_n and γ_e are the gyromagnetic ratios of the nucleus and of the free electron respectively, g is the Lande's factor, μ_B is the Bohr magneton, k_B is the Boltzmann constant. The coefficients $A^{\pm}$ (q) and A^z(q) are the Fourier transforms of the spherical components of the product of two dipole-interaction tensors describing the hyperfine coupling of a given proton to the magnetic moments whereby the symbols $\pm$ and z refer to the components of the electron spins transverse and longitudinal with respect to the quantization direction which is here the external magnetic field. The collective q-dependent spin correlation function is written as the product of the static response function times a normalized relaxation function $f_q^{\pm,z}$ (ω).

At high temperature ($k_B T\gg J$) one can neglect in Eq.11 the q-dependence of the generalized susceptibility χ^{α}(q) and of the spectral density function $f_q^{\alpha}(\omega)$. If one assumes an isotropic response function 1/2 $\chi^{\pm}$(q) = χ^z(q) = χ(q=0) and one takes a q-independent average value for the dipolar hyperfine interaction of the protons with the local moment of the electronic spins: $A^{\pm}$(q) = $A^{\perp}$; A^z(q) = A^z in units of cm^{-6}, then Eq.11 reduces in this high temperature limit to: $1/T_1 = (h\gamma_n\gamma_e)^2/(4\pi g^2\mu_B^2)$ k_B T χ(q=0)[1/2 $A^{\pm}\Phi^{\pm}(\omega_e)+A^z$ $\Phi^z(\omega_n)$]. This is indeed the expression for the nuclear relaxation rate which was utilized to analyze the high temperature data in molecular nanomagnets in the previous lecture.

By decreasing the temperature at values such that $k_B T$ becomes comparable to J one expects that the nuclear spin lattice relaxation rate displays a characteristic temperature dependence related to the correlated spin dynamics. As shown in Fig.12, measurements of proton T_1 as a function of temperature in a number of antiferromagnetic molecular rings has shown a surprisingly large enhancement of the relaxation rate at low temperatures resulting in a field dependent peak of T_1^{-1} centered at a temperature of the order of the magnetic exchange constant J/k_B. The results are shown for three different rings. The systems investigated are: Cr8 (s=5/2, J$\sim$17.2K);

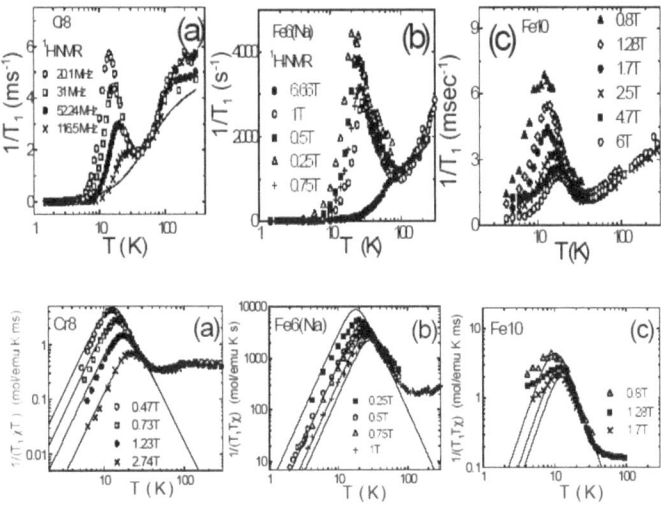

Fig. 12.

Fe6(Na) (s=5/2, J~ 28.2K); Fe10 (s=5/2, J~ 13.8K) (for the last two compounds the respective complete formulas are: [NaFe$_6$(μ_2-OMe)$_{12}$(dbm)$_6$]Cl and [Fe(OMe)$_2$(O$_2$CCH$_2$Cl)]$_{10}$). In all three samples the ground state is non magnetic with total spin S$_{Total}$=0 and the energies of the lowest lying exchange multiplets can be described at first approximation by Lande's interval rule E$_S$=2JS(S+1)/N where N is the number of spins in the ring. The main feature in the temperature dependence of T$_1^{-1}$ is the strong enhancement at low T and the presence of a maximum at a temperature T$_0$ for each of the samples investigated. For T<T$_0$, T$_1^{-1}$ decreases approaching at low T an exponential drop due to the "condensation" into the S$_{total}$=0 singlet ground state as discussed later on. It should be noted that the behavior of the relaxation rate is different than the behavior of the uniform magnetic susceptibility. The latter, when plotted as χT vs T, shows a continuous decrease with an exponential drop at very low temperature consistent with what expected for an AFM system with a singlet ground state.

The plot of $1/T_1\chi T$ also shown in Fig.12 indicates that the enhancement of the relaxation rate is of purely dynamical origin. As shown in the same figure the peak in the relaxation rate is depressed by the application of an external magnetic field and the position of the maximum moves at higher temperature on increasing the external field. The maximum of the relaxation rate divided by χT can be fitted by a simple expression of the type τ $(1+\omega_N^2\tau^2)$ where the correlation time τ is temperature dependent but field independent. The simple expression which fits the data is known as Bloembergen,Purcell and Pound (BPP) expression and it normally describes the nuclear relaxation driven by a random hyperfine field fluctuations which follows an exponential CF with a unique correlation time. A further remarkable finding is that the renormalized relaxation rate plotted as a function of the temperature normalized to the temperature of the peak T_0 is given by a universal function:

$$1/(T_1 T \chi)/(1/(T_1 T \chi))_{max} = 2t^n/(1 + t^{2n}) \tag{12}$$

with $t = T/T_0$ and a fitting exponent n close to 3 as shown in Fig.13: It is noted that the universal function Eq.12 is a direct consequence of the BPP -type fitting expression but only if the correlation time is assumed to have a power law T dependence i.e. $\tau^{-1} \propto T^n$. This BPP -type maximum in the

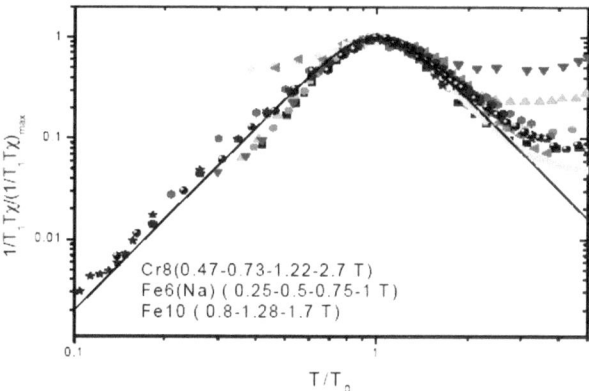

Fig. 13.

relaxation rate has been observed also in other molecular nanomagnets and it appears now to be a very general feature of finite size magnetic systems at a temperature comparable with the magnetic exchange interaction J. For example in the AF cluster Fe30 which has a S=0 ground state and J = 1.57 K, the peak is observed at much lower temperature compared with the AF rings as shown in Fig.14. Even in the more complex molecular ferrimagnets, the peak in the proton $1/T_1$ can be observed although the analysis of the data is complicated by the loss of NMR signal in the region of the peak for low applied magnetic fields. The data for different clusters together with the fitting curves are shown in Fig.15. Two different but probably related interpretations of the characteristic maximum in the proton $1/T_1$ in molecular nanomagnets have been given and will be briefly outlined in the following. In the first interpretation one starts from the quantum expression for the relaxation rate (Eq.9) which we rewrite below for the case of a dominant dipolar nuclear-electron hyperfine interaction.

$$T_1^{-1} = \frac{1}{2(\gamma_e \gamma_N \hbar)^2} \sum_{i,j=1}^{N} \frac{1}{(r_i r_j)^3} \frac{1}{\sum_\mu e^{-\beta E_\mu}} \sum_{\mu,\nu} e^{-\beta E_\mu} \pi [\delta(\omega_N + \omega_{\mu\nu}) + \delta(\omega_N - \omega_{\mu\nu})].$$

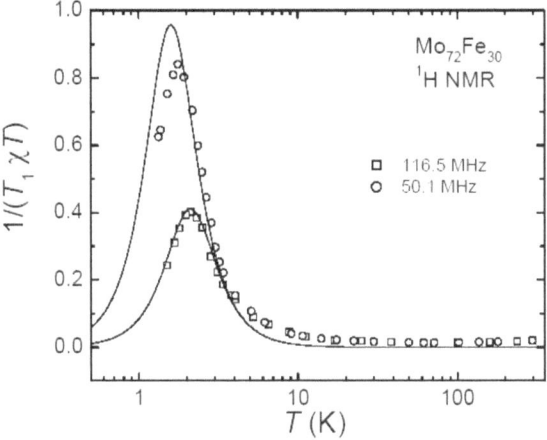

Fig. 14.

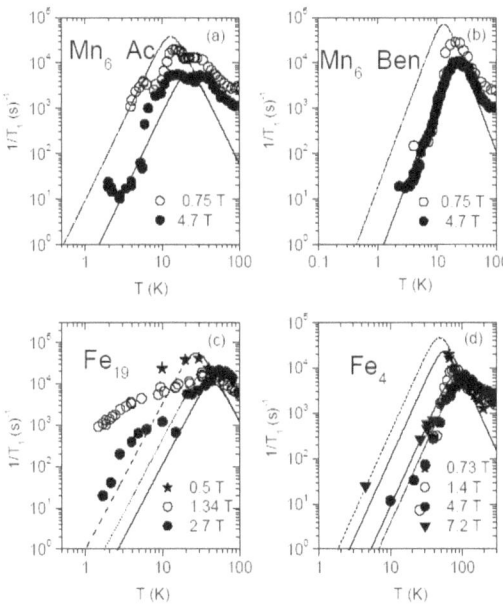

Fig. 15.

$$\cdot[\alpha_{ij}\langle\mu|s_i^z|\nu\rangle\langle\nu|s_j^z|\mu\rangle + \frac{1}{4}\beta_{ij}(\langle\mu|s_i^-|\nu\rangle\langle\nu|s_j^+|\mu\rangle + \langle\mu|s_i^+|\nu\rangle\langle\nu|s_j^-|\mu\rangle))], \quad (13)$$

If the energy levels are well separated in energy no direct relaxation transition is possible between different magnetic states since $\omega_N \ll \omega_{\mu\nu}$.

On the other hand the term which depends on the longitudinal spin component s_z can give rise to relaxation transitions if the energy levels have some finite broadening bigger than ω_N. Thus the BPP -type maximum observed in finite size magnetic systems can be simply explained as arising form the longitudinal fluctuations of the local spin via lifetime broadening of the magnetic levels. The expression which fits the data i.e. $1/T_1 = \chi T A \tau / (1 + \omega_N^2 \tau^2)$ can be obtained from the above expression if one identifies $\chi T = \exp(-\beta E_n)/\sum_n \exp(-\beta E_n)$ i.e. the population of the magnetic states, the hyperfine constant A with the average coupling coefficients times the matrix elements between the total spin states of the local electron spin operator and the Lorenzian $\tau/(1 + \omega_N^2 \tau^2)$ with the broadened delta function. The correlation frequency $\tau^{-1} \equiv \omega_c$ is thus identified here as an average lifetime broadening of the magnetic states. The characteristic frequency ω_c appears to have a power law T dependence and to scale with the gap Δ between ground state and first excited state i.e. $\tau^{-1} \equiv \omega_c = DT^3$ with $D \propto \Delta^{-2}$ or Δ^{-3}.

These findings are illustrated in the two figures 16 and 17. Both the power

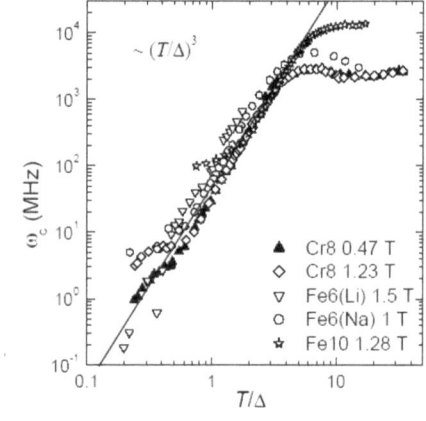

Fig. 16.

law temperature dependence and the dependence on the gap parameter for the lifetime broadening are in qualitative agreement with a lifetime broadening of the magnetic states dominated by spin-phonon interaction although no detailed theory has been yet generated.

The alternative explanation starts by considering the q-dependent expression for the relaxation rate Eq.10 or 11 valid in presence of SRO and strong correlation among interacting electron spins. Under the approximations $f_q^{\pm}(\omega_e) \ll f_q^z(\omega_N)$; $\chi^z(q) \approx 1/N \chi$ then Eq.11 takes the form of the BPP expression $1/T1 = \chi T A \tau / (1 + \omega_N^2 \tau^2)$ where now the correlation time τ can be interpreted as a dominant relaxation time for the macroscopic magnetization. This interpretation has been based on the results by P. Santini et al. of first principle calculations of the time autocorrelation of the molecular magnetization M(t)

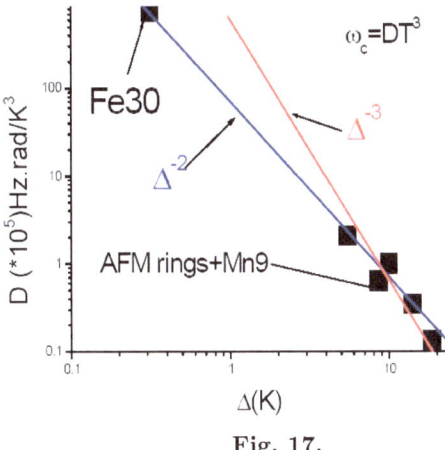

Fig. 17.

which corresponds to the q=0 component of the spin CF in Eq.10. It was found that the exponential decay of the fluctuations of M(t) is characterized by a single characteristic time $\tau(T, B)$ which can be identified with the characteristic correlation time entering the expression which fits the NMR relaxation rate. The calculation of the T and B dependence of τ yielded good agreement with the NMR data. We refer to the original literature for further details.

4 NMR at low temperatures ($k_B T \ll J$). Nuclear spin lattice relaxation in gapped molecular antiferromagnets and effects at level crossing

When the temperature is much lower than the exchange interaction among magnetic moments in the molecule the system is mostly in its collective quantum ground state characterized by a total spin S. We have to distinguish the two cases of singlet ground state S=0 and of high spin ground state S>0. In this section we will treat the first case while the second case will be discussed in the next section. In the case of singlet S=0 ground state, which pertains to AFM rings, the residual weak magnetism of the molecule at low temperature is due to the thermal population of the first excited state which is normally a triplet S=1 state.

4.1 Determination of the energy gap of AFM rings in the magnetic ground state

AFM rings such as Fe10, Fe6, Cr8 are characterized by a single nearest neighbor exchange interaction J which generates a singlet ground state of total spin S=0 separated by an energy gap Δ from the first excited triplet state S=1. From simple Lande's interval rule one has E(S)= 2J/N S(S+1). Thus in absence of crystal field anisotropy the gap is Δ= 4J/N where N is the number of magnetic moments in the ring. In presence of crystal field anisotropy with axial symmetry characterized by the parameter D the gap between S=0 and S=1, M=$\pm$1 is 4J/N +D/3 for the case of positive axial anisotropy.

Measurements of T_1^{-1} versus temperature in a T range where the molecule is mostly in the non magnetic S=0 ground state can allow the determination of the energy gap Δ which separates the singlet ground state from the first magnetic excited state. We give first the general theoretical argument and then we give a specific example by choosing the cluster V12 which can be assimilated to a square of V^{4+} magnetic ions. The electronic spin correlation function entering the expression of T_1^{-1} is defined as:

$$G_{ij}^{\alpha}(r,t) = \Sigma_n \Sigma_m \langle n|S_i^{\alpha}|m\rangle exp(-\beta E_n + iE_n t/\hbar - iE_m t/\hbar)\langle m|S_j^{\alpha}|n\rangle \quad (14)$$

where n,m number the eigenstates, E_n, E_m are the energy eigenvalues, $\beta=1/k_B T$, $S_{i(j)}^{\alpha}$ are the spin operators of the ith (j^{th}) spin and α=x,y,z. For a finite system the energy difference between eigenstates is very large compared to the Larmor Zeeman energy. Therefore for a direct process only the matrix elements with n=m in Eq.14 need to be considered and a broadening of the energy levels has to be introduced in order to fulfill energy conservation (i.e., in order to have some spectral density of the fluctuations at ω_e and ω_L). It should be noted that an alternative approach is to describe the nuclear relaxation in terms of a Raman process. Even in this case one needs to have a broadening of the levels or a spin wave band. We have not explored this possibility since the direct relaxation process appears to be able to explain the experimental data.

As a consequence of the presence of the Boltzman factors in Eq.13 the NSLR at very low temperature will be simply proportional to the population of the excited states. For temperatures less than the energy gap Δ one has approximately:

$$T_1^{-1} = Aexp(-\Delta/k_B T)/(1 + 3exp(-\Delta/k_B T)) \quad (15)$$

where A is a fitting constant which contains the hyperfine coupling constants and the matrix elements in Eq.13. The gap Δ depends on the applied magnetic field as a result of the Zeeman splitting of the excited triplet state of the AFM rings. The validity of the simple prediction of Eq.15 was tested in several AFM rings. Here we show the results in V12 which is a magnetic cluster which contains three squares of V^{4+} ions. Two of the squares are magnetically mute since they are in a S=0 singlet ground state even at high T due to the huge magnetic exchange coupling J. The middle square instead has a J coupling of only 17.6 K and thus can be treated as a spin tetramer with Hamiltonian H = J (s_1+s_3). (s_2+s_4) + gμ_B **B**.S_{tot}.

The resulting low lying energy levels are shown in Fig.18 below: Part (a) is from inelastic neutron scattering and part (b) is from the solution of the isotropic Heisenberg Hamiltonian written above. For T≪J the nuclear spin

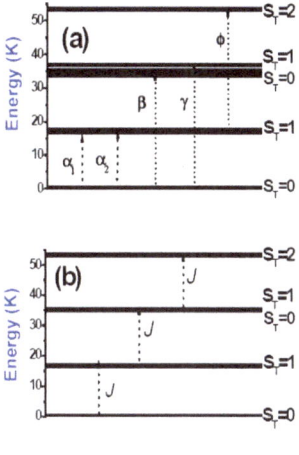

Fig. 18.

lattice relaxation rate (NSLR) is proportional to an exponential function of the gap between the ground and first excited state as predicted by Eq.15. The experimental measurements of proton $1/T_1$ at low temperature display an exponential decrease in agreement with Eq.15 as shown in the Fig.19. From

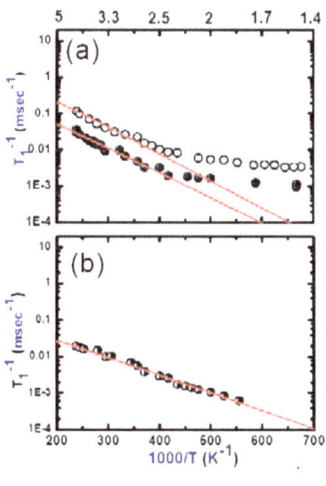

Fig. 19.

the slope of the semi-log plot one can derive the gap between the singlet ground state and the first excited state which turns out in good agreement with the theory and the INS data shown in the above figure. By applying an external magnetic field in the direction of the anisotropy axis the gap is reduced.

As long as the applied field is small enough so that one is far from the level crossing conditions, the reduction of the gap Δ between the ground state $S_{tot}=0$ and first excited state $S_{tot}=1$, Ms $=+1$ is expected to be linear in the applied field as the result of the Zeeman splitting of the excited state i.e. $\Delta(H) = \Delta_0 - g\mu_B B$. This has been confirmed in V12 from proton relaxation measurements as shown in Fig.20.

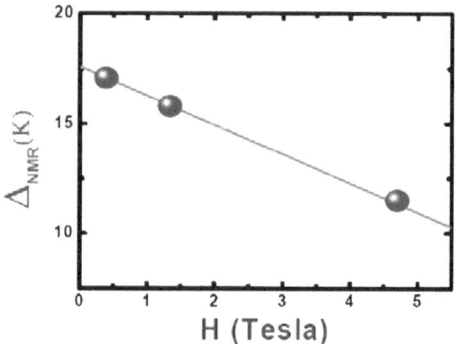

Fig. 20.

4.2 Level crossing in AFM rings

In a AFM ring of the kind discussed above with a S=0 singlet ground state and a S=1 triplet excited state, an external magnetic field H removes the residual Kramers degeneracy of the triplet state and induces multiple level-crossings (LCs) at specific magnetic field values H_{ci}, whereupon the ground state of the molecule changes from S = 0 to S = 1 (H_{c1}), from S = 1 to S = 2 (H_{c2}), and so on. Because of magnetic anisotropy, the values of H_{ci} depend on the angle between H and the molecular axis z. When the energy separation between the ground state and the first excited state decreases in proximity of the level crossing condition energy conserving transitions involving a nuclear flip accompanied by a magnetic transition of the molecule can become possible. In other words matrix elements of the local transverse spin operators between different energy states are no longer zero in Eq.9. This equation can be rewritten for dipolar hyperfine interactions as:

$$T_1^{-1} = \frac{1}{2N^2}\gamma_n^2\gamma_e^2\hbar^2[2\pi\delta(\omega_n)\frac{1}{1+e^{-\beta\hbar\omega_{eg}}}(S+1)^2]G_\alpha+$$

$$\pi[\alpha(\omega_n+\omega_{eg})+\delta(\omega_n-\omega_{eg})]\frac{(S*1)[(Ns+1)^2-(S+1)^2]}{2(2S+3)}G_\beta] \qquad (16)$$

S is the total spin value in the ground state i.e S=0 for the first crossing. G_α and G_β are the geometrical dipolar coupling coefficients and ω_{eg} is the frequency separation between the ground state and the excited state. The first term which arises from the longitudinal spin components (with respect to the field H) will be named quasi-elastic term since it does not involve any change of magnetic state of the molecular cluster. The second term is the inelastic term which is present only close to the level crossing condition. The quasi elastic term is proportional to the thermal population of the excited state while the inelastic term has a maximum when the energy separation of the two magnetic levels is equal to the nuclear Zeeman splitting i.e. $\omega_L \approx \omega_{eg}$.

The first example of the effect of level crossing on the nuclear relaxation was obtained by proton NMR in the AF ring Fe10. A picture of the Fe10 ring and of the field dependence of the magnetic energy levels is shown in Fig.21. Three level crossings are accessible for available magnetic fields. The

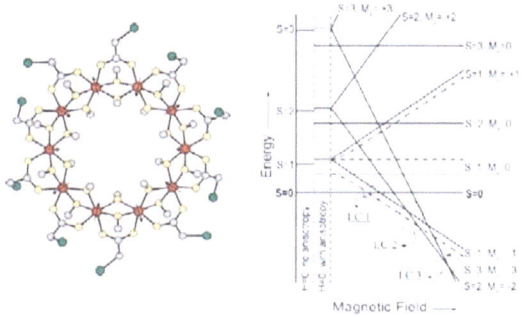

Fig. 21.

enhancement of the proton relaxation rate around the critical fields for level crossing in Fe10 is shown in Fig.22. One crucial issue in the study of LC in AFM rings is the structure of the magnetic levels at the critical field where the S=0 state becomes degenerate with the S=1, M_S=1 state (first level crossing) and similarly for the other LC's. If the magnetic Hamiltonian does not contain terms that admix the two degenerate levels one can have in principle a true LC. If on the other hand the Hamiltonian contains terms which strongly admix the levels one expects a gap at the critical field resulting in what

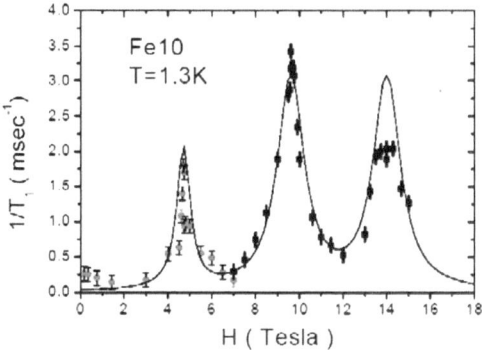

Fig. 22.

we call level anticrossing (LAC). The analysis of the width and shape of the peak of NSLR at the critical field can in principle give information about the structure of the magnetic levels near the critical field. To analyze the data one can use a phenomenological expression obtained from the theoretical expression discussed above where the delta functions are replaced by finite width Lorenzian functions.

$$T_1^{-1} = \frac{e^{-\frac{\Delta_1}{k_B T}}}{1 + e^{-\frac{\Delta_1}{k_B T}}} A \frac{\Gamma_1(T, H)}{\Gamma_1^2(T, H) + \omega_n^2} + B \frac{\Gamma_2(T, H)}{\Gamma_2^2(T, H) + (\omega_n - \Delta_1)^2} \qquad (17)$$

Both the first term (quasi elastic) and the second term (inelastic) display a maximum at the LC or LAC point. The maximum of the quasi elastic term reflects the rapid change of the magnetization at level crossing. In fact one has approximately $1/T_1 = A \, \delta \langle M^2 \rangle f(\omega_N) = A' \, T \, dM/dH \, f(\omega_N)$. The change of magnetization at level crossing is shown in Fig.23 for the case of the Fe6 AF ring. The maximum of the inelastic term is instead due to the overlap of the broadened energy levels of the ground state and excited state in the vicinity of the LC or LAC condition which makes the scattering of a nucleus by a magnetic excitation possible. The width of the nuclear relaxation peak at the level crossing can thus give an indication of the anticrossing gap. This is illustrated by the results in two AF rings: Cr8 and Fe6(Li). In the first case the crossing is almost a pure LC while in the second case there is a sizeable LAC gap. The data are shown in Fig.24. The expression utilized to describe the field dependence of the gap is $\Delta_1 = \{[g\mu_B(H_{c1} - H)]^2 + \Delta_{01}\}^{1/2}$ where the LAC gap is Δ_{01}. From the fit of the inelastic peak in Cr8 one obtains an anticrossing gap: $\Delta_{01} = 0.1$K much smaller than the one in Fe6(Li) i.e. $\Delta_{01} = 1$K.

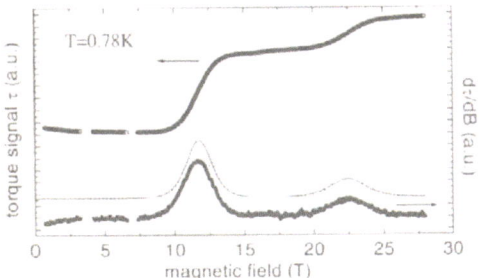

Fig. 23.

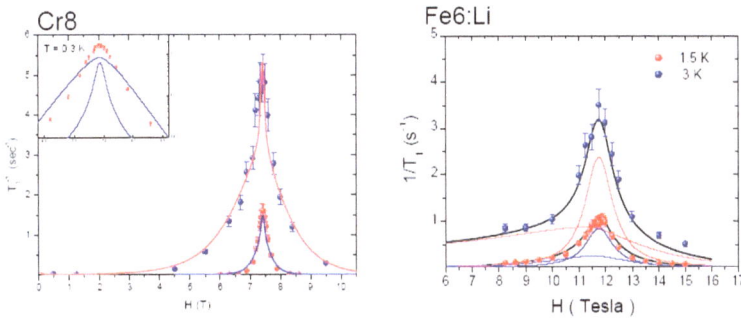

Fig. 24.

5 NMR relaxation in single molecule magnets (SMM) in the ground state. Detection of quantum tunneling of the magnetization by NMR

We consider now the case of molecular nanomagnets with a high spin ground state which at low temperature behaves like a nanomagnet with a spontaneous magnetization proportional to the value of the ground state spin S. If there is no anisotropy the molecule acts like a soft nanomagnet with a magnetization which can be aligned by an external magnetic field with no hysteresis in the magnetization cycle. This is the case of Cu6 FM ring. In presence of an anisotropy the molecule behaves as a hard nanomagnet with hysteresis in the magnetization cycle. However, since there is no long range order each molecule acts as a superparamagnetic particle. At temperatures much lower than the anisotropy barrier the relaxation of the magnetization can be dominated by quantum tunneling. NMR can give interesting information in this low temperature regime. The best examples of single molecule magnets with superparamagnetic behavior are Mn12 and Fe8. As a result of crystal field anisotropy the total S ground state is split into 2S+1 doubly degenerate magnetic levels which can be labelled with the magnetic quantum number M_S. The two lowest $M_S = \pm 10$ levels are separated by an energy barrier which depends on the anisotropy (for Mn12 is 60K while for Fe8 is about 30K).

The NSLR at low temperature i.e. when the SMM is in its total S ground state has interesting features and can give useful information on the collective spin dynamics of the molecule. There are two regimes which can be treated separately. The first regime is at a temperature comparable to the energy barrier kT≈ ΔE. In this case the nuclei sense essentially the fluctuations of the magnetization due to thermal transitions among different M_S states induced by spin phonon coupling. At temperatures much lower than the barrier, kT≪ ΔE the SMM is in the highest Ms state(lowest energy) and the dominant fluctuation is due to quantum tunneling of the magnetization. We are going to discuss below the two cases separately:

5.1 NMR relaxation due to thermal fluctuations of the magnetization.

Proton relaxation in the ground state of SMM is driven by fluctuations of the magnetization due to transitions of among different crystal field M_S substates induced by spin phonon coupling. The transitions among different magnetic levels generates a fluctuation in the local hyperfine field at the nuclear site which is responsible for the nuclear relaxation. The theoretical treatment of this process in terms either of local spin correlation functions or in terms of matrix elements of the local spin among the total magnetic eigenstates which we used before for high T and low T respectively is not very practical in this case. A model has been proposed which describes the nuclear relaxation in terms of the correlation function of the hyperfine field itself. One starts from the usual Moriya expression Eq.3. One then assumes an exponential form for the CF of the hyperfine field i.e. $\langle \delta H\pm(t)\ \delta H\pm(0)\rangle = h^2_{eff}$ exp (-t/τ_m). One makes the further assumption that h^2_{eff} is an average square value of the change of hyperfine field when the SMM undergoes a transition between different magnetic M_S states (corresponding classically to different orientations of the magnetization with respect to the anisotropy axis). The correlation time τ_m corresponds then to the lifetime of the corresponding m state. The NSLR can be written by summing the contributions of the different magnetic states:

$$1/T_1 = \sum_{m=+10,-10} h^2_{eff} exp(-E_m/k_B T)/Z \int exp(-t/\tau_m) exp(-i\omega t) dt =$$

$$A/Z \sum_{m=+10,-10} \tau_m \exp(-E_m/kT)/(1+\omega_N^2\tau_m^2) \quad (18)$$

where $A = \gamma_N^2 h_{eff}^2$. The lifetime τ_m for each individual m state is determined by the probability of a transition from m to m $\pm$ 1, $W_{m,m\pm1}$,
plus the probability for a transition with $\Delta m=\pm2$, $W_{m,m\pm2}$ i.e: $1/\tau_m =$
$W_{m\rightarrow m+1}+W_{m\rightarrow m-1}+W_{m\rightarrow m+2}+W_{m\rightarrow m-2}$. The transition probabilities are
due to spin-phonon interaction and can be expressed in terms of the energy
level differences as:

$$W_{m\rightarrow m\pm1} = W_{\pm1} = C's_{\pm1}\frac{(E_{m\pm1}-E_m)^3}{\exp[\beta(E_{m\pm1}-E_m)]-1}$$
$$W_{m\rightarrow m\pm2} = W_{\pm2} = 1.06C's_{\pm2}\frac{(E_{m\pm2}-E_m)^3}{\exp[\beta(E_{m\pm2}-E_m)]-1} \quad (19)$$

where $s_{\pm1} = (s\mp m)(s\pm m+1)(2m\pm1)^2$ and $s_{\pm2} = (s\mp m)(s\pm m+1)(s\mp m-1)(s\pm m+2)$ The spin-phonon parameter C' is given by C'= $D'^2/(12\pi\rho v^5 h^4)$
with ρ the mass density and v the sound velocity and D' a constant related
to the crystal field anisotropy. The energy levels E_m in the above equations
can be obtained from the Hamiltonian of the molecules expressed in terms of
the total spin S:

$$H = -DS_z^2 - BS_z^4 + E(S_x^2 - S_y^2) + g\mu_B S_z H \quad (20)$$

For Mn12 one has D=0.55 K, B=1.2x10^{-3} K, and E=0 while for Fe8 one has
D=0.27 K, B=0 and E= 0.046 K. Some representative experimental results

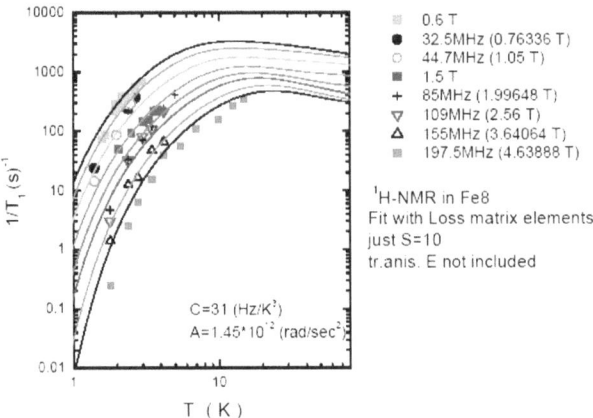

Fig. 25.

of proton NSLR are shown in the figures 25 and 26. For Fe8 we show the

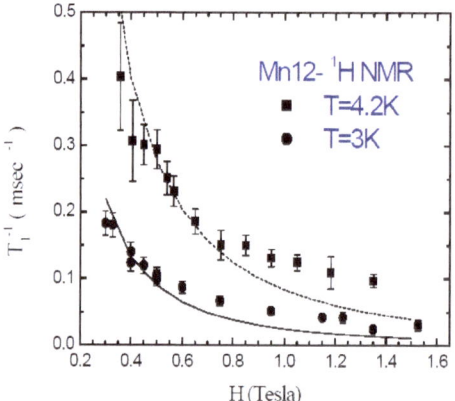

Fig. 26.

temperature dependence at different fields and for Mn12 the field dependence at different temperatures. The curves are best fit curves according to the theoretical expressions discussed above with A and C′ treated as adjustable parameters. From the fit of the proton relaxation in Fe8 one obtains the parameters: C′ = 31 Hz/K^3 and A= 1.02x10^{12} (rad/sec)^2. The fit of the proton relaxation data in Mn12 was obtained with a somewhat larger value of C′ and a smaller value of A=0.45x10^{12} (rad/sec)^2. The coupling constant A represents the average hyperfine interaction squared between protons and transition metal magnetic moments. The value found for Fe8 is larger than the value obtained in Mn12 indicating that in Fe8 the protons are subject to non negligible hyperfine interaction due to contact terms in addition to the dipolar interaction. From the knowledge of the spin-phonon coupling parameter C′ one can estimate the lifetime of the m sublevels. For a detailed discussion we refer to the original papers.

5.2 NMR relaxation due to incoherent quantum tunneling

At very low temperature, when the nanomagnet occupies mostly the magnetic ground state the thermal fluctuations of the magnetization become vanishingly small. In this temperature range the spin dynamics is dominated by quantum fluctuations. In this section we will concentrate on the issue of what information NMR can give about the phenomenon of quantum tunneling of the magnetization (QTM). Let us first briefly summarize the QTM phenomenon. The ground state of both Mn12 and Fe8 clusters is a high total spin ground state i.e. S=10. The S=10 ground state is split into eleven sublevels by a strong easy axis anisotropy. The remaining Kramers degeneracy is removed by an external magnetic field directed along the z easy axis. The energy levels for H // z are obtained from Hamiltonian Eq.20 as:

$$E_m = -Dm^2 - Bm^4 + g\mu_B Hm \qquad (21)$$

Assuming g=2 one has $g\mu_B H = 1.33H$ (K) for H in Tesla. For Mn12 one has D=0.55K, B=1.2x10^{-3} K; for Fe8 one has D= 0.27K, B=0 and in the total Hamiltonian H = -DS$_z$²-BS$_z$⁴+gBH·S, the rhombic term E(S$_x$²-S$_y$²), with E=0.046K, must be added. This term modifies the Eq.21 of the energy levels in a non-simple way. Below liquid helium temperature the clusters occupy mostly the $M_S = \pm 10$ states and the reorientation of the magnetization between these two states becomes extremely long (about one day for Mn12 at 2.4 K) due to the anisotropy barrier giving rise to a pronounced superparamagnetic behavior. When the relaxation rate of the magnetization is measured in response to a varying magnetic field H$_z$ along the easy axis peaks are observed which have been interpreted as a manifestation of resonant tunneling of the magnetization. The qualitative explanation is that the relaxation rate of the magnetization is maximum at zero field and at field values where the total spin states become pairwise degenerate again. The longitudinal field at which this occurs can be easily calculated from Eq.21 with the parameters given for Mn12 and Fe8 respectively. It is this degeneracy which increases the tunneling probability and thus shortens the relaxation time. The size of the effect depends on terms not shown in the Hamiltonian Eq.20, terms which couple the pairwise degenerate states. In particular a transverse magnetic field component can greatly enhance the tunneling splitting of the degenerate levels and thus the QTM. On the other hand the QTM is reduced by the smearing out of the energy levels of the spin states due to spin-phonon coupling, intermolecular interactions and/or hyperfine interactions with the nuclei. It should be stressed that the QTM which has been detected up to now in molecular nanomagnets is an incoherent tunneling probability. In other words the decoherence due to the coupling of the magnetic levels with the thermal bath (phonons and nuclear spins) is always sufficiently strong to overdamp the Rabi oscillations at the frequency given by the tunneling splitting. The incoherent tunneling probability can be calculated in this simple way. Let us consider first the quantum oscillations in time of the probability of occupation of one of the two degenerate $M_S = \pm 10$ states in presence of a small offset ΔE of the energies of the two quasi degenerate magnetic states:

$$|C_{-10}(t)|^2 = \Delta_T^2/(\Delta E^2 + \Delta_T^2) \, \text{sen}^2 \, (\Delta E^2 + \Delta_T^2)^{1/2} \, t$$

If the two quasi degenerate states have a finite broadening W which describes the decoherence effects, the incoherent tunneling probability can be defined as the average over the offset levels whose broadening is described by a Lorenzian function of width W:

$$\Gamma = 1/t \int |C_{-10}(t)|^2 \, W/ \, (W^2 + (E - \Delta E)^2) \, dE$$

the result for the incoherent tunneling probability is

$$\Gamma = \Delta_T^2 W/(\Delta E^2 + W^2) \qquad (22)$$

Quantum tunneling of the magnetization produces a fluctuation of the hyperfine field at the nuclear site and thus is expected to be a source of NSLR. The problem is to discriminate between the nuclear relaxation due to thermal fluctuations from the one due to QTM. We give in the following two representative cases where the contribution to NSLR coming from QTM can be sorted out without ambiguity.

5.3 Proton NSLR in Fe8 due to QTM in transverse applied field.

In zero applied field the tunneling splitting $\Delta_T$of the ground state is very small. On the other hand by applying a magnetic field perpendicular to the easy axis (transverse field), one can increase Δ_T of all levels while leaving the symmetry of the double well potential intact. An increase of the tunneling splitting corresponds to an increase of the tunneling frequency and thus of the incoherent tunneling probability (see Eq.22). For H>1T, in Fe8, the relaxation (fluctuation) of the magnetization driven by tunneling becomes so fast that it falls within the characteristic frequency domain (MHz) of a NMR experiment as shown in the Fig.27. Therefore when the magnetic field is ap-

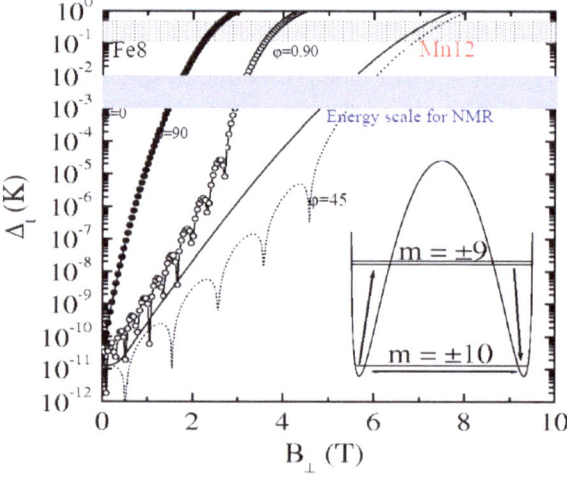

Fig. 27.

plied perpendicular to the main easy axis z (transverse field) a pronounced peak in the proton spin-lattice relaxation rate, $1/T_1$, of protons in a single crystal of Fe8 as a function of external magnetic field can be observed at 1.5K as shown in Fig.28. The effect is well explained by considering that by increasing the transverse field the incoherent tunneling probability which is proportional to the square of the tunneling splitting Δ_T becomes faster and

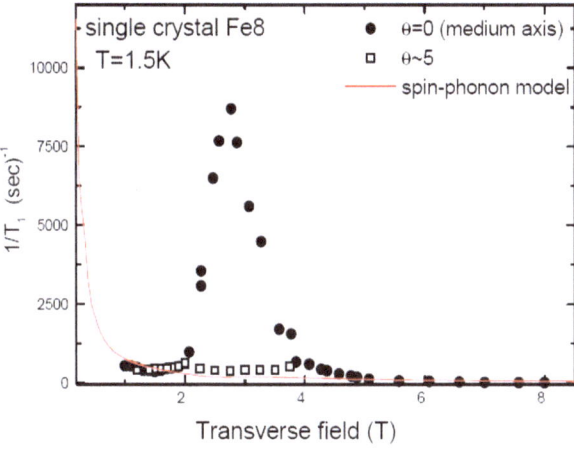

Fig. 28.

faster and at a certain value of the transverse field it can match the Larmor frequency. When the applied field goes through this condition the fluctuation rate of the magnetization is most effective in driving the nuclear relaxation and a maximum appears in $1/T_1$. The peak disappears when a parallel field component is introduced in addition to the transverse field, by tilting the single crystal about 5 degrees in yz plane. Since the parallel field component removes the degeneracy of the $\pm M_S$ magnetic states and consequently the possibility of tunneling it is clear that the peak of $1/T_1$ must be related to a contribution to the nuclear relaxation rate from the tunneling dynamics. It should be stressed that the theoretical description in this case follows the "weak collision"approximation i.e. a perturbation approach and one can use the formula's used previously to describe the relaxation due to thermal fluctuations. In particular one can use Eq.18 where the correlation frequency τ^{-1} is replaced with the tunneling probability Γ. In fact the reversal of the magnetization due to QTM is a small perturbation at the proton site in Fe8 since the hyperfine field is the weak dipolar coupling with the Fe^{3+} moments which is much smaller than the proton Zeeman energy in the high applied field. A detailed quantitative description can be found in the original literature.

5.4 ^{57}Fe NMR relaxation in zero field in Fe8 due to QTM: a case of "strong collision".

As we have shown in the first section the ^{57}Fe NMR signal can be detected in isotopically enriched Fe8 at low temperature in absence of external field.

The NSLR of anyone of the eight NMR lines of the ^{57}Fe spectrum (see Fig.4) is dominated at very low T by QTM. This can be inferred by the fact that the NSLR becomes temperature independent as shown in Fig.29. In this T independent region the "weak collision"approximation cannot be applied.

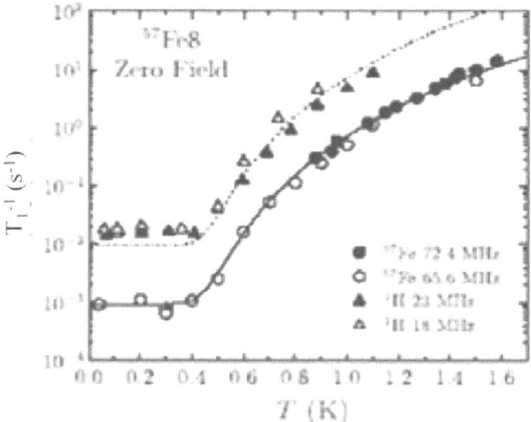

Fig. 29.

In fact the ^{57}Fe nuclei are quantized in the internal hyperfine field (of the order of 40T!!!). When a tunneling transition takes place the direction of the quantization axis changes by 180^0 degrees. Obviously this fluctuation cannot be treated as a perturbation. As a matter of fact a tunneling transition corresponds to interchanging the high energy nuclear Zeeeman level with the low energy one. This is equivalent to a relaxation transition for the nuclear spin system. A detailed treatment of this "strong collision" approximation can be found in the original literature. Here we limit ourselves to the qualitative argument that identifies (apart from a numerical constant of order of 1) the nuclear relaxation probability with the tunneling probability i.e. $1/T_1 \approx \Gamma$. To convince ourselves that the NSLR at the low T plateau is a direct measurement of the incoherent tunneling probability Γ one can refer to two experiments. In the first one a small longitudinal (along the anisotropy axis z) magnetic field was applied in a single crystal of Fe8. The longitudinal field introduces an offset in the energies of the two otherwise degenerate $M_S = \pm 10$ states and thus destroys the possibility of tunneling. As a consequence the NSLR decreases abruptly as shown in Fig.30. Since the experiment was performed at a temperature not sufficiently low a sizeable contribution due to thermal fluctuations remains at high magnetic fields when no tunneling is present. In the second experiment the ^{57}Fe NMR in a isotopically enriched single crystal of Fe8 with almost complete deuteration was measured The replacement of hydrogen nuclei with the lower gyromagnetic ratio deuterium nuclei should not affect the ^{57}Fe NSLR if this were due to thermal fluctuations of the magnetization. On the other hand the deuteration, by changing the hyperfine interactions with the Fe^{3+} moments can affect the broadening of the magnetic levels and thus the incoherent tunneling probability. Since the $1/T_1$ at the temperature plateau is a direct measure of the tunneling probability we

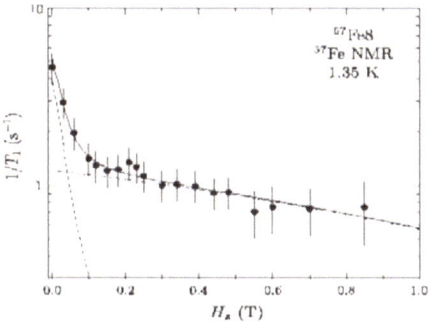

Fig. 30.

observe a dramatic change by almost one order of magnitude of the ^{57}Fe NSLR as shown in Fig.31 which corresponds to an equivalent change in the incoherent tunneling probability. Before closing this section we would like to recall

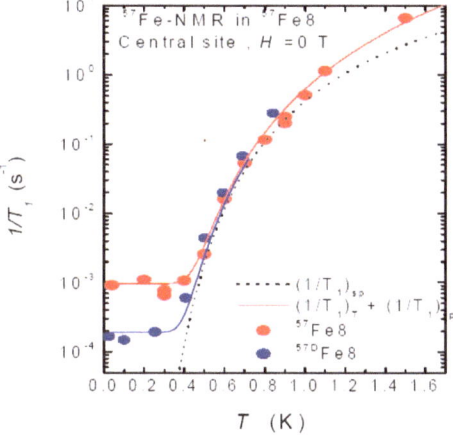

Fig. 31.

that one can measure the relaxation rate of the magnetization by monitoring the NMR signal intensity in off-equilibrium state. The idea is quite simple. We have seen above that in the NMR spectrum at low temperature in Mn12 and Fe8 the position of the resonance lines depend upon the internal field due to the magnetization of the molecule. Thus if an external field is also applied the position of the line depends on the vector sum of the external field plus the internal field the latter being directed along the magnetization of the molecule. If the direction of the external field is suddenly reversed (or the sample is flipped by $180°$) the position of the NMR line changes in the new off-equilibrium situation. The intensity of that particular line starts from zero and it grows back to the full intensity as the magnetization of the molecule relaxes back to equilibrium along the applied field. The method was first described for proton NMR in Mn12 by looking at the echo intensity at the Larmor frequency when the external field is turned on adiabatically at low temperature. A more straightforward implementation was later applied to the signal intensity of shifted proton lines in Mn12. The detailed results for proton NMR in Mn12 are discussed in the original literature. The method has been confirmed by measuring the ^{55}Mn NMR spectrum in Mn12 with results very similar to the ones obtained with proton NMR and it has been extended to very low temperature to observe the avalanche effect of the spin reversal in the magnetization recovery in Mn12. Also it has been applied to proton NMR in Fe8 at very low temperature to obtain information about Landau-Zener transition as the field is swept through a level crossing condition.

Conclusions

In these lectures we have shown how NMR can give information about the microscopic static and dynamic magnetic properties of molecular nanomagnets. Regarding static properties NMR spectra give access to the hyperfine interactions between nuclei and magnetic ions. Measurements in external fields and in zero field at low temperature allow the microscopic determination of the local spin structure of the ground state and its modifications as a function of temperature and field. Regarding the dynamic properties the NMR relaxation measurements (NSLR) give a clear picture of the evolution of electronic spin dynamics from the uncorrelated high temperature regime to the quantum spin dynamics in the low temperature ground state. In particular it is shown that at high temperature the spin dynamics is characterized by the persistence at long time of the decay of the electronic spin correlation function due to the restricted dimensionality of the nanomagnet. At intermediate temperature the spin dynamics is surprisingly dominated by a uniform collective fluctuation characterized by a single correlation time which increases fast as the temperature decreases. This result appears to be robust and universal for all kind of nanomagnets. At temperatures lower than the exchange interaction energy J/k the spin dynamics is dominated by the fluctuations of the total magnetization. In AFM rings and in clusters with a singlet $S=0$ total spin ground state the fluctuations are due to thermal transitions from the non magnetic ground state to the first magnetic excited state. At magnetic fields small compared to the energy gap between these two states the NMR relaxation measurements allow the determination of the energy gap and of its dependence on the applied magnetic field. For large applied fields one can observe a peak in the NSLR in correspondence to the critical field for which the lowest lying magnetic states cross. From the width of the peak one can obtain information about the nature of the crossing i.e. the degree of level repulsion. In molecular clusters with a magnetic high spin ground state one can distinguish in the NSLR a temperature region where thermal excitations among the crystal field sublevels of the high spin ground state are dominant (superparamagnetic region) and a very low temperature region where the spin dynamics is dominated by incoherent quantum tunneling. In this T region it is shown that, under favorable circumstances, the NSLR can measure directly the incoherent tunneling probability.

Acknowledgements: The work reviewed in these lectures is largely the result of research financed by FIRB National project and European project Quemolna and MagmaNet with many collaborators whom I would like to thank. In particular I would like to thank E.Micotti for help in editing the manuscript.

References

1. **Molecular nanomagnets**
   - D. Gatteschi and R. Sessoli, *Magnetism: Molecules to Materials III*, ed by J. S. Miller and M. Drillon (Wiley-VHC, Verlag GmbH, Weinheim 2002)
   - D. Gatteschi et al: Science **265**, 1055 (1994)
   - A. Muller et al: Chem. Phys. Chem. **2**, 517 (2001)
   - K. Wieghardt, K. Pohl, H. Jibril and G. Huttner: Angew. Chem. Int. Ed. Engl. **23**, 77 (1984)
   - T.Lis: Acta Cryst. **B36**, 2042 (1980)
   - D.Gatteschi, A.Caneschi, R.Sessoli and A.Cornia: Chem. Soc. Rev. **2**, 101 (1996); K.L. Taft, C.D. Delfs, G.C, Papaefthymoiu, S. Foner, D. Gatteschi and S.J. Lippard: J. Am. Chem. Soc. **116**, 823 (1994)
   - G.Cristou, D.Gatteschi, D.N.Hendrickson, and R.Sessoli: MRS Bull. **25**, 66 (2000)
   - O.Kahn, *Molecular Magnetism* (Wiley-VCH, Weinheim, Germany 1993)

2. **Review paper on NMR in molecular nanomagnets**
   - F.Borsa, A.Lascialfari and Y.Furukawa: NMR in magnetic molecular rings and clusters. In: *New developments in Magnetic Resonance*, (Springer & Verlag 2006 and cond-mat/0404378,0404379)

3. **NMR and relaxation in magnetic systems**
   - A. Abragam: *The principles of Nuclear Magnetism* (Clarendon Press, Oxford 1961)
   - T. Moriya: Prog. Theor. Phys. **16**, 23 (1956), ibidem **28**, 371 (1962)
   - F.Borsa and A.Rigamonti. In: *Magnetic Resonance of Phase Transitions*, (Academic Press 1979)

4. **NMR relaxation in one dimensional paramagnets**
   - F. Borsa and M. Mali: Phys. Rev. B **9**, 2215 (1974)
   - J. P. Boucher, M. Ahmed Bakheid, M. Nechschtein, G. Bonera, M. Villa, and F. Borsa: Phys. Rev. **13**, 4098 (1976)

5. **Theory of spin dynamics in molecular nanomagmets at high temperature**
   - A. Lascialfari, D. Gatteschi, A. Cornia, U. Balucani, M.G. Pini and A. Rettori: Phys. Rev. B **57**, 1115 (1998)
   - J. H. Luscombe and M. Luban: J. Phys. Condens. Matter **9**, 6913 (1997)
   - J. Tang, S. N. Dikshit, and J. R. Norris: J. Chem. Phys. **103**, 2873 (1995)
   - J.H. Luscombe, M. Luban and F. Borsa: J. Chem. Phys. **108**, 7266 (1998)

6. **NMR and relaxation in molecular nanomagnets at high temperature**
   - A. Lascialfari, Z.H. Jang, F. Borsa, D. Gatteschi and A. Cornia: J. Appl. Phys. **83**, 6946 (1998)
   - A.Lascialfari, D.Gatteschi, F.Borsa and A.Cornia: Phys. Rev. B **55**, 14341 (1997)
   - D.Procissi, B.J.Suh, J.K.Jung, P.Kogerler, R.Vincent and F.Borsa: J. of Applied Physics **93**, 7810 (2003)
   - A.Lascialfari, Z.H.Jang, F.Borsa, A.Cornia, D.Rovai, A.Caneschi and P.Carretta: Phys.Rev.B **61**, 6839 (2000)

7. **Static magnetic properties in molecular nanomagnets by NMR**
   - Y. Furukawa, K. Kumagai, A. Lascialfari, S. Aldrovandi, F. Borsa, R. Sessoli and D. Gatteschi: Phys. Rev. B **64**, 094439 (2001)
   - T. Goto, T. Koshiba, T. Kubo, and K. Awaga, Phys. Rev. B **67**, 104408 (2003)
   - T. Goto, T. Kubo, T. Koshiba, Y. Fuji, A. Oyamada, J. Arai, T. Takeda and K. Awaga: Physica B **284-288**, 1277 (2000)
   - Y. Furukawa, K. Watanabe, K. Kumagai, F. Borsa and D. Gatteschi: Phys. Rev. B **64**, 104401 (2001)
   - Y. Furukawa, S. Kawakami, K. Kumagai, S-H. Baek and F. Borsa: Phys. Rev. B **68**, 180405(R) (2003)
   - Y. Furukawa, K. Watanabe, K. Kumagai, F. Borsa and D. Gatteschi: Physica B **329-333**, 1146 (2003)
   - S.H.Baek, S.Kawakami, Y.Furukawa, B.J.Suh, F.Borsa, K.Kumagai, and A.Cornia: Journal of Magnetism and Magnetic Materials **272-276**, 771 (2004)
   - E.Micotti, Y.Furukawa, K.Kumagai, S.Caretta, A.Lascialfari, F.Borsa, G.A.Timco and R.E.P. Winpenny: Phys. Rev. Lett. **97**, 267204 (2006)
   - T.Kubo,T.Goto,T.Koshiba,K.Takeda and K.Agawa: Phys. Rev.B **65**, 224425 (2002)

8. **NMR Relaxation in molecular nanomagnets at intermediate temperatures**
   - A.Lascialfari, D.Gatteschi, F.Borsa and A.Cornia: Phys. Rev. B **55**, 14341 (1997)
   - J.K.Jung, D.Procissi, R.Vincent, B.J.Suh , F.Borsa, C.Schroder, M.Luban, P.Kogerler and A.Muller: J.of Applied Physics **91**, 7388 (2002)

- D.Procissi, B.J.Suh, J.K.Jung, P.Kogerler, R.Vincent and F.Borsa: J. of Applied Physics **93**, 7810 (2003)
- B.J.Suh, D.Procissi, K.J.Jung, S.Budko, W.S.Jeon, Y.J.Kim and D.Y.Jung: J. of Applied Physics **93**, 7098 (2003)
- S.H.Baek, M.Luban, A.Lascialfari, E.Micotti, Y.Furukawa, F.Borsa, J.van Slageren and A.Cornia: Phys. Rev. B **70**, 134434 (2004)
- M.Belesi, A.Lascialfari, D.Procissi, Z.H.Jang and F.Borsa: Phys.Rev.B **72**, 014440 (2005)
- P.Santini, S.Carretta, E.Liviotti, G.Amoretti, P.Carretta, M.Filibian, A.Lascialfari and E.Micotti: Phys. Rev. Lett. **94**, 077203 (2005)
- M.Corti,M.Filibian,P.Carretta,L.Zhao and L.K.Thompson: Phys. Rev B **72**, 064402 (2005)

9. **NMR at level crossing in AF rings**
   - M.H. Julien, Z.H. Jang, A. Lascialfari, F. Borsa, H. Horvatic, A. Caneschi and D. Gatteschi: Phys. Rev. Lett. **83**, 227 (1999)
   - M. Affronte, A. Cornia, A. Lascialfari, F. Borsa, D. Gatteschi, J. Hinderer, M. Horvatic, A.G.M. Jansen, and M.-H. Julien: Phys. Rev. Lett. **88**, 167201 (2002)
   - E.Micotti, A.Lascialfari, F.Borsa, M.-H.Julien, C.Berthier, M.Horvatic, J.Van Slageren and D.Gatteschi: Phys. Rev. B **72**, 020405(R) (2005)

10. **NMR and the energy gap of AFM rings in the magnetic ground state**
    - A. Lascialfari, F. Tabak, G. L. Abbati, F. Borsa, M. Corti and D. Gatteschi: Journal of Applied Physics **85**, 4539 (1999)
    - A. Lascialfari, Z.H.Jang, F.Borsa, A.Cornia, D.Rovai, A.Caneschi and P.Carretta: Phys.Rev.B **61**, 6839 (2000)
    - D.Procissi, A.Shastri, I.Rousochadsakis, M.Al Rifai, P.Kogerler, M.Luban, B.J.Suh and F.Borsa: Phys. Rev.B **69**, 094436 (2004)
    - S. Maegawa and M.Ueda: Physica B **329**, 1144 (2003); T.Yamasaki, M.Ueda and S. Maegawa: Physica B **329**, 1187 (2003)

11. **NSLR due to thermal fluctuations**
    - A.Lascialfari, Z.H.Jang, F.Borsa, P.Carretta and D.Gatteschi: Phys. Rev. Lett. **81**, 3773 (1998)
    - S. Yamamoto and T. Nakanishi: Phys. Rev. Lett. **89**, 157603 (2002); H.Hori and S.Yamamoto: Phys.Rev. B **68**, 054409 (2003)
    - Goto, T. Koshiba, T. Kubo, and K. Awaga: Phys. Rev. B **67**, 104408 (2003)
    - Y.Furukawa, K.Watanabe, K.Kumagai, F.Borsa and D.Gatteschi: Phys.Rev.B **64**, 104401 (2001)
    - Y.Furukawa, K.Kumagai, A.Lascialfari, S.Aldrovandi, F.Borsa, R.Sessoli and D.Gatteschi: Phys.Rev.B **64**, 094439 (2001)

12. **Theory of spin-phonon coupling in molecular nanomagnets**
    - M.N. Leuenberger and D. Loss: Phys. Rev. B **61**, 1286 (2000)
    - J.Villain, F. Hartmann-Boutron, R. Sessoli, and A. Rettori: Europhys. Lett. **27**, 537 (1994); F. Hartmann-Boutron, P. Politi, and J. Villain: Int. J. Mod. Phys. **10**, 2577 (1996)

13. **Theory and experiments on QTM in molecular nanomagnets**
    - I.Tupitsyn and B.Barbara, *Magnetism: Molecules to Materials III*, ed by J. S. Miller and M. Drillon (Wiley-VHC, Verlag GmbH, Weinheim 2002)
    - M. Luis, F.L. Mettes and L. Jos de Jongh, *Magnetism: Molecules to Materials III*, ed by J. S. Miller and M. Drillon (Wiley-VHC, Verlag GmbH, Weinheim 2002)

14. **NMR detection of QTM**
    - M. Ueda, S. Maegawa and S. Kitagawa: Phys. Rev. B **66**, 073309 (2002)
    - A.Morello, O.N.Bakharev, H.B.Brom and L.J.de Jongh: Polyhedron **22**, 1745 (2003)
    - A.Morello, O.N.Bakharev, H.B.Brom, R.Sessoli and L.J.de Jongh: Phys. Rev. Lett. **93**, 197202 (2004)
    - Y.Furukawa, K. Aizawa, K.Kumagai, R.Ullu, A.Lascialfari and F.Borsa: J. Appl. Phys. **93**, 7813 (2003)see also Phys. Rev B **69**, 014405 (2004)
    - S.H.Baek, F.Borsa, Y.Furukawa, Y.Hatanaka, S.Kawakami, K.Kumagai, B.J.Suh and A.Cornia: Phys.Rev. B **71**, 214436 (2005)
    - Y.Furukawa, K.Watanabe, K.Kumagai, Z.H.Jang, A.Lascialfari, F.Borsa and D.Gatteschi: Phys.Rev B **62**, 14246 (2000).

Structure, synthesis and characterization of contrast agents for magnetic resonance molecular imaging

Sophie Laurent, Luce Vander Elst, Alain Roch, Robert N. Muller

Department of General, Organic and Biomedical Chemistry, NMR and Molecular Imaging Laboratory, University of Mons-Hainaut, B-7000 Mons, Belgium
robert.muller@umh.ac.be

1 Introduction

Magnetic resonance imaging (MRI) is a non invasive method that, in a medical context, allows the identification and characterization of pathological tissues and lesions. The principle of this technique is based on the application of magnetic field gradients in three dimensions. The image results from the spatial identification of hydrogen nuclei. This image is constituted of elements, "pixels", where grey levels represent the signal intensity emitted by corresponding volume elements, "voxels". As the signal depends on the concentration in protons and on nuclear relaxation times, T_1 and T_2, modulations of image intensity are observed.

To make a precise medical diagnosis, contrast agents facilitating the distinction between pathological and healthy tissue are administered. Their role is to modify the nuclear density and nuclear relaxation times. Instrumental parameters such as the choice of the sequence can also influence the contrast. Among the characteristics of these substances, the efficiency and safety for the patient must be determined before considering medical application. The contrast agents therefore have to be stable, non toxic, biocompatible and efficient at weak doses.

The intensities on MRI image are mainly determined by water proton relaxation. The process of relaxation is a function of the water proton environment and varies according to the tissues. The contrast can be modulated by two phenomena governing the relaxation of protons: the T_1 effect or longitudinal relaxation, which increases the signal intensity (positive agent), or the T_2 effect or tranverse relaxation, which decreases the signal intensity (negative agent). Images will then be weighted in T_1 or T_2.

2 Contrast agents

To increase the signal intensity and image quality, contrast agents are used
[1,2,3] (fig. 1). Their key property is their "relaxivity", defined as the aptitude
to increase the water proton relaxation rate by one mmol l^{-1} of paramagnetic
center. These substances play the role of a real catalyst in relaxation and
are classified into 3 categories according to their magnetism: diamagnetic,
paramagnetic or superparamagnetic compounds. Contrast agents decrease the
proton relaxation time and, thus, reduce the acquisition time of the image.
More of the signal can be accumulated, which increases the sensitivity and
the image contrast.

a)

b)

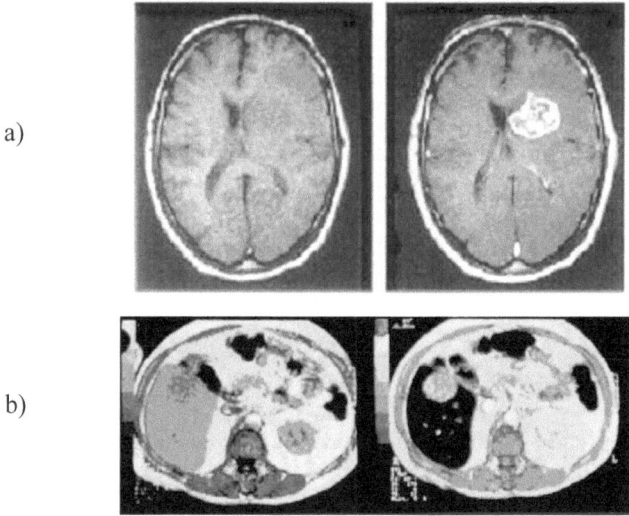

Fig. 1. MRI before injection of the contrast agent and after injection with the
positive contrast agent (a) [4] and with a negative contrast agent (b).

Diamagnetic materials have a weak negative susceptibility in an external
magnetic field. Their effect on signals is very weak.

Paramagnetic compounds include oxygen, nitroxydes, transition metal
ions and lanthanides. They possess one or several free electrons reponsible
for the presence of a dipolar magnetic field during the application of an external field B_0. This magnetic moment is 657 times higher than that of the
proton, and the paramagnetic effect is therefore superior to the diamagnetic
effect. Gadolinium is a very toxic metal in the hydrated form $[Gd(H_2O)_8]^{3+}$.
To avoid toxicity while maintaining the magnetic and electronic properties
of this ion, paramagnetic gadolinium complexes have been used. In vivo, the
free Gd is in competition with calcium-dependent systems and blocks the
reticulo-endothelial system. To avoid all in-vivo toxicity, it has to be used

in the form of a thermodynamically stable inert complex. Further, the ligand which complexes the metal has to leave free coordination sites such that one or several molecules of water can link to the metal, thus increasing the relaxivity. The first gadolinium complexes commercially available were Magnevist® (Gd-DTPA) and Dotarem® (Gd-DOTA). These compounds are accompanied by counter-ions that increase the osmolality of the injected solution. Two other "neutral" complexes exist, Omniscan® (Gd-DTPA-BMA) and Prohance® (Gd-HP-DO3A) [5]. These 4 paramagnetic complexes (fig. 2) disseminate freely in the extracellular space and present no specificity.

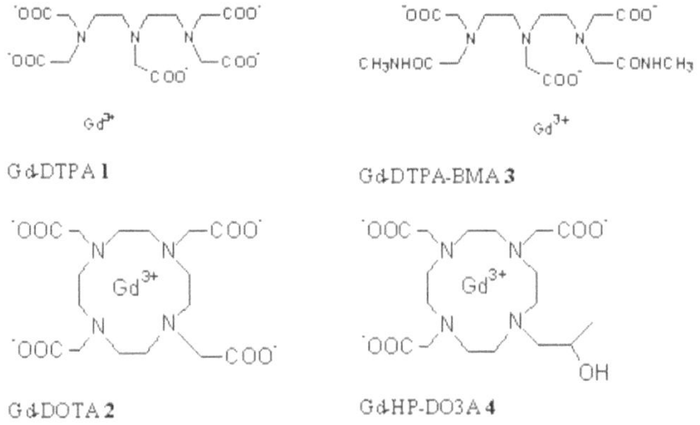

Fig. 2. Structure of commercial paramagnetic contrast agents.

The addition of lipophilic groups onto the skeleton of Gd-DTPA has improved the tissular tropism. When these derivatives accumulate at the level of some pathological tissues, they can greatly increase the signal intensity. Eovist® (Gd-EOB-DTPA), for example, targets hepatocytes (fig. 3). Another strategy is to covalently couple these gadolinium complexes to macromolecules such as human seric albumin (HSA) and polysaccharides (dextran), or to synthetic polymers such as polylysine, polyethyleneglycol, dendrimers, etc. An alternative for bettering the efficiency of contrast agents is the non-covalent interaction with HSA. Aryl moieties are grafted onto the contrast agent, which are able to recognize the hydrophobic sites of the protein.

The recent development in contrast agent research is the targeting of a sickness or its manifestations, for example the early detection of tumors. This strategy is based on the principle of recognition of specific receptors. Their ligands, i.e. peptides, antibodies, folate,... are grafted onto Gd-DTPA. Another class of smart agents modulates their efficiency according to the biological environment (the presence of enzymes, pO2, pH, and so on). For example, pH-sensitive agents are promising for pathologies that acidify the environment (in tumors, the pH is ∼6.8, while in a healthy extracellular medium, the

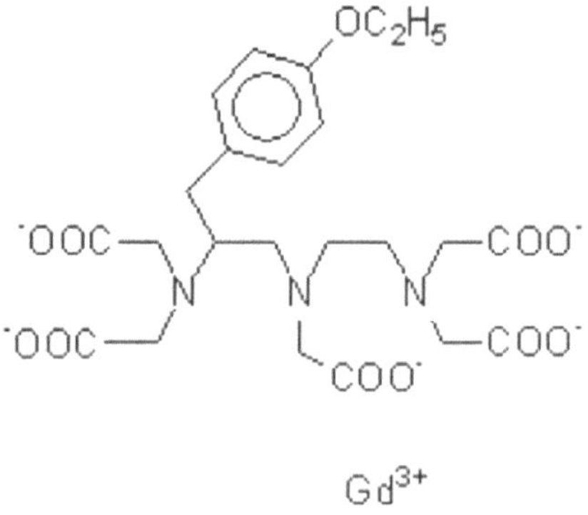

Fig. 3. Structure of Gd-EOB-DTPA.

pH is ~7.4).

Contrast compounds for angiography are notably characterized by prolonged vascular remanence time, decreased extravasation, biological inertia, efficient elimination, etc. To increase the vascular residence time, various strategies have been tried: imitating sanguine cellular structures (liposomes, micelles) or increasing the molecular mass by covalent bond to the macromolecules. However, these approaches have disadvantages that limit their application because, among other problems, they leave the bloodstream rapidly. Thus, the liposomes accumulate in the liver and spleen, the plasma protein mimetics are opsonized and recognized by the reticulo-endothelial system. A new strategy consists in grafting molecules of glucose onto the Gd-DTPA. Prolongation of the half-life of glucosylated Gd-DTPA derivatives is hypothesized to be due to a possible interaction with the renal glucose carrier, which can entail delayed renal excretion.

The last category is that of superparamagnetic nanoparticles (fig. 4), which are composed of an iron oxide nucleus of 5 nm diameter (SPIO, SuperParamagnetic Iron Oxide). To increase their stability in aqueous medium, particles are coated with polymers (dextran) and form colloidal solutions. The diameter of these particles is on the order of 10 - 20 nm. After intravenous injection, particles are trapped by cells of the reticulo-endothelial system (i.e. Kupffer cells in the liver, the spleen). They induce a diminution of the signal in these organs and are used for hepatic tumor detection. Current experiments aim to graft these contrast agents onto different molecules that target specific cellular receptors of a given pathology. MRI applications for these superparamagnetic compounds are quite various and go from angiography to tumoral diagnosis and atherosclerotic pathology. The progress of molecular biology in recent years has led to the development of new, increasingly specific methods of diagnostic imaging. Among these, molecular imaging has burgeoned thanks to the utilization of the magnetic marker and to a better spatial resolution. The

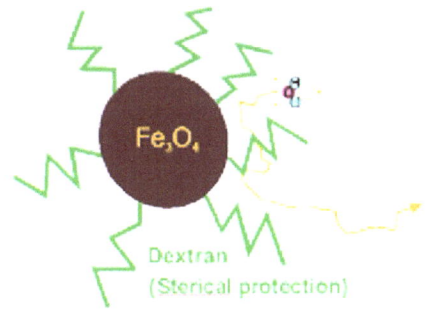

Fig. 4. Structure of iron oxide nanoparticles.

recent studies on cellular labeling in MRI have used the specific targeting of some cellular receptors with ligands grafted to SPIO (superparamagnetic iron oxide) or USPIO (ultrasmall superparamagnetic iron oxide) particles.

The most common method for producing magnetic nanoparticles involves coprecipition of ferrous and ferric salts in an alkaline medium. This synthesis approach can be used in the absence or presence of surface complexing agents, such as PEG, dextran, synthetic polymer, to name a few. For the method without surface complexing agents, the nanoparticles precipitated are isolated through magnetic decantation or centrifugation. The precipitate is then treated with nitric acid, centrifuged and peptized in water to produce a stable acidic magnetic sol. Alkaline magnetic sols can be obtained using tetramethylammonium. A variety of factors can be adjusted in the synthesis protocol of the iron oxide particles in order to control size, magnetic characteristics, stability in solution or surface properties. Although coprecipitation methods are used for their simplicity, the nanoparticles produced are fairly polydisperse. Consequently, several other techniques are currently being developed to obtain nanoparticles with more uniform dimensions.

3 Relaxation mechanims

3.1 Nuclear relaxation

Nuclear relaxation corresponds to the return to thermodynamic equilibrium of a spin system excited by the energy absorbed during the application of an electromagnetic field of appropriate frequency.

The interaction of magnetic moments of the excited spins with the environment creates fluctuating local microscopic magnetic fields. These magnetic fluctuations are linked to molecular movement. Longitudinal (R_1) and transverse (R_2) relaxation rates are therefore modulated by the phenomenon of molecular distribution and are expressed by the generic relationship:

$$\frac{1}{T_i} = R_i = KE_c^2 f(\tau_c) \tag{1}$$

where:

$i = 1, 2$

K is a constant

E_c is the amplitude of the interaction responsible for the relaxation

$f(\tau_c)$ is a function of the correlation time, τ_c, that modulates the interaction.

3.2 Paramagnetic relaxation

The presence of paramagnetic ions entails the increase of the observed relaxation rate of water protons. This equation can be written as:

$$\frac{1}{T_i^{obs}} = \frac{1}{T_i^{dia}} + \frac{1}{T_i^P} \tag{2}$$

where $1/T_i^{dia}$ is the diamagnetic relaxation rate of water protons without paramagnetic contribution and $1/T_i^P$ is the paramagnetic relaxation rate.

The paramagnetic center influences the relaxation rate of the water molecule which interacts directly with it and the neighboring molecules.

The efficiency of contrast compounds is linked to molecular movements but also to intrinsic properties of the nuclei (magnetic moment, gyromagnetic ratio, spin). The paramagnetic relaxation $R_1{}^P$ is characterized by two contributions: the contribution of internal sphere $R_1{}^{is}$ ("inner sphere", IS, fig. 5) and external sphere $R_1{}^{os}$ ("outer sphere", OS, fig. 6). The principle of "inner sphere" relaxation is a chemical exchange during which one or several water molecules are in contact with the electronic spin; after leaving the first sphere of coordination of the paramagnetic center, they are replaced by other molecules. This mechanism allows the propagation of the paramagnetic effect to the totality of the solvent and constitutes a situation where the water molecule is exchanged between two sites (the interior and the exterior of the first coordination sphere). The IS model has been described by the Solomon-Bloembergen-Morgan theory (SBM) [6, 7].

The inner sphere contribution is given by:

$$R_1^{is} = fq \frac{1}{T_{1M} + \tau_M} \tag{3}$$

$$\frac{1}{T_{1M}} = \frac{2}{15} \left(\frac{\mu_0}{4\pi}\right)^2 \gamma_H^2 \gamma_S^2 \hbar^2 S(S+1) \frac{1}{r^6} \left[\frac{7\tau_{c2}}{1+(\omega_S\tau_{c2})^2} + \frac{3\tau_{c1}}{1+(\omega_H\tau_{c1})^2}\right] \tag{4}$$

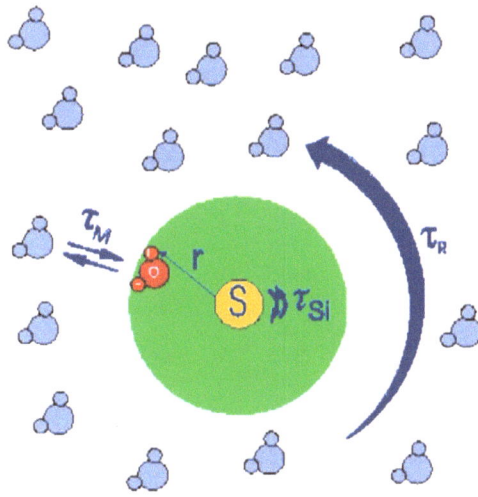

Fig. 5. Schematic representation of the inner sphere theory (IS).

$$\frac{1}{\tau_{ci}} = \frac{1}{\tau_R} + \frac{1}{\tau_M} + \frac{1}{\tau_{si}} \tag{5}$$

$$\frac{1}{\tau_{S1}} = \frac{1}{5\tau_{SO}} \left[\frac{1}{1+\omega_S^2\tau_V^2} + \frac{4}{1+4\omega_S^2\tau_V^2} \right] \tag{6}$$

$$\frac{1}{\tau_{S2}} = \frac{1}{10\tau_{SO}} \left[3 + \frac{5}{1+\omega_S^2\tau_V^2} + \frac{2}{1+4\omega_S^2\tau_V^2} \right] \tag{7}$$

where f is the relative concentration of the paramagnetic complex and of the water molecules; q is the number of water molecules in the first coordination sphere; τ_M is the water residence time; γ_S and γ_H are the gyromagnetic ratios of the electron (S) and of the proton (H), respectively; $\omega_{S,H}$ are the angular frequencies of the electron and of the proton; r is the distance between coordinated water protons and the unpaired electron spin; $\tau_{c1,2}$, the correlation times modulating the interaction, are defined by Eq. 5 – where τ_R is the rotational correlation time of the hydrated complex and $\tau_{s1,2}$ are the longitudinal and transverse relaxation times of the electron. These latter parameters are field-dependent (Eqs. 6 and 7). τ_{SO} is the value of $\tau_{s1,2}$ at zero field and τ_v is the correlation time characteristic of the electronic relaxation times.

The second contribution to paramagnetic relaxation is "outer sphere" relaxation. It is explained by the dipolar interaction at long-distance between the spin of the paramagnetic substance and the nuclear spin. This intramolecular mechanism is modulated by the translational correlation time (τ_D) that takes into account the relative molecular diffusion constant (D) between the paramagnetic center and the solvent molecule, as well as their distance of closest approach (d). The OS model has been described by Freed [8].

Fig. 6. Schematic representation of outer sphere theory (OS).

The outer sphere contribution is given by:

$$R_1^{os} = \frac{6400\pi}{81} \left(\frac{\mu_0}{4\pi}\right)^2 \gamma_H^2 \gamma_S^2 \hbar^2 S(S+1) N A \frac{[C]}{dD} [7j(\omega_S \tau_D) + 3j(\omega_H \tau_D)] \quad (8)$$

$$j(\omega\tau_D) = Re \left[\frac{1 + \frac{1}{4}\left(i\omega\tau_D + \frac{\tau_D}{\tau_{S1}}\right)^{1/2}}{1 + \left(i\omega\tau_D + \frac{\tau_D}{\tau_{S1}}\right)^{1/2} + \frac{4}{9}\left(i\omega\tau_D + \frac{\tau_D}{\tau_{S1}}\right) + \frac{1}{9}\left(i\omega\tau_D + \frac{\tau_D}{\tau_{S1}}\right)^{3/2}} \right]$$
$$(9)$$

[C] is the molar concentration of the paramagnetic ion and $\tau_D = d^2/D$ is the translational correlation time.

4 Physico-chemical characterization of contrast agents

4.1 Paramagnetic complexes

The complexity of the equations describing the relaxation rate justifies the large number of parameters describing the relaxation IS and OS (8 parameters: τ_M, q, τ_R, D, r, d, τ_V, τ_{S0}). Considering the high number of parameters introduced by the equations, the estimation of all parameters using the technique of field cycling is often ambiguous. Thus, the determination of some parameters using independent methods increases the overall reliability of the theoretical adjustment of the NMRD curve (fig. 7) [9, 10, 11, 12, 13].

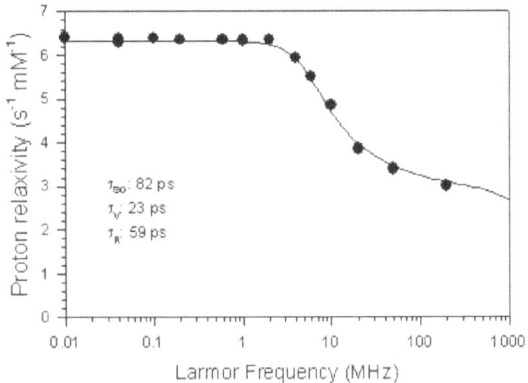

Fig. 7. NMRD profile of Gd-DTPA.

Rotational correlation time

The rotational correlation time (τ_R) characterizes the reorientation of the spin-nucleus vector, such as the Gd^{3+}-proton vector of the water molecule in the case of a gadolinium complex. Generally, τ_R limits the complex relaxivity of low molecular weight complexes in the fields range used in imaging. This characteristic can be exploited advantageously. The rotational correlation time can be obtained through several methods: analysis of the longitudinal relaxation of ^{17}O on the Gd complex, EPR measurement of the vanadyl complex, measurement of the longitudinal relaxation rate by ^{13}C NMR, fluorescence polarization spectroscopy or by deuterium relaxation rate.

In small diamagnetic molecules, the longitudinal relaxation rate of deuterium nuclei is given by equation 10, where the quantity in brackets is the quadrupolar coupling constant ($e^2qQ/h = 170$ kHz for D-C_{sp^3}).

$$R_1 = \frac{1}{T_1} = \frac{3}{8} \left(\frac{e^2qQ}{\hbar} \right)^2 \tau_R \qquad (10)$$

Electronic relaxation times

Longitudinal and transversal electronic relaxation times (τ_{S1} and τ_{S2}) describe the process of return to equilibrium of the magnetization associated with electrons undergoing transitions between the electronic levels of the Gd ion. These transitions produce fluctuations that enable the relaxation of protons. The degeneration of the higher electronic level of gadolinium ions occupied by single electrons is attributed to the Zeeman effect. It seems that the origin of electronic transition is represented by the collisions between solvent molecules and the complex that induces distortions in symmetry and leads to transitory ZFS of electronic levels.

Electronic relaxation rates depend on the field. In the case of Gd complexes, electronic relaxation rates are quite long ($\sim 10^{-10}$ s at low field and $\sim 10^{-8}$-10^{-7} s at high field). Consequently, the influence of electronic relaxation is especially evident at fields inferior to 0.2 Tesla.

Number of water molecules coordinated to metal

The number of water molecules strongly influences the IS contribution because the relation is directly proportional. For complexes like Gd-DTPA, an increase in the number of water molecules from 1 to 2 enhances relaxivity by about 30%, but nearly all derivatives of Gd-DTPA are q=1. Two types of measurements can be distinguished: q in solid phase and q in solution. Measurements in solid phase are obtained by diffraction (X rays, neutron diffraction). Although there is generally a good correlation between measurements obtained in these two different states of matter, by definition, measurement in solid phase does not take into account the dynamic state of the complex. It is therefore useful to confirm these measurements by doing studies in solution. The fluorescence of Eu or Tb complexes can be induced by laser. The disexcitation rate by fluorescence of the two lanthanides follows an exponential decrease and is proportional to the number of water molecules coordinated to the metal. This disexcitation rate differs according to whether the complexes are coordinated to H_2O or D_2O. The measurement in these two media and their subtraction allows the calculation of number q. Another possibility is the LIS (Lanthanide Induced Shift) method that makes it possible to determine the number of water molecules coordinated by ^{17}O NMR.

Proton-to-metal distance

For protonic relaxation in the presence of paramagnetic centers, the IS contribution is manifested by dipolar interactions. The efficiency of the dipolar mechanism comprises the term in $1/r^6$, where r is the metal-to-proton distance. One understands therefore that even a slight reduction in this distance will have a noticeable impact on the relaxivity of Gd-DTPA complex.

This reduction could be the cause of the higher relaxivity of Gd-EOB-DTPA (5.5 mM^{-1} s^{-1} at 310 K, 20 MHz) with respect to Gd-DTPA (3.9 mM^{-1} s^{-1} at 310 K, 20 MHz). While the exact distance between the gadolinium and the oxygen in water is well defined thanks to cristallographic structures, the Gd-water-proton distance is only an estimate because few experimental methods allow this measurement.

Coordinated water exchange rate

The mechanism of IS relaxation is based on an exchange between water molecules surrounding the complex and the water molecule(s) coordinated to the lanthanide. Consequently, for agents of paramagnetic relaxation, the exchange rate (kex= $1/\tau_M$) is an essential parameter for the transmission of the "relaxing" effect to the solvent surrounding the complex. The principle of measurement is based on Swift and Connick's works on diluted paramagnetic solutions. It consists of analyzing the transverse relaxation rate of ^{17}O as a function of the temperature.

The paramagnetic transverse relaxation rate of water in solutions of gadolinium complexes is given by equation 11 where T_{2M}, the transverse relaxation rate of the oxygen atom of the bound water, results from scalar interaction between the electron and the oxygen nucleus (Eq. 12) and $\Delta\omega_M$, the chemical shift of the oxygen in this water molecule is given by Eq. 13. The outer sphere contribution is neglected.

$$\frac{1}{T_2^{is}} = fq\frac{1}{\tau_M}\frac{\frac{1}{T_{2M}^2} + \frac{1}{\tau_M T_{2M}} + \Delta\omega_M^2}{\left(\frac{1}{\tau_M} + \frac{1}{T_{2M}}\right)^2 + \Delta\omega_M^2} \tag{11}$$

$$\frac{1}{T_{2M}} = \frac{1}{3}S(S+1)\left(\frac{A}{\hbar}\right)^2\left[\tau_{e1} + \frac{\tau_{e2}}{1+\omega_S^2\tau_{e2}^2}\right] \tag{12}$$

$$\Delta\omega_M = \frac{g_L\mu_B S\,(S+1)\,B_o}{3k_B T}\frac{A}{\hbar} \tag{13}$$

$A/\hbar$ is the hyperfine or scalar coupling constant between oxygen and Gd^{3+}; τ_{ei} are given by $[\tau_M\,^{-1} + \tau_{Si}\,^{-1}]^{-1}$; g_L, the Landé factor, is equal to 2.0 for Gd^{3+}; μ_B is the Bohr magneton; B_0 is the external magnetic field.

The temperature dependence of τ_M and τ_V can be described by Eqs. 14 and 15, respectively.

$$\frac{1}{\tau_M} = \frac{k_B T}{h}\exp\left(\frac{\Delta S^{\neq}}{R} - \frac{\Delta H^{\neq}}{RT}\right) \tag{14}$$

$$\tau_V = \tau_V^{298}\exp\left(\frac{E_v}{R}\left(\frac{1}{T} - \frac{1}{298.15}\right)\right) \tag{15}$$

$\Delta S^{\neq}$ and $\Delta H^{\neq}$ are the entropy and the enthalpy of activation for the exchange process, τ_V^{298} is the correlation time at 298.15 K and E_v is the activation energy for this process.

4.2 Superparamagnetic nanoparticles

Proton relaxation in superparamagnetic colloids occurs because of the fluctuations of dipolar magnetic coupling between nanocrystal magnetization and proton spin. The relaxation is described by an outer sphere model where the dipolar interaction fluctuates because of both the translational diffusion process and the Néel relaxation process.

The simplest model [14] is derived when the anisotropy energy of the crystal is great enough to prevent any precession of its magnetic moment. In this high anisotropy condition, the magnetization of the crystal is locked along the easy axes. The magnetic fluctuations then arise from the jumps of the moment between different easy directions according to the Néel relaxation process.

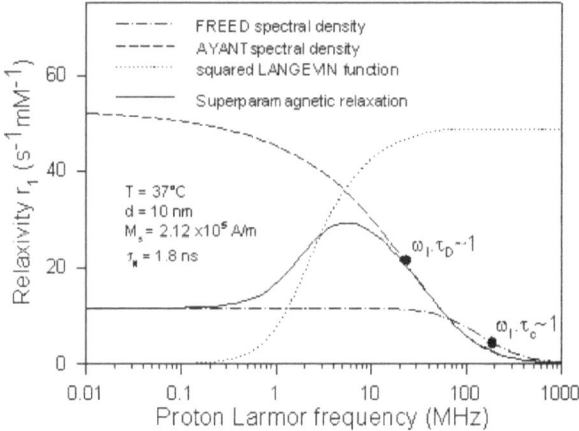

Fig. 8. Illustration of the superparamagnetic model.

At low fields (Fig. 8), the longitudinal relaxation rate of the protons is obtained by introducing in the outer sphere equations the limitation of the precession as mentioned above: the electron Larmor precession frequency is set to zero. The spectral density function determining this component of the relaxation is then characterized by a global correlation time depending on τ_N and τ_D. τ_N and τ_D are respectively the Néel relaxation time and the translation correlation time. Figure 8 shows the dispersion of this spectral density function, called the Freed function.

For very small crystals, the assumption of a complete locking of the magnetization along the easy axes, assuming an infinite anisotropy energy, becomes less and less valid. Subsequently, the orientation of the magnetization vector out of the easy axes becomes more probable [15]. This results in the presence of low field dispersion (Fig. 9). The evaluation of the amplitude of the low field component requires a more complete and difficult theory that takes into account the anisotropy. Note that the low field dispersion is always smaller than that predicted by the classical paramagnetic outer sphere theory. Indeed, the classical theory would be valid only for superparamagnetic colloids characterized by a null anisotropy [16].

At high fields (Fig. 8), the magnetic vector is locked along the external field B_o, and Curie relaxation dominates. The corresponding relaxation rates are

given by an outer sphere model assuming a stationary magnetization component in the B_o direction and, therefore, an infinite value of the Néel relaxation
time. The dispersion of this spectral density (named the Ayant function) occurs when $\omega_I.\tau_D \sim 1$.

At intermediate fields (Fig. 8), the relaxation rates are combinations of
the high and low field contributions, weighted by factors depending on the
Langevin function, which gives the average magnetization of the sample.

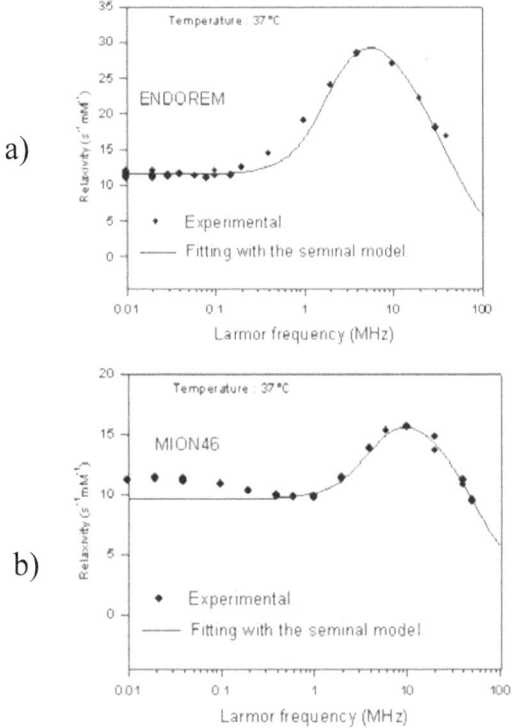

Fig. 9. Comparison between NMRD curves of particles with high (a) and low (b)
anisotropy. As the anisotropy energy is proportional to the crystal volume, curve A
corresponds to larger crystals.

The analysis of the proton NMRD profiles thus gives:
1. *the average radius (r)*: at high magnetic fields, the relaxation rate depends
only on τ_D and the inflection point corresponds to the condition $\omega_I.\tau_D \sim 1$
(Fig. 10). As shown in Eq. 16, the determination of τ_D gives the crystal size
r. r, D and ω_I are the average radius of the superparamagnetic crystals, the
relative diffusion coefficient, and the proton Larmor pulsation, respectively.

$$\tau_D = \frac{r^2}{D} \qquad (16)$$

2. *the specific magnetization* (M_s): at high fields, M_s can be obtained from the equation $M_s \sim C.(R_{max}/\tau_D)^{1/2}$, where C is a constant and R_{max} the maximal relaxation rate.

3. *the crystal anisotropy energy* (E_a): the absence or the presence of dispersion at low fields informs about the magnitude of the anisotropy energy. For crystals characterized by a high E_a value as compared to the thermal agitation, the low field dispersion disappears. This was confirmed in a previous work with cobalt ferrites [17], known to have high anisotropy energy.

4. *the Néel relaxation time* (τ_N): the relaxation rate at very low field R_0 is governed by a "zero magnetic field" correlation time τ_{C0} which is equal to τ_N if $\tau_N \ll \tau_D$. However, this situation is not met often, so τ_N is frequently reported as qualitative information additional to the crystal size and the specific magnetization.

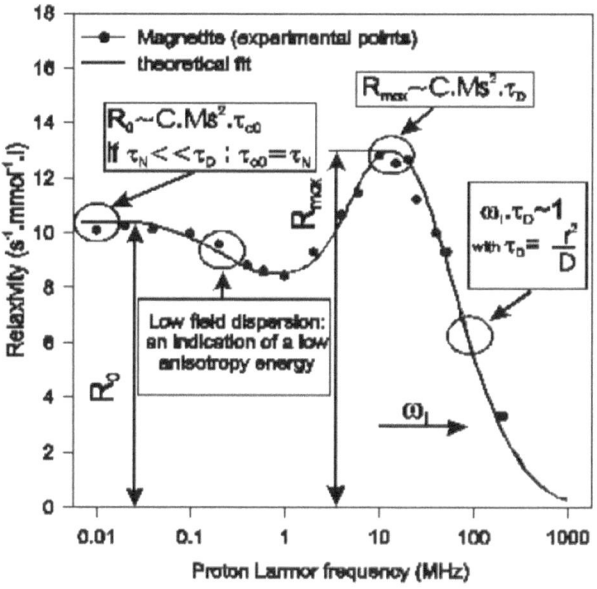

Fig. 10. NMRD profile of magnetite particles in colloidal solution.

5 Magnetic applications: MRI, cellular targeting

One possible way of extending the vascular residence time of contrast agents resides in the renal reabsorption mechanisms of some molecules such as glucose. Thus, various small-molecular-weight glucosyl derivatives of gadolinium diethylenetriaminepentaacetic (Gd-DTPA) were synthesized and their vascular half-life was studied. The sugar moieties linked to Gd-DTPA efficiently reduce the renal excretion of some derivatives. The interaction with the renal carrier was not clearly demonstrated, nor was any interaction with blood components observed. New glucosylated derivatives of Gd-DTPA (Cd-DTPA-BC$_2$-beta-cellobionA and Gd-DTPA-BC$_4$-beta-glucosylA) have been proposed as blood-pool MR contrast agents, considering their vascular remanence [18].

Apoptosis is a physiological process that becomes pathologic either by overactivity or inhibition. A dedicated contrast agent evidencing pathologies where apoptosis takes place would be useful for monitoring antitumor therapies. Phage display is a powerful new method used to select peptides with high affinity for a given target - phosphatidylserine (PS) in this study. Subsequent coupling of the selected peptides with a magnetically active species produces selective MRI contrast agents. The efficacy of this new contrast agent was tested on cells in culture [19].

Targeting of the endothelial inflammatory adhesion molecule E-selectin by MRI can be performed with a paramagnetic or superparamagnetic contrast agent in the context of in vitro and in vivo models of inflammation [20, 21]. The specific contrast agent was obtained by grafting a synthetic mimetic of sialyl Lewis^x (sLe^x), a natural ligand of E-selectin expressed on leukocytes, on the DTPA-bisanhydride or on the dextran coating of ultrasmall particles of iron oxide (USPIO).

Bulté has used MRI to provide information on the location and migration of cells after transplantation or transfusion. This approach requires magnetic prelabeling of the cells. With the magnetic labeling methods currently available, it is anticipated that cellular MRI will find applications in biology and medicine [22].

Magnetic hyperthermia involves dispersing nanoparticles into the targeted tissue and then applying a magnetic field to heat the particles. This heat is conducted into the immediately surrounding diseased tissue. Hyperthermia treatment of cancers is based on the finding that some cancer cells are more sensitive to temperatures in excess of 41°C than are their normal healthy counterparts [23, 24]. Magnetic nanoparticles encapsulated in polymer matrix beads have been used successfully for the treatment of macroscopic liver tumor.

Conclusions

MRI plays an important role in modern medicine and is at the interface of several scientific domains such as chemistry, biology and the physical sciences. Superparamagnetic particles present a lot of advantages in terms of molecular imaging because: (i) they constitute very powerful transverse relaxophores and therefore are very efficient for use as a negative contrast agent. (ii) super-paramagnetic crystals allow the fixation of several thousand active iron atoms, instead of only one in the case of paramagnetic complexes, to a single cell receptor. (iii) iron oxide particles are absolutely non toxic and bio-compatible. (iv) as shown in this review, there are many ways to graft the molecular entities that are the most suitable for targeting the desirable receptor. (v) the new means of synthesis recently discovered would optimize the size distribution of the crystal and thus the performance of the contrastophore. The challenge for chemists is, hence, to increase the relaxivity of their "contrastophores".

References

1. P. Caravan, J.J. Ellison, T.J. Mc Murry, R.B. Lauffer, Chem.Rev, 99, 2293 (1999)
2. P.A. Rinck in *Magnetic Resonance in Medicine*, 4th edition, (Blackwell Wissenschafts, Verlag, Berlin-Vienna, 2001)
3. R.N. Muller, in *Contrast Agents in whole body MR: operating mechanisms*, Encyclopedia of NMR, (Wiley, New-York (1996)) p. 1438
4. A. Alaux, in *L'imagerie par Résonance Magnétique* (Éditions Sauramp Médical, 1994)
5. S. Laurent, L. Vander Elst, R.N. Muller, Contrast Med. Mol. Imaging, 1(3), 128 (2006)
6. I. Solomon, Phys. Rev. 99, 559 (1955)
7. N.J. Bloembergen, Chem. Phys. 27, 572 (1957)
8. J.H. Freed, J. Chem. Phys. 68, 4034 (1978)
9. L. Vander Elst, F. Maton, S. Laurent, F. Seghi, F. Chapelle, R.N. Muller, Magn. Reson. Med. 38, 604 (1997)
10. R.N. Muller, B. Raduchel, S. Laurent, J. Platzek, C. Piérart, P. Mareski, L. Vander Elst, Eur. J. Inorg. Chem. 1949 (1999);
11. S. Laurent, L. Vander Elst, S. Houzé, N. Guérit, R.N. Muller, Helv. Chim. Acta 83, 394 (2000)
12. S. Laurent, F. Botteman, L. Vander Elst, R.N. Muller, Magn. Reson. Mater. Phys. Biol. Med. 16(5), 235 (2004)
13. S. Laurent, F. Botteman, L. Vander Elst, R.N. Muller, Helv Chem Acta 87, 1077 (2004)
14. A. Roch, R.N. Muller in *Longitudinal relaxation of water protons in colloidal suspensions of superparamagnetic crystals* Proceedings of the 11th Annual Meeting of the Society of Magnetic Resonance in Medicine 11, 1447 (1992).
15. A. Roch, R.N. Muller, P. Gillis, J Chem Phys. 110, 5403 (1999).
16. A. Roch, R.N. Muller, P. Gillis, J Magn Reson Imaging 14, 94 (2001).
17. A. Roch, P. Gillis, A. Ouakssim, R.N. Muller, J Magn Magn Mater. 201, 77 (1999)
18. C. Burtea, S. Laurent, J-M. Colet, L. Vander Elst, R.N. Muller, Invest. Radiol. 38(6), 320 (2003)
19. C. Laumonier, J. Segers, S. Laurent, A. Michel, F. Coppée, A. Belayew, L. Vander Elst, R.N. Muller, J. Biomol. Screening 11(5), 537 (2006)

20. S. Boutry, C. Burtea, S. Laurent, L. Vander Elst, R. Muller, Magn. Reson. Med. 53(4), 800 (2005)
21. S. Boutry, S. Laurent, L. Vander Elst, R.N. Muller, Contrast Med. Mol. Imaging 1(1), 15 (2006)
22. J.W. Bulté, Methods Mol. Med. 124, 419 (2006)
23. A. Jordan, R. Scholz, P. Wust, H. Fähling, R. Felix, J. Magn. Magn. Mater. 210, 413 (1999)
24. P. Moroz, S.K. Jones, C. Metcalf, B.N. Gray, Int. J. Hyperthermia 19, 23 (2003)
    See also the following WebSites:
    - http://www.umh.ac.be/~nmrlab/
    - http://www.ami-imaging.org
    - http://www.molecularimaging.org
    - http://www.emrf.org
    - http://www.ismrm.com
    - http://www.esmrmb.com
    - https://matar.ciril.fr/JSTIM/

Basic concepts of Magnetic Resonance Imaging

Alessandro Lascialfari[1,2], Maurizio Corti[1]

1 Department of Physics "A. Volta", University of Pavia, and CNR-INFM research unit, Via A. Bassi 6, I-27100 Pavia, Italy; 2 S3-CNR-INFM, Modena (Italy) and Institute of General Physiology and Biological Chemistry "G. Esposito", University of Milano, Via Trentacoste 2, I-20134 Milano (Italy) `corti@fisicavolta.unipv.it`

Fundamental aspects of the Magnetic Resonance Imaging (MRI) technique which is one of the preferred modalities for non-invasive clinical applications and "in vivo" medical and biological research, are presented. The physical principles of MRI and protocols recently introduced to investigate the microscopic details of pathological cerebral damages are introduced. The main results obtained from the analysis of T_2 relaxation curves are briefly summarized.

1 Introduction

The techniques available for medical diagnosis of internal tissues and organs of human body can be grossly divided into non-invasive and invasive ones. X-ray computer assisted tomography (CT X-rays), SPET (single photon tomography) and PET (Positron Emission Tomography) are part of the invasive techniques as they make large use of ionizing radiations. On the other hand Magnetic Resonance Imaging (MRI) is a non-invasive investigation tool (together with Diagnostic Ultrasounds) that has had a very fast technological development in the last 20 years, resulting in a substantial improvement of diagnostic images and protocols. Particularly, it should be noted that for some sectors and/or pathologies MRI reached the highest resolution in images and the fastest acquisition times (e.g. in the case of internal cerebral damages to which, in some cases, CT is blind). In the present work, before going into some details of recent results obtained by means of MRI in models of cerebral ischemia, we briefly describe the principles of MRI and some potentialities of this technique.

2 MRI physical principles and examples

The Magnetic Resonance Imaging is the last application of the Nuclear Magnetic Resonance (NMR) technique which is a powerful tool in many research fields: in chemical and pharmaceutical analysis for the investigation of molecular structure and molecular motion in solids and liquids, in solid state physics to study physical properties at the atomic level [1,2], in biomedicine to obtain information about organs, cells and tissues [3,4].

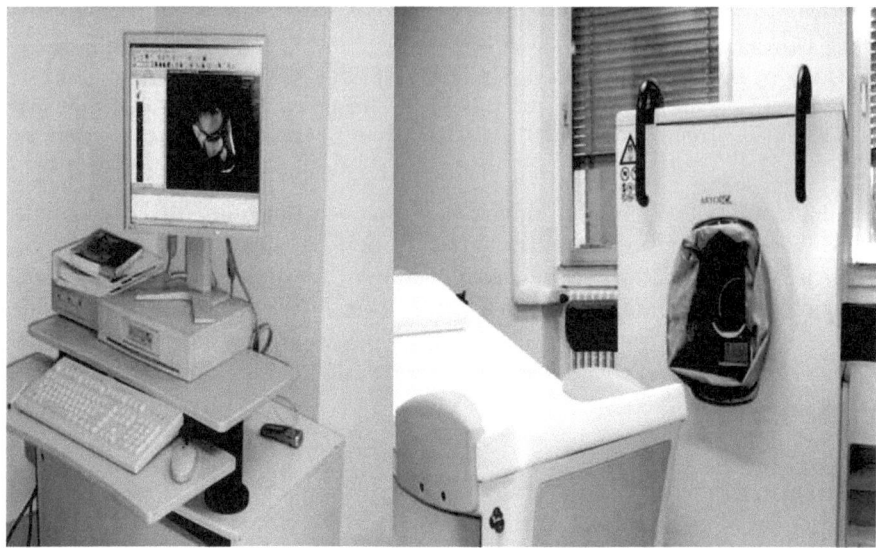

Fig. 1. MRI Artoscan (Esaote SPA Genova, Italy) Imager dedicated to the investigation of articular diseases. The system in the picture is placed in the NMR laboratory of the Department of Physics "A. Volta", University of Pavia, Italy.

The process of acquiring 2D and 3D images by NMR was first illustrated by Lauterbur and Mansfield in 1973 (awarded by the Nobel Prize in Medicine in 2003) and, over the last 20 years, the development of this technique was accelerated by Fourier transform image processing [3,4]. Nowadays MRI has become one of the most important techniques in the diagnosis of many diseases. As said above, contrary to CT which uses ionizing radiations MRI is based on the magnetic properties of the nuclei, excited by (non-invasive) radiofrequency radiation.

The NMR experiment is based on the precession of a nuclear spin I when it is immersed in a static magnetic field. For the sake of simplicity in the following we are only focusing on the nuclear spin of the hydrogen nucleus, ^{1}H, the most relevant one for MRI. When the patient is positioned within the magnetic field B_0 of the MRI apparatus (see Fig.1), the spins I_i of the atomic nuclei inside the human tissues are forced to align along B_0, thus giving a total nuclear magnetization $M = \sum_i I_i/V$. As shown in Fig. 2, in a magnetic field spinning nuclei have lower energy when aligned to the magnetic direction than when they are opposed to it.

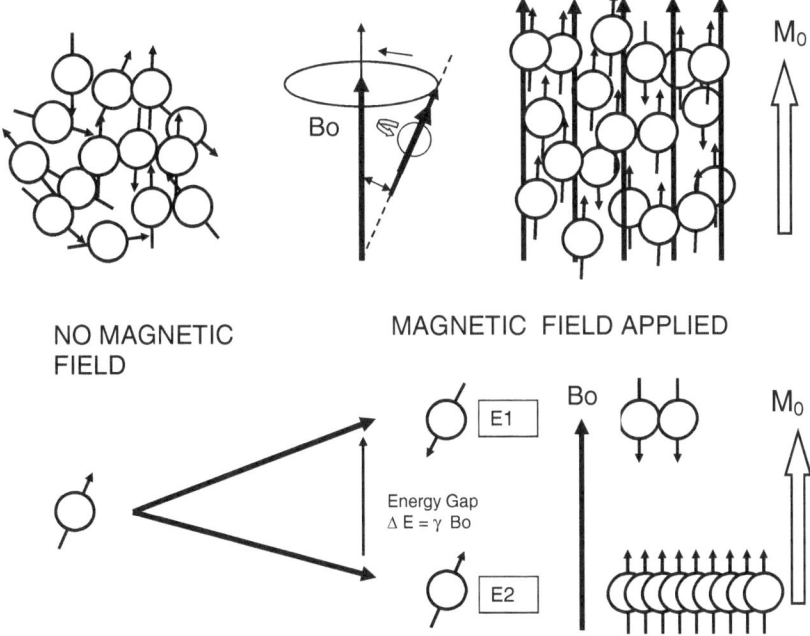

Fig. 2. The basis of NMR in a semi-classical picture: a nuclear spin precesses around the external magnetic field B_0. For example the ^{1}H nuclei that possess a nuclear spin, can be simplified by visualizing them as small magnets. This simplification allows one to explain the alignment of these spins in the magnetic field of the MR system and how they generate nuclear magnetization in the patient's body. The two possible directions of the spin vectors determine two energy levels E_1 and E_2. The frequency difference, $\Delta \nu = (E_1 - E_2)/h$, is in the range of radio-frequencies.

The energy difference between the two levels generated by the Zeeman effect, $\Delta E = \gamma B_0$, corresponds to a certain frequency $\nu = \Delta E/h$ which is in the range of radio frequencies. The energy of the nuclei can be increased if they absorb radio waves with the same frequency (condition of resonance). After each stimulation by a RF pulse, the atomic nuclei return to their previous energy level emitting radio waves [1-4]. The detected NMR response signal contains all the information related to the recovery of the magnetization toward the equilibrium state (i.e. the relaxation). The relaxation process of the nuclear magnetization is described mainly by two microscopic time constants: the longitudinal one (T_1) called nuclear spin-lattice relaxation time and the transversal one (T_2) called nuclear spin-spin relaxation time. The fascinating and extraordinary characteristic of the NMR experiment in general and in particular of the MRI experiment is that microscopic nuclear parameters like T_1 and T_2 are material-dependent, thus giving the possibility of distinguishing different tissues/diseases with a proper analysis of the received RF signal.

2.1 Spatial localization principle: Fourier imaging

In the classical NMR experiment the signal contains information about all the resonant nuclei in the whole body; thus, no information about the spatial distribution of nuclei belonging to different tissues can be obtained. In order to generate an image showing spatial structures in scale of gray intensities, a method able to spatially differentiate the NMR signal must be developed.

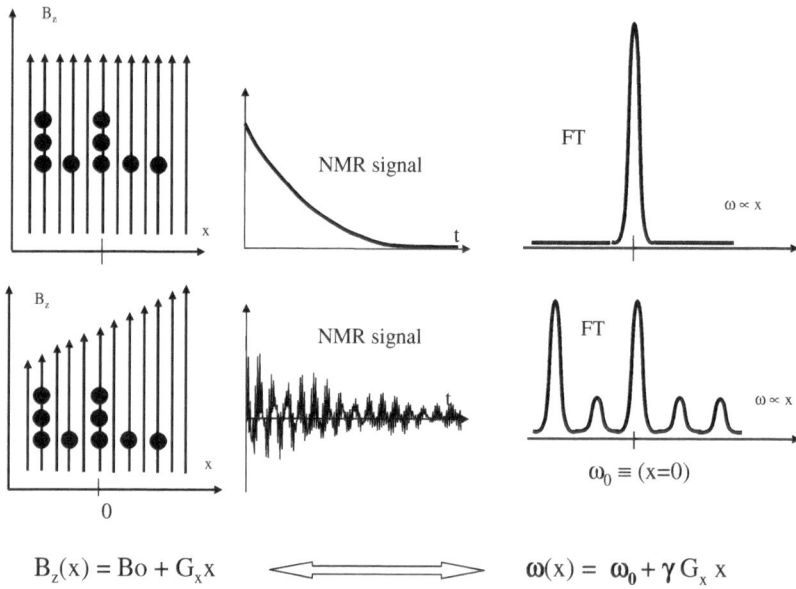

$$B_z(x) = Bo + G_x x \qquad \Longleftrightarrow \qquad \omega(x) = \omega_0 + \gamma G_x x$$

Fig. 3. Resonance spectrum (1D-FT of the acquired NMR signal) of a 9 capillaries water phantom which form a "F" letter (the height of each column is proportional to the number of resonating ^1H nuclei contained inside), in a presence of a uniform magnetic field (upper part) and in a presence of a field gradient G_x superimposed to the uniform one (lower part). The field gradient produces a linear distribution of resonance frequencies of the proton (i.e. ^1H nuclei) density along the gradient direction.

In a uniform magnetic field the proton resonance frequency is the same everywhere. On the other hand, in the presence of a field gradient superimposed to the uniform magnetic field the resonance frequency become site dependent opening the way to identify the different nuclei position in the human body.

The used technique is well illustrated in Fig. 3 where we consider a phantom composed by 9 capillaries. By taking the Fourier Transform (FT) of the acquired NMR signal in presence of a linear spatial (site) dependent magnetic field, that follows the expression $B_z(x) = G_x x = \frac{\delta B_z}{\delta x} x$, instead of a single line at the Larmor frequency, one observes a series of resonance lines which describe the proton intensity of the different isochromates. A frequency encoding along the direction of the field gradient is thus obtained (1D imaging).

In the case of a single magnetic field gradient not all the power of the FT technique is applied. FT in fact is a complex function that can be represented by real and imaginary part or, in equivalent notation, by magnitude and phase. With one field gradient only the magnitude of proton density along the direction of the gradient could be detected. In order to completely characterize a two-dimensional region a further information related to the second spatial dimension must be take into account. This is achieved by using a further field gradient perpendicular to the one that produces the frequency encoding. In particular, if a gradient G_y is applied for a fixed time t_y, the magnetization vectors along the y axis are forced to precess at a frequencies $\omega = \omega_0 + \gamma G_y y$. When G_y is turned off, the accumulate phase of each local magnetic vector along the y axis will be $\phi_y = \omega_y t$ and a complete characterization of the 2D domain is obtained. Generally in the imaging sequences the encoding procedure is done by first applying the phase encoding gradient G_y which is followed by the frequency encoding gradient G_x. This last gradient is maintained during the acquisition of the whole NMR signals. If one considers that the MRI spectrometers are equipped with a double phase sensitive demodulator detector with two reference signals at frequency ω_{rf} in quadrature (see Fig. 10), all the information giving a complete characterization of a 2D region can be obtained.

Following the presented excitation and acquisition procedure, the local NMR signal can be described by the expression

$$s_{nmr} = I(x,y)exp[j((\omega(x,y) - \omega_{rf})t + \phi(x,y) - \beta)] \tag{1}$$

where $I(x,y)$ take into account that the local magnetization is related to the proton density ρ and to the local relaxation times (T_1 and T_2) of the different human tissues. More generally each pixel gives a contribution to the demodulated NMR signal of the form

$$s_{nmr} = I(x,y)exp(j[\Omega(x,y)t + \Phi(x,y)] = I(x,y)exp[j(\gamma G_x t_x x + \gamma G_y t_y y)] \tag{2}$$

The detected NMR signal, by introducing the related variables $k_x = \gamma G_x t_x / 2\pi$ and $k_y = \gamma G_y t_y / 2\pi$ (here t_x represent the time where the NMR signal is acquired and t_y is the time length of the encoding phase gradient) and by considering all the pixels of the interested region becomes

$$S_{nmr}(k_x, k_y) = \int \int I(x,y)exp[j2\pi(k_x x + k_y y)]dxdy \tag{3}$$

The form assumed by the $S_{nmr}(k_x, k_y)$ expression shows that the 2D proton density image can be obtained by the Fourier Transform of the detected NMR signal. More precisely, the MRI apparatus uses the digital Fast Fourier Transform (FFT) to obtain a two dimensional proton image with (NxN) elements; for example to get a 2D image with 128x128 pixels one has to acquire a matrix composed of 128 different values of the encoding phase gradient (i.e. $G_y(i) = ig$ with $g = G_y(max)/128$ and i that spans from 1 to 128, each one followed by 128 samples of the NMR signals obtained in presence of the same frequency encoding gradient G_x applied along the transverse x direction.

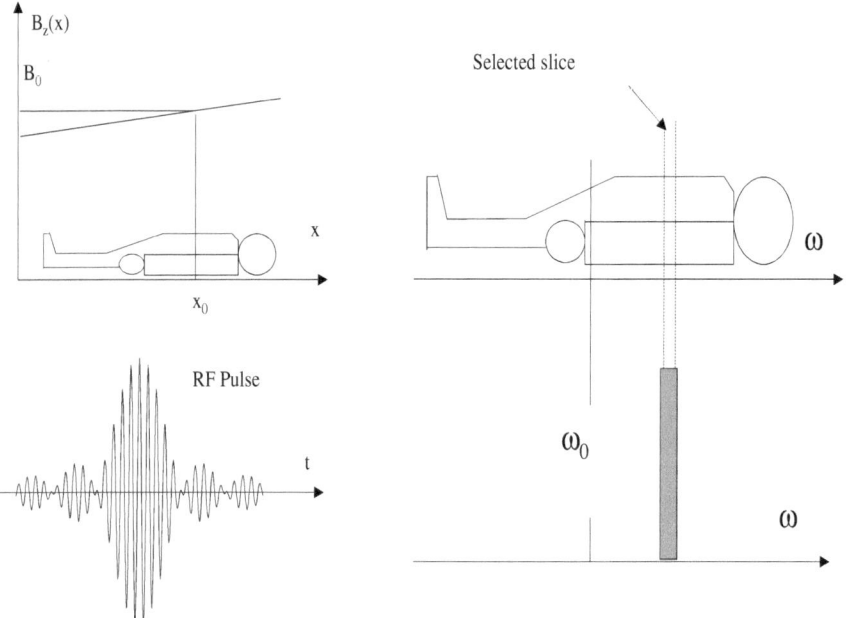

Fig. 4. In a uniform magnetic field, the nuclear magnetic resonance condition is the same everywhere. By using a magnetic field gradient and a sinc(t) shaped RF pulse it is possible to irradiate the nuclei contained in a thin slice leaving the outside nuclei at the rest conditions. The gradients allow one to position slices at will.

The proton density image is then obtained by applying the 2D-DFFT (Discrete Fast Fourier Transform) to the acquired (NxN) matrix. It is now easy to understand that by applying a further gradient (slice-selecting gradient) along the third space direction (z) it is possible to select a small volume of sample, generally of the order of 1 mm^3 (voxel), as expected in the conventional MRI technique. The time required to obtain the image in this case becomes so long that it is not applicable for the patients that should have to stay immobile for more than half an hour. With the aim of reaching a good compromise between image resolution and acquisition time, the investigation along the third space dimension, by using particular RF pulses in conjunction with a magnetic field gradient, is usually limited to a set of slices with thickness of the order of 1 mm. The slice selection technique is realized by using pulses of the form $sinc(x) = sin(x)/x$ whose FT is the rectangular function. By applying a $sinc(t)$ amplitude modulated RF pulse in a presence of a linear magnetic field gradient, a rectangular window of exciting nuclear resonance frequencies is obtained (see Fig. 4). By modifying the gradient intensity, the RF strength, the argument of the $sinc(t)$ function and the frequency carrier of the RF pulse, it is possible to select more than one slice during the same image acquisition. The slice selection is generally the first implemented operation in a MRI pulse sequence and it is followed by the phase and the frequency encoding. It is now clear the importance of the magnetic field gradients that, in conjunction with particular RF excitations sequences, allow to distinguish the classical NMR from MRI and to obtain beautiful images of the internal tissues of living subjects, like the ones reported in Fig. 5.

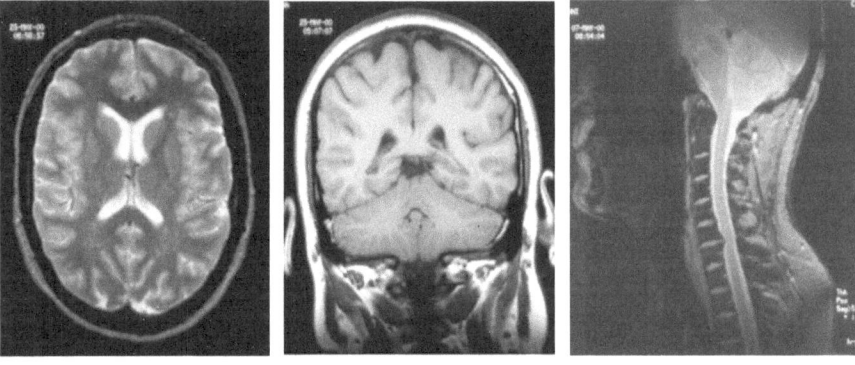

Fig. 5. Images of the human head and rachis with different kinds of contrast: (a) SE T_2 -weighted images; (b) SE T_1 -weighted images; (c) GE T_2 -weighted images. These different images are collected to visualize different contrasts between white matter, gray matter, rachis and cerebrospinal fluid. They all reveal excellent anatomic details.

2.2 Pulse sequences

The spatial selection techniques introduced in the previous section can be used as the guide line to implement the acquisition sequences with the aim of obtaining a good quality of the tomographic image in the shortest time. The most general and traditional image sequences are based on the Hahn-Echo technique that generates the so call Spin-Echo signal (see Chapter 1 by P. Carretta). This is the typical pulse sequence used in the traditional NMR experiment to measure the spin-spin relaxation time. The "short" echo dephasing time of the NMR signal observed when a gradient is applied after the exciting RF pulse, forces the image acquisition process to the detection of the echo signal. In fact with the echo sequence it becomes possible to observe the NMR signal till to time value of the order of T_2. The most important echo-based imaging sequences are the Spin-Echo (SE) sequence, the Inversion Recovery (IR) and the Multi-SE (MSE) sequence. More recently due to the relatively long acquisition time of the spin-echo based sequences, in order to speed up the image acquisition, a new sequence able to generate an echo signal with a simple inversion of the pulsed gradient was implemented; this is a very useful MRI sequence that takes the name of Gradient-Echo (GE) sequence. Other sequences that couple the SE and the GE sequences i.e. MPRAGE, GRASE, FLASH, etc., belong to the same family. Other image sequences that reduce the phase encoding gradient configurations (Fast-Spin-Echo sequence) or allow to acquire more than one phase encoding gradient during the same echo dephasing time (Turbo-Spin-Echo sequence) were also developed. Particular attention is devoted to the last image acquisition based on the Ultra-Fast image sequences that are able to detect angiography images and to study the encephalon in times of the order of tenth of second [4,5]. The most relevant MRI sequences will be analyzed in detail in the following subsections.

Spin-Echo sequence

In order to well understand how an image acquisition sequence works, in Fig. 6 the Spin Echo Sequence is represented. As one can see from the figure, to generate the echo signal of the selected slice the SE sequence uses two pulse gradients aligned along the z axis, G_z, in conjunction with a 90° followed by a 180° selective pulses, separated by a time $TE/2$. Between the two RF pulses a phase encoding gradient G_y is applied along the y axis for a time t_y. After the second RF pulse, symmetric with respect to the time echo position TE, the frequency encoding gradient G_x is applied along the x axis for the total acquisition time of the NMR echo signal. To obtain an image of 256x 256 pixels one has to memorize 256 echo signals acquired in the presence of 256 different values of the phase encoding gradient G_y. Generally all the gradient configurations are obtained by using the expression $G_y = ng_y$ with g_y gradient step increment and n spanning between -127 and 128. It is possible to prove that the intensity of the acquired echo signal intensity (SI) follows the expression

$$SI \propto \rho[1 - exp(-TR/T_1)]exp(-TE/T_2) \qquad (4)$$

where ρ represents the proton density, TR is the repetition time (the time between two excitation sequences) and TE the echo time. As one can see from Eq. (4), the echo intensity depends on both the relaxation processes T_1 and T_2. As a consequence the produced image will be also T_1- and T_2- dependent. If one considers that the values of the relaxation times in the human tissue can change by orders of magnitude between the normal and pathological tissues, the possibility to change the weight of one relaxation process with respect to the other plays a crucial role in the quality of the final image and the differentiation of tissues.

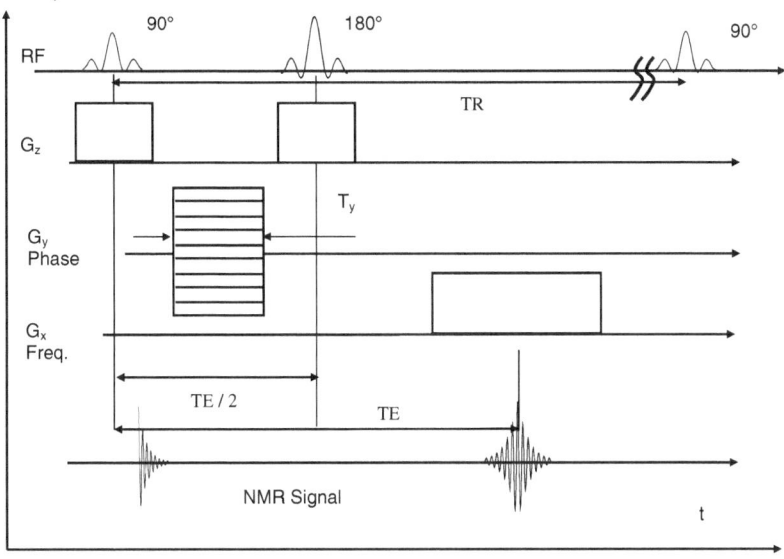

Fig. 6. Typical sequence diagram for a 2D spin echo (SE) imaging sequence. The RF pulses are modulated by a sinc(t) function to irradiate only the selected slice. The phase encoding gradient G_y is pictured as a series of horizontal lines to indicate that it is regularly increased in intensity during the different repetition periods TR.

One can easily observe that with $TR >> T_1$ and $TE << T_2$ an image of proton density of the selected slice is obtained, with $TR \approx T_1$ and $TE << T_2$ a T_1-weighted image is observed and by choosing $TR >> T_1$ and $TE \geq T_2$ a T_2-weighted image is detected. It is important to note that, although the SE sequence is the slowest imaging technique, it gives the best image quality. The variant of the SE sequence that acquires more than one echo signal during the same TR (Multi-SE sequence) allows to obtain concurrently more than one T_2-weighted images of the selected slice.

Inversion Recovery sequence

The Inversion Recovery (IR) Imaging sequence is very similar to the SE. As in classical IR-T_1 measuring sequence a selectively inversion 180° RF pulse that reverses the magnetization of the selected slice is applied before the echo generating pulses (see Fig.7). By indicating with TI the time interval between the inversion pulse of the magnetization and the 90° pulse of the echo pulses sequence and with ρ the proton density, it is possible to obtain an echo signal intensity, acquired with $TE \ll T_2$, that follows the expression

$$SI \propto \rho[1 - 2exp(-TI/T_1)] \times [1 - exp(-TR/T_1)] \tag{5}$$

This sequence is able to detect a T_1-weighted image with a contrast higher than the one that it is possible to obtain with a SE technique with short TR and TE.

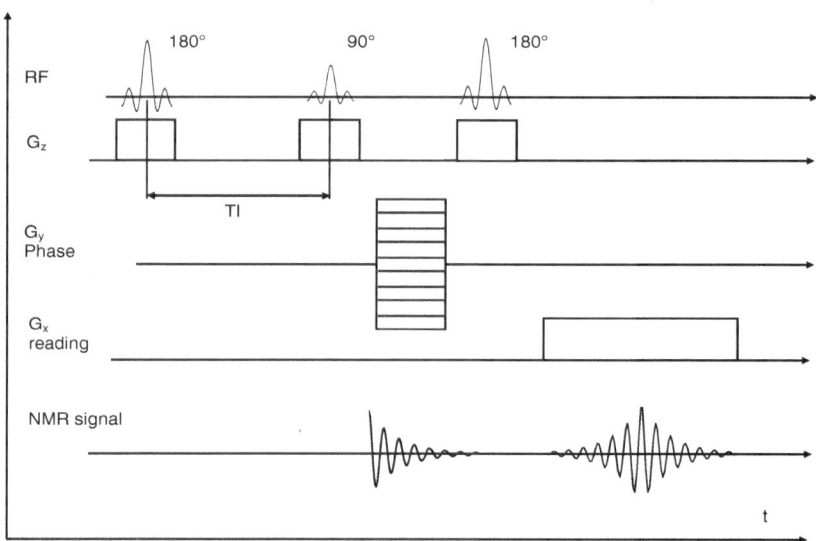

Fig. 7. Inversion recovery (IR) sequence diagram. By applying a selective 90° RF pulse at the beginning of the SE sequence it is possible to eliminate the signal of particular tissues, e.g. the fat signal.

The peculiarity of the IR sequence is related to the ability to remove a signal coming from a particular tissue when the inversion time TI is set equal to $T_1 ln2 \simeq 0.69 T_1$, being T_1 the relaxation time of the tissue whose signal must be removed. For example if we are interested in removing the fat proton signal of the selected slice, we can use a IR sequence with an inversion time $TI \approx 150$ ms. The IR sequence specific for the fat suppression take the name of STIR (Short-Tau Inversion Recovery).

Gradient Echo sequence

In this sequence the echo signal is generated without the refocusing $180°$ RF pulse. In fact within the GE sequence the echo signal is produced by applying, after the $90°$ RF selective pulse, a magnetic field gradient that is inverted after a chosen time τ.

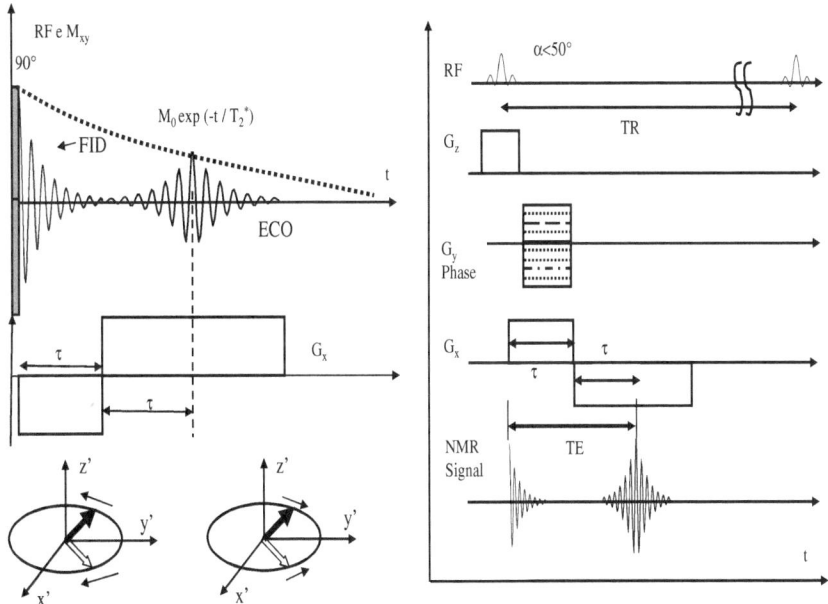

Fig. 8. Standard 2D gradient eco (GE) sequence diagram.

As explained above, a field gradient G_x produces a distribution of resonance frequencies along the x direction determining a controlled dephasing of the transverse magnetization. If after a time τ the field gradient is reversed, after a further time τ one observes the refocusing of the magnetization with an echo signal development. The gradient echo generation process can be easily pictured by comparing the nuclear spin to the runners on a race. If at time τ they turn around and run back toward the starting point, as the slower runners are leading, they will reach the starting point at exactly the same time (at time 2τ). The lack of the refocusing pulse paves the way for the decreasing of the repetition time of the RF pulses (TR) and consequently of the image acquisition time. A shorted TR however produces a worsening of the quality of the image, due to the incomplete recovery of the longitudinal magnetization. To partially overcome this problem a selective RF pulse with a flip-angle $\alpha < 90^o$ is used, as suggested by Ernst [6] for a rapid determination of T_1 in an NMR experiment with progressive saturation pulse sequence. The typical excitation procedure is shown in Fig. 8. As one can see after the excitation of the slice with a selective RF pulse the pulsed phase encoding field gradient G_y is applied, followed by the frequency encoding gradient G_x that is inverted after a time τ in order to acquire the full echo signal. By using small flip-angles ($10^o - 30^o$ RF pulse) and relatively short TR (50-100 ms) a T_2^*-weighted images with a contrast comparable to the ones obtained with a SE sequence with long TR can be acquired. By applying the mathematical expressions that control the NMR signal generation introduced in the first chapter of this book, it is possible to obtain for the echo signal intensity (SI) the equation

$$SI \propto \rho \frac{1 - exp(-TR/T_1)}{1 - cos(\alpha)exp(-TR/T_1)} sin(\alpha)exp(-TE/T_2^*) \qquad (6)$$

The remarkable reduction of the acquisition time in the GE sequence have led to strong attention for 3D imaging techniques, for cardiac and cerebrospinal fluid (CSF) dynamic-MRI, for magnetic resonance angiography (MRA) investigation and for functional magnetic resonance imaging (f-MRI).

Imaging flow sequences

Flow-related effects are used in magnetic resonance angiography where vascular structures can be investigated without the use of contrast agent. One of these techniques uses the NMR phase shift that is accumulated by the moving volume element. In fact if a nuclear spin is moving in a magnetic field gradient, it acquires an extra phase with respect to the spin at rest which is proportional to the spin speed. This is the effect used in the phase-contrast angiography imaging. The second angiography technique uses the time employed by the moving element to pass through the selected slice: this is called time of flight or TOF technique. The time of flight sequence uses a large flip angle $\alpha \simeq 90^o$ RF pulse with a short repetition time T_R. In these conditions while the stationary tissue saturates, the signal of the nuclei flowing into the excited vessels slice is enhanced. The increasing intensity can be explained by the flow, into the selected image slice, of nuclei not irradiated by the previous RF excitations (see Fig. 9). The two previously illustrated techniques cannot be applied for the imaging of large volumes. In such cases, "blood pool" contrast agents must be used. Here the shortening of the spin-lattice relaxation time T_1 produced by the blood pool contrast agent generates a very strong signal of the vessels with respect to the other tissues.

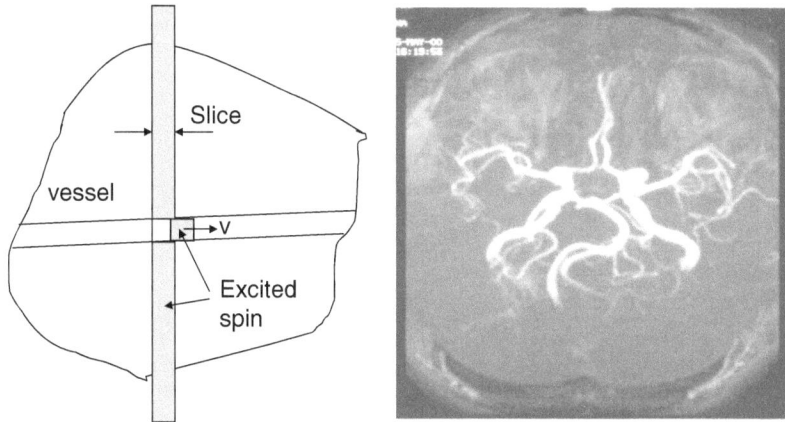

Fig. 9. Schematic representation of the TOF technique, on the left. On the right is reported a 3D image of intracranial circulation vessels obtained by means of TOF and the Maximum Intensity projection algorithm (MR angiography).

MRI instrumentation

One of the basic part of a MRI apparatus is represented by a superconducting (or permanent or electro) magnet that generates the external magnetic field B_0 required to perform an NMR experiment. The investigated sample (or animal or patient) is positioned within the force lines of the "homogeneous" magnetic field generated by the magnet (see Fig. 10). The second main part is represented by the NMR pulse spectrometer where one has electronic devices in charge of generating and transmitting to the patient the exciting RF pulsed sequences and receiving from the sample the NMR generated signals. The pulsed RF signal is transmitted through proper "coils" that create an oscillating B_1 magnetic field perpendicular to the static magnetic field B_0

where a system of coils, designed to create a spatial variation of the magnetic field (field gradient) along the 3 orthogonal directions, is added. Finally a computerized system able to analyze (acquire, average and process by Fourier Transform, etc.) the RF signal coming from the sample is also included. The block diagram of a MRI standard apparatus where all the main electronic units are evidenced, is reported in Fig. 10. As already introduced, the MR system has three pairs of gradient coils along the orthogonal axes x, y, z and, by switching them, it is possible to acquire images from different "slices", without moving the patient (depending on the slices, i.e. "sections" of the patient, the images are called coronal, sagittal, axial or oblique). Thus the MR images display the internal structure of the patient under investigation in two or three dimensions. These images show the spatial distribution of the excited nuclei within the object.

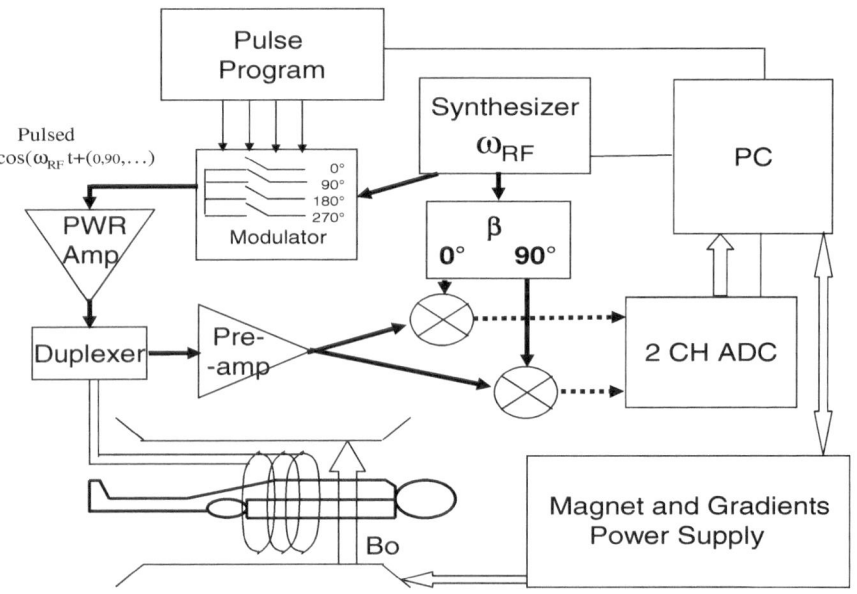

Fig. 10. Block diagram of a standard MRI apparatus.

MRI signal contrast

The contrast of the images depends primarily on a combinations of the following parameters: a) nuclear spin density (SD) or, equivalently for ^1H nuclei, proton spin density (PD); b) nuclear spin-lattice relaxation time T_1; c) nuclear spin-spin relaxation time T_2; d) apparent diffusion coefficient (ADC);

e) chemical shift; f) magnetic field (B_0) strength; g) concentration (and kind) of contrast agent (CA).

The ADC contains information on the nuclei's mobility inside different kinds of tissues and in the framework of different clinical situations (in case e.g. of edema the ADC changes remarkably with respect to normal situation). The chemical shift represents the shift of the NMR resonance line due to different chemical environment of the investigated nuclei. This is a typical situation for ^1H nuclei in the human body, belonging to different molecular structures. As the magnetic field is concerned, it should be remarked that, if all the other software and electronic parameters are kept to a constant efficiency, by increasing the magnetic field the quality of the images is improved. Finally, MRI contrast agents [7,8] are paramagnetic or superparamagnetic compounds (generally dissolved in solution) that can be injected inside the human body to highlight pathological tissues or diseases, eventually by the so-called "dynamic-MRI" (here the CA uptaking is followed as a function of time; the modalities of uptaking depend on the current state of the disease; see Fig. 11). They are distinguished in non-specific CA and specific-CA. The magnetic properties of non-specific CA tend to decrease T_1 and T_2 of all the nuclei in the human body, with different quantitative reduction depending on the CA.

In most cases they can be distinguished in intra-vascular and extra-vascular CA. The specific CA were deeply investigated in recent years. In fact, they are synthesized in form of molecules able to link to proteins, macrophages or other molecules. In this way it is possible to realize the so-called "Molecular Imaging" that targets specific molecules during the development of an investigated disease (for example transferrin and related gene's expressions in cancer diagnosis). This field of research is currently in rapid evolution and it is one of the most promising experimental protocols to investigated targeted pathologies.

Before going on, we would like to remark the importance of functional MRI (f-MRI), a very recent field of MRI research. With f-MRI, the evolution of brain activities during normal life actions can be followed in real time. A typical example is the brain activity when a rotating black and white disk (the MRI paradigm) is showed to a primate for some seconds. Within this time interval the MR responses of the brain's ^1H nuclei change, thus giving evolving (i.e. different with time) MR Images. Effects of brain's memory can be also evidenced [9].

Finally, we remark that the Diffusion weighted MRI (DWI) is the only non-invasive method able to give diffusion imaging and quantification [3,4,10]. As a particular applicative example we report in Fig. 12 some images highlighting the potential of DWI as a clinical tool able to provide the earliest detectable clinical signs of brain ischemia, i.e. cytotoxic edema [11,12]. These images will be discussed more deeply in the next paragraph.

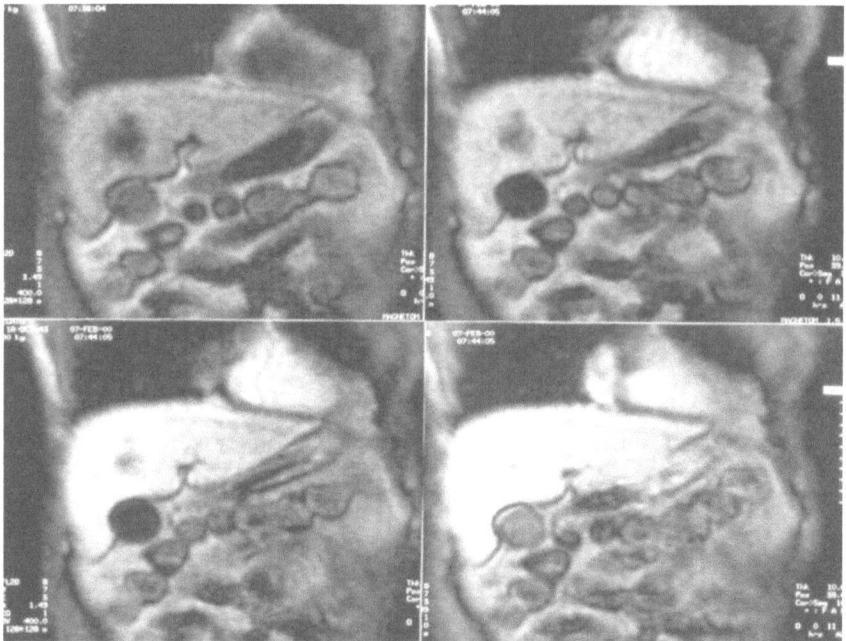

Fig. 11. A dynamic-MRI sequential study of a hepatic lesion by TURBO-FLASH sequence. The uptaking of a Gd-DTPA paramagnetic contrast agent is followed with time evolving.

3 Recent results on cerebral ischemia

In the last years, the researchers of the Department of Pharmacological Sciences of the University of Milan partly in collaboration with the Department of Physics of the University and INFM Unit of Pavia, studied the brain damages occurring in Spontaneously Hypertensive Stroke-Prone (SHR-SP) Rats and in the rat's model of induced ischemia by Occlusion of Middle Cerebral Arteria (MCAO rats) [11-14] (see also ref. [15]). These two animal models are thought to approximate the situation of cerebral ischemia in humans. In recent years, it was shown how it is possible to discriminate between the two investigated animal models by monitoring the temporal behavior of the apparent diffusion coefficient (ADC) and the nuclear spin-spin relaxation time T_2 of the water's ^{1}H nuclei [9,10]. The main finding is that the pattern of ADC variation in SHR-SP has different features from those reported in MCAO (see Fig. 12 for T_2-weighted and diffusion-weighted images). In particular, the decrease of ADC in MCAO rats is interpreted as a consequence of cytotoxic edema (e.g., the transfer of water from the extracellular space into the cells due to cellular energy failure): since ADC is a weighted average between the intracellular (assumed to be lower) and extracellular diffusion coefficients (assumed to be higher), its decrease reflects the changes in the ratio of intracellular and extracellular volume. As a further information, the increase of T_2 occurs at times later than 24 hrs. In MCAO rats it was shown that vasogenic edema sets in as a result of the increase in absolute extracellular water content due to the increase in vascular permeability. In SHR-SP, the phase of cytotoxic edema is missing. Data obtained by means of different techniques (like immunohistologic analysis) have revealed the spread of plasma constituents into the brain, presence of anomalies in arteries and alterations of the Blood Brain Barrier (BBB) [12].

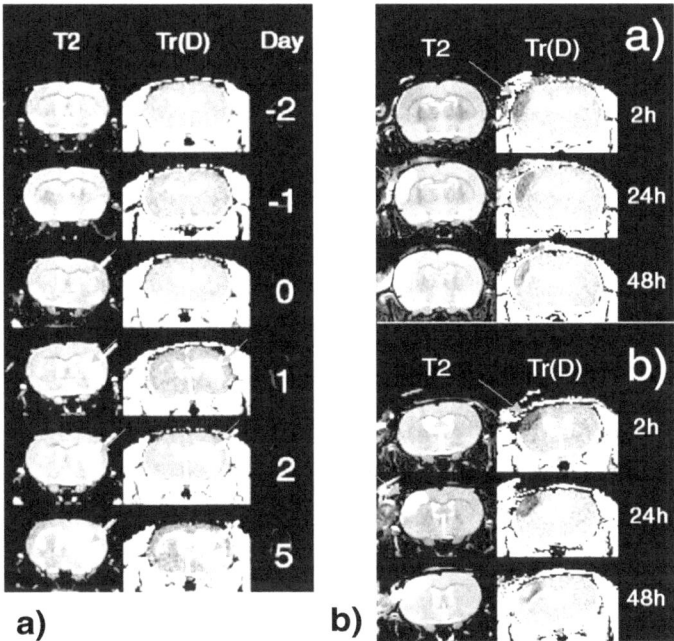

Fig. 12. Images of the brain of : (a) SHR-SP rats. Coronal sections of the same slice imaged on different days are shown: T_2-Weighted images and maps of the diffusion tensor trace (Tr(D)). Arrows show the main area of injury: note that both T_2 and D increase after the beginning of the damage, thus showing its vasogenic origin. (b) MCAO rats. Coronal sections of the same slice imaged at different times after the surgery are shown: T_2-Weighted images and maps of the diffusion tensor trace. Arrows show the main area of tissue injury: note that the decrease of the diffusion coefficient evidences early the beginning of the damage, thus showing its cytotoxic edema [12].

It should be remarked however that these MRI protocols of investigation are not enough to understand the details of the microscopic situation created by brain damages. For this reason , T_{1W} (Fig. 13) and T_{2W} images modified by the introduction of paramagnetic and superparamagnetic contrast agents [13,14,16] and multiexponential T_2 relaxation curves were collected and analyzed. In the following, a very brief summary of the experimental results obtained in ref. [13,14 and 16] is given. For experimental details the reader is referred to the cited papers.

The analysis of magnetic resonance imaging (MRI) images in the absence and presence of contrast agents is useful for the investigation of pathological states [17-20] in humans and in animal models [21-25]. In particular, contrast agents can be used to assess increased vascularity, blood-brain barrier breakdown, etc. The images obtained by means of T_2-weighted (T_{2W}) MRI are generally used to obtain qualitative information concerning the extent of brain damage in biological subjects, whereas the possibilities offered by a more detailed analysis of nuclear relaxation curves in different in vivo experimental models are often ignored [26-28].

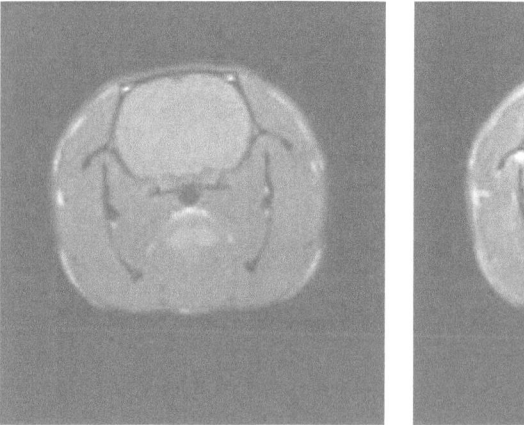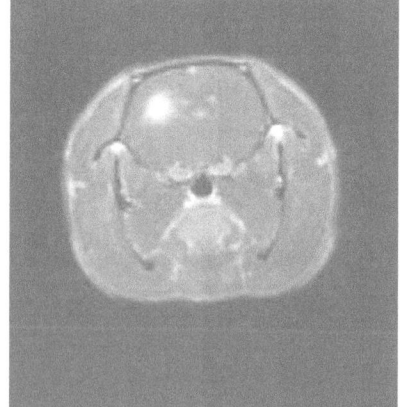

Fig. 13. SHR-SP with cerebral damages. T_{1W} MR images of the coronal section of the same slice imaged before (a) and after (b) gadolinium injection show leakage through the Blood Brain Barrier [13].

In T_{1W} and T_{2W} images, relaxation curves are generally analyzed using a single exponential component, which leads to conventional images in which the "gray" intensity depends on the local T_2 (T_1) value. However, in many cases, studies of transversal (T_2) nuclear magnetization by means of the well-known Carr-Purcell-Meiboom-Gill (CPMG) technique reveal a behavior that is not monoexponential (see fig. 14), which can be explained by the presence of many "non-equivalent" hydrogen nuclei in biological tissues that have different nuclear relaxation mechanisms (rotational, diffusional, etc.) also due to different environments (i.e. the different molecules causing the relaxation). The use of MRI contrast agents can be of further help in understanding the evolution of brain damage [7,29]. It is well known that the gadolinium-based paramagnetic contrast agents (CA) reduce the T_1 and T_2 relaxation times (see [7] and references therein). Given the small size of the molecule (mean diameter ≈ 0.35 nm), it should be possible to highlight variations in the permeability of the membranes separating different tissue compartments. On the other hand, the superparamagnetic USPIO (ultra-small superparamagnetic iron-oxide particles) contrast agent is formed by molecules with dimensions of the order of 5/10 nm and gives extraordinary variations in magnetic susceptibility, thus leading to very well imaged loss of MRI signal.

Despite the considerable interest in brain pathologies and MRI contrast agents, no published study has compared the T_2 times extracted from multiexponential relaxation without and with contrast agent in the presence of lesions in rat models. The experimental CPMG data in MCAO and SHR-SP rats, without and with Multihance (Bracco Imaging SpA, Milan, Italy) paramagnetic contrast agent, reveal two components (at intermediate times) in the exponential relaxation of the T_2-curves that reflect the behavior of the protons belonging to different molecular groups [14]. Different T_2 values can be attributed to a well-defined water molecular group belonging to a specific compartment. Concentrating on the two intermediate relaxation times T_{2A} (relative weight A) and T_{2B} (relative weight B), the data suggest the following interpretations:

1) If A is interpreted as the relative weight of the protons in the intracellular compartment, and B the relative weight of the protons in the extracellular compartment, it can be observed that the qualitative temporal behavior of A and B is the same in SHR-SP at times around the damage onset, and in the first 24 hrs. in MCAO rats.

2) In MCAO rats, the data obtained two hours after surgery suggest that the ADC decreases whereas the monoexponential T_2 used in T_{2W} images does not change, as has previously been verified in the case of cytotoxic edema.

3) The increase in B at the time of the damage onset in SHR-SP is sudden and marked, and less marked in MCAO rats 24 hrs after surgery.

In conclusion, the analysis of multiexponential relaxation curves confirmed the dynamics of edema development in models of cerebral ischemia.

For further investigation on cerebral damages, data with the use of USPIO contrast agent were collected. T_2-weighted (T_{2W}) images obtained on MCAO and SHR-SP rats after injection of USPIO (Sinerem, Guerbet Group, Roissy, France), evidence very clearly the damaged zone thus confirming the Blood

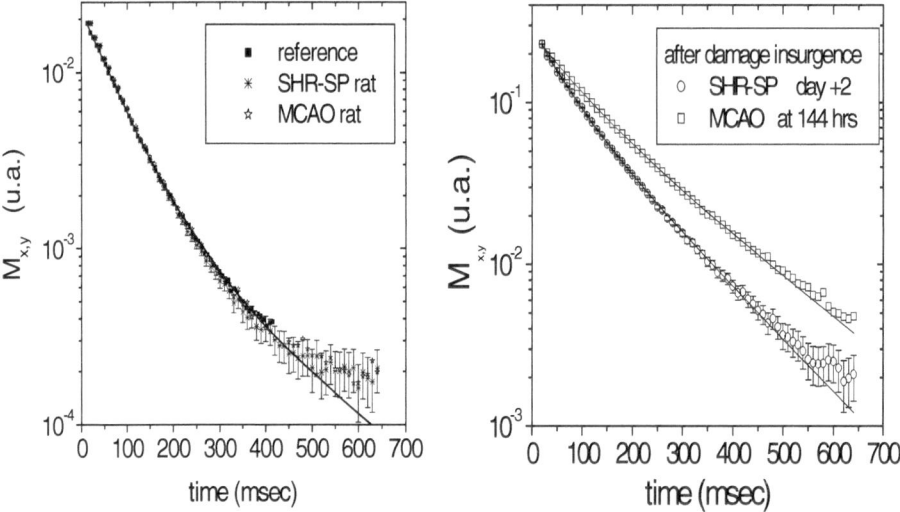

Fig. 14. Relaxation of transverse nuclear magnetization. (a) No damage and no contrast agent (Multihance) data in MCAO rats before artery occlusion and in SHR-SP rats well before damage onset; the results obtained in one reference rat are also shown. (b) The same MCAO and SHR-SP rats with damage at different times from onset. The solid lines are fits to the experimental data in the range $20 < t < 450$ ms [14]. In (a), only the fit to the reference rat data is shown; the other two fits are not shown for reasons of clarity. The error bars correspond to the standard deviation of uniform distribution of width equal to the experimental accuracy.

Brain Barrier disruption [16]. The region of interest appears darker in the central core of the damage (because of T_2^* effect generated by high concentration of USPIO) and more brilliant in the border zones. The multiecho T_{2W} images obtained with USPIO were analyzed in terms of multiexponential T_2-relaxation and compared with the case of Multihance. The relaxation curves resulted almost identical for both CA (data not reported) and the presence of two T_2 values in the range 0.03-1 s with very different weights, named T_{2A} and T_{2B}, was again evidenced.

References

1. A. Abragam, "the principles of Nuclear Magnetism", Clarendon Press (Oxford, 1961)
2. for an introduction to NMR in liquids see e.g. R. V. Parish, "NMR; NQR, EPR and Mossbauer spectroscopy in inorganic chemistry", Ellis Horwood (Chichester, 1990)

3. E. Mark Haacke, Robert W. Brown, Michael R. Thompson, Ramesh Venkatesan. "Magnetic Resonance Imaging. Physical Principles and Sequence Design", Wiley-Liss (New York, 1999)

4. Paul T. Callaghan, "Principles of Nuclear Magnetic Resonance Microscopy", Oxford University Press (Oxford, 1994)

5. L. Cei, A. La Fianza, C. Baluce, "Tecniche di Tomografia Computerizzata e Risonanza Magnetica nella Diagnostica per Immagin", Societa' Editrice Universo (Roma, 2001)

6. R.R. Ernst, W.A. Anderser, Rev. Sci. Intrum. 37, 93 (1966)

7. R.N. Muller. Contrast agents in whole body magnetic resonance: operating mechanisms, Encyclopedia of nuclear magnetic resonance, eds. Wiley J. and Sons; 1438-1444 (N.Y. 1996)

8. S. Aime, M.Botta et al., Chem.Soc.Rev. 27, 19 (1998); P. Caravan, J.J.Ellison et al., Chem.Soc.Rev. 99, 2293 (1999)

9. Kamil Ugurbil, Wei Chen, Xiaoping Hu, Seong-Gi Kim, Xiao-Hung Zhu. Functional MRI at high fields: practice and utility.

10. see e.g., Rohde GK, Barnett AS, Basser PJ, Marenco S, Pierpaoli C, Comprehensive approach for correction of motion and distortion in diffusion-weighted MRI, Magn. Res. Med. 51, 103-114 (2004), and references therin

11. U. Guerrini, L. Sironi, E. Tremoli, M. Cimino, B. Pollo, A.M. Calvio, R. Paoletti, M. Asdente. New insights into brain damage in stroke-prone rats: a nuclear magnetic imaging study. Sroke 33, 825-830 (2002)

12. L. Sironi, E. Tremoli, I. Miller, U. Guerrini, A.M. Calvio, I. Eberini , M. Gemeiner, M. Asdente, R. Paoletti , E. Gianazza. Acute-phase proteins before cerebral ischemia in stroke-prone rats. Identification by proteomics. Stoke 32, 753-760 (2001)

13. L. Sironi, U. Guerrini, E. Tremoli, I. Miller, P. Gelosa, A. Lascialfari, I. Zucca, I. Eberini, M. Gemeiner, R. Paoletti, E. Gianazza. Analysis of patological events at the onset of brain damage in stroke-prone rats: a proteomics and magnetic resonance imaging approach. Journal of Neuroscience Research 78, 115-122 (2004)

14. A. Lascialfari, I. Zucca, M. Cimino, U. Guerrini, R. Paoletti, L. Sironi, E. Tremoli, V. Lorusso and M. Asdente. Multiexponential analysis of MRI T2-relaxation in cerebral damaged rats with and without a paramagnetic Gd-based contrast agent. Mag. Res. Med., 53, 1326 (2005)

15. M. Takahashi, B. Fritz-Zieroth, T. Chikugo, H. Ogawa. Differentiation of chronic lesions after stroke in stroke-prone spontaneously hypertensive rats using diffusion weighted MRI. NMR 30, 485-488 (1993).

16. A. Lascialfari, I. Zucca, M. Cimino, U. Guerrini, R. Paoletti, L. Sironi, E. Tremoli, "Guerbet Group", work in progress

17. M.R. Crain, W.T.C. Yuh, G.M. Greene, D.J. Loes, T.J.Ryals, Y. Sato, M.N.Hart. Cerebral ischemia: evaluation with contrast-enhanced MR imaging. AJNR Am J Neuroradiol 12, 631-639 (1991)

18. R.O. Lindsey, Z.F. Yetkin, R. Prost, V.M. Haughton. Effect of dose and field strength on enhancement with paramagnetic contrast media. AJNR Am J Neuroradiol. 15, 1849-1852 (1994)

19. V. P. Mathews, L.H. Monsein, C.A. Pardo, R.N. Bryan. Histologic abnormalities associated with gadolinium enhancement on MR in the initial hours of experimental cerebral infarction. AJNR Am J Neuroradiol 15, 573-579 (1994)

20. S. Imakita, T. Nishimura, N. Yamada, H. Naito, M. Takamiya, Y. Yamada, J. Minamikawa, H. Kikuchi, M. Nakamura, T. Sawada, J. Choki, T. Yamaguchi. Magnetic resonance imaging of cerebral infarction: time course of Gd-DTPA enhancement and CT comparison. Neuroradiology **30**, 372-378 (1988)
21. R.A. Knight, P.B. Barker, S.C. Fagan, Y. Li, M.A. Jacobs, K.M.A. Welch. Prediction of impending hemorrhagic transformation in ischemic stoke using magnetic resonance imaging in rats. Stroke **29**, 144-151 (1998)
22. W. Reith, M. Forsting, H. Vogler, S. Heiland, K. Sartor, M.R. Early. Detection of experimentally induced cerebral ischemia using magnetic susceptibility contrast agents: comparison between gadopentetate dimeglumine and iron oxide particles. AJNR Am J Neuroradiol. **16**, 53-60 (1995)
23. O. Haraldseth, R.A. Jones, T.B. Muller, A.K. Fahlvik, A.N. Oksendal. Comparison of dysprosium DTPA BMA and superparamagnetic iron oxide particles as susceptibility contrast agents for perfusion imaging of regional cerebral ischemia in the rat. J. Magn. Reson. Imag. **16**, 714-717 (1996)
24. A. Doerfler, T. Engelhorn, S. Heiland, M. Knauth, I. Wanke, M. Forsting. MR contrast agents in acute experimental cerebral ischemia: potential adverse impacts on neurologic outcome and infarction size. J .Magn. Reson. Imag. **11**, 418-424 (2000)
25. M. Takahashi, B. Fritz-Zieroth, T. Chikugo, H. Ogawa. Differentiation of chronic lesions after stroke in stroke-prone spontaneously hypertensive rats using diffusion weighted MRI. MRM. **30**, 485-488 (1993)
26. R.M. Kroeker, R.M. Henkelman. Analysis of biological NMR relaxation data with continuous distributions of relaxation times. J. Magn. Reson. **69**, 218-235 (1986)
27. K.P. Whittall, M.J. Bronskill, R.M. Henkelman. Investigation of analysis techniques for complicated NMR relaxation data. J. Magn. Reson. **95**, 221-234 (1991)
28. K.P. Whittall, A.L. MacKay. Quantitative interpretation of NMR relaxation data. J. Magn. Reson. **84**, 134-152 (1989)
29. A.D. Watson, S.M. Rocklage. Theory and mechanisms of contrast-enhancing agents. Magnetic Resonance Imaging of the Body. Second Edition, Raven Press, New York, (1992); 1257-1287.

Phases, phase transitions and spin dynamics in strongly correlated electron systems, from antiferromagnets to HT_C superconductors: NMR-NQR insights

A. Rigamonti[1] and F. Tedoldi[2]

[1] Department of Physics "A. Volta" Via Bassi 6, I-27100 Pavia (Italy)
  rigamonti@fisicavolta.unipv.it
[2] Bruker BioSpin S.r.l. Via Pascoli 70/3, I-20133 Milano (Italy)
  fabio.tedoldi@bruker.it

1 Basic aspects of phase transitions and critical dynamics as reflected in NMR-NQR relaxation.

1.1 Introduction and free energy for homogeneous systems

Phase transitions (PT) are typical collective phenomena occurring in many-body systems with interparticle interactions. A rather general description of the PT's and of the accompanying "critical dynamics" can be given in the framework of Landau-type statistical theory, which includes more specific theories such as the Weiss mean field theory for magnetic systems and the Van der Waals theory for fluids. This approach is suited for second order or slightly first-order PT[3].

---

[3] At the transition one observe "anomalous" behaviour (such as divergencies or tendencies to zero) of the *response functions*, namely the derivatives of the *thermodynamical densities* (e.g. magnetization, particle density, entropy) with respect to the *conjugate fields* (magnetic or electric fields, pressure, temperature). That behaviour reflects the onset of correlation and slowing down of the fluctuations around the equilibrium values, fluctuations in turn driven by some microscopic critical dynamics. The *order of a transition* is related to the order of the derivative of the thermodynamical potentials which display discontinuity at the critical point. *First order*, when the first derivatives with respect to the conjugate variable display discontinuity. Tipically this classification refers to the *order parameter*, namely the density going toward zero when the critical point is approached along the coexistence curve. For *second order transition* only the second derivatives of the thermodinamical potential have discontinuity at the critical point. At variance, the order parameter goes to zero with continuity on approaching the critical point.

In that framework one defines a generalized site-dependent order parameter $m(\mathbf{r})$ having a field h as thermodinamically conjugated (we shall use scalar quantities for simplicity). A proper average of $m(\mathbf{r})$ is defined as the *thermodinamic density*, going gradually to zero on moving along the coexistence curve towards the critical point. From a microscopic point of view there is a certain correspondence of m to a local critical variable (e.g. local spins to the magnetization), as we shall see.

The general principle is to expand the free energy density $f(m(\mathbf{r}, T))$ in powers of $m(\mathbf{r})$, with coefficients that depend on $(T - T_c)$ (T_c critical temperature), the total free energy being

$$F = \int f(m(\mathbf{r}, T)) \, d\mathbf{r}$$

By taking into account that f cannot depend from the sign of m and must increase when a gradient of m occurs (and $\boldsymbol{\nabla} m$ must be involved to the second power for isotropy reasons) one writes[4]

$$f(m(\mathbf{r}), T) = f_0(T) + \alpha(T)m^2(\mathbf{r}) + \frac{1}{2}\beta(T)m^4(\mathbf{r}) + \gamma(T)|\boldsymbol{\nabla} m(\mathbf{r})|^2 \qquad (1)$$

Then $\alpha(T)$ and $\beta(T)$ can be expanded in series of $(T - T_c)$. For homogeneous system $\boldsymbol{\nabla} m = 0$ and by assuming $\beta(T) = b_0$ temperature independent, Eq.(1) is rewritten

$$f(m, T) = f_0(T) + a_0(T - T_c)m^2 + \frac{b_0}{2}m^4 \qquad (2)$$

Thus for $T > T_c$ the minimum of F is at $m = 0$. For $T < T_c$ the minimum is at the equilibrium value

$$\overline{m} = \left[-\left(\frac{a_0}{b_0}\right)(T - T_c)\right]^{\frac{1}{2}} = m_0(-\epsilon)^{\frac{1}{2}} \qquad (3)$$

with $\epsilon = \frac{(T-T_c)}{T_c}$. Therefore according to (3) the *critical exponent* describing the temperature dependence of the order parameter for T-decreasing below T_c is $\beta = \frac{1}{2}$, often indicated as "classical" or mean-field exponent.

NMR-NQR spectra (through splittings or shifts of some resonance lines) or μSR (from precessional frequencies) are among the best experimental tools to study the temperature behaviour of the order parameter and of other related "static" quantities (such, for instance, expectation values of local spin operator through magnetic hyperfine fields or atomic displacements at structural phase transitions, through the electric field gradients at the nuclear sites).

---

[4] In this expansion the odd terms are not present because the form of the forces cannot change on reversing the sign of m. The term $\boldsymbol{\nabla} m$ has the role to damp out the spatial variation in m (for $\gamma(T) > 0$), f having a minimum for m constant. The expansion does not converge but it is customary to suppose that the singularities involve the coefficients of the terms which are not explicitly considered.

From the equation of state yielding $m = m(T, h)$ (as for instance the equation for the magnetization in the Weiss mean field theory, or for the inverse density $\rho^{-1} \equiv V$ for the isotherms of real gases) one can find the response of m to the conjugate field, namely the *generalized susceptibility* $\chi \equiv \frac{\partial \overline{m}}{\partial h}$ tipically going around T_c as $\chi \propto (T - T_c)^{-1} \propto \epsilon^{-1}$, with classical critical exponent $\gamma = 1$. From the free energy density $f(T)$ as in Eq.(2), the specific entropy $s = -\left(\frac{\partial f}{\partial T}\right)$ and the specific heat $c = T\left(\frac{\partial s}{\partial T}\right)$ (the response function to a thermal stimulus) can be directly derived.

It should be remarked that for the superconducting transition the role of the free energy expansion requires further detail and arguments. This happens because the order parameter $\Psi(\mathbf{r})$ in that case is complex, i.e. $\Psi(\mathbf{r}) = \sqrt{n_c}e^{i\theta(\mathbf{r})}$ (with n_c = density of Cooper pairs, and $\theta(\mathbf{r})$ = phase) and it cannot be considered site-independent, as we shall see.

1.2 Non-homogeneous systems and fluctuations

The assumption of homogeneity for $m(\mathbf{r})$ is inappropriate in the neighborhood of a phase transition, where the fluctuations around the equilibrium value $\overline{m}$ play a dominant role, since the curvature of f vs. m is going to zero for $T \to T_c$, according to Eq(2). Actually, one could say that is just the enhancement of the fluctuations for $T \to T_c^+$ that drives the transition to the low-temperature phase.

The probability that a thermodynamic variable yields a fluctuation implying an increase of free energy ΔF is given by[5]

$$W \propto e^{-\Delta F/k_B T} \qquad (4)$$

Then, in the presence of fluctuations one has to go back to Eq.(1), by considering $\gamma \neq 0$, often assumed temperature independent. The term in m^4 is neglected[6].

---

[5] The probability that at constant temperature a fluctuation occurs is given by $w(m) = A \, exp[S(m)/k_B]$, where S is the entropy of the system in correspondence to the value m of the density. As usual $S(m)$ has a gaussian distribution around the mean value $S(\overline{m})$ and thus the probability of a fluctuation is given by Eq.(4). When the transition is approached, the free energy being a slowly varying function of m, large fluctuations occur. The mean square value $\langle|m|^2\rangle$ is related to the response function χ by the fluctuation-dissipation theorem. Therefore the divergency of χ at the critical point, for a second-order PT, again evidences the divergency of the fluctuations.

[6] This is known as *gaussian approximation* (or first-order fluctuation correction) and it breaks down in the so called *critical region*. In most systems this region corresponds to a very narrow temperature range, where ϵ is close to zero. Only in particular cases one cannot neglect the term in m^4, as for instance in restricted dimensionality or for the superconducting transition in small size grains, where the range of the critical region is expanded.

Furthermore one has to expand $m(\mathbf{r})$ in Fourier series

$$m(\mathbf{r}) - \overline{m} = \sum_{\mathbf{q}} m_{\mathbf{q}} e^{-i\mathbf{q}\cdot\mathbf{r}} \tag{5}$$

and then

$$\frac{\Delta F}{V} = \frac{1}{V} \int (f - f_0)dV = \sum_{\mathbf{q}} |m_{\mathbf{q}}|^2 \alpha(T) \left[1 + \frac{\gamma q^2}{\alpha(T)}\right] \tag{6}$$

One has to take the thermodynamic average of $|m_{\mathbf{q}}|^2$ over all the possible values of the fluctuating order parameter, in accordance to Eqs.(4) and (6):

$$\langle |m_{\mathbf{q}}|^2 \rangle = \frac{\int |m_{\mathbf{q}}|^2 e^{-\Delta F/k_B T} dm_{\mathbf{q}}}{\int e^{-\Delta F/k_B T} dm_{\mathbf{q}}} \quad ,$$

or

$$\langle |m_{\mathbf{q}}|^2 \rangle = \frac{k_B T}{2V\alpha(T)\left[1 + \frac{\gamma}{\alpha(T)}q^2\right]} \tag{7}$$

which corresponds to the energy $k_B T$ for each orthogonal mode of the system, namely for each $\mathbf{q}$ value. From the second term in Eq.(2) $\alpha(T) = a_0(T - T_c)$, Eq.(7) can be rewritten

$$\langle |m_{\mathbf{q}}|^2 \rangle \propto \frac{1}{\left[(T - T_c) + k_0^{-2}q^2\right]} \tag{8}$$

with k_0 constant wave vector.

For homogeneous fluctuation (i.e. $\mathbf{q} = 0$) one notes the divergence of $\langle |m_{\mathbf{q}=0}|^2 \rangle$ as in the static generalized susceptibility $\chi(\mathbf{q} = 0, \omega = 0)$, with critical exponent $\gamma = 1$ (see note 5). This divergence describes the *enhancement of the fluctuations* on approaching the transition. Here $\mathbf{q} = 0$ can be defined the *critical wave vector*.

One can also write Eq.(8) in the form

$$\langle |m_{\mathbf{q}}|^2 \rangle = \frac{\langle |m_{\mathbf{q}=0}|^2 \rangle}{1 + \xi^2 q^2} \tag{9}$$

where

$$\xi(T) \equiv \left(\frac{\gamma}{\alpha}\right)^{\frac{1}{2}} = \frac{1}{k_0 \epsilon^{\nu}} \equiv \xi_0 \epsilon^{-\frac{1}{2}} \tag{10}$$

is the *correlation length*, diverging with the critical exponent $\nu = 1/2$ for $T \to T_c$.[7] The role of ξ as correlation length can be better understood looking at the spatial correlation function for the fluctuations of $m(\mathbf{r})$.

---

[7] The critical exponents in most cases are different from the "classical" values derived in mean field-like theories. In particular that can happen when the temperature range of the critical region is expanded or when the interactions are short-range or when the dimensionality of the system is reduced.

By referring for simplicity to the case $T > T_c$ where $\overline{m} = 0$, the correlation function is

$$g(\mathbf{r}, \mathbf{r}') = \langle m(\mathbf{r})m(\mathbf{r}') \rangle_0 = \sum_{\mathbf{q}} \langle |m_{\mathbf{q}}|^2 \rangle e^{-i\mathbf{q}\cdot(\mathbf{r}-\mathbf{r}')} \tag{11}$$

For $\mathbf{R} = \mathbf{r} - \mathbf{r}'$, from Eq.(7) and $\sum_{\mathbf{q}} \rightarrow \frac{1}{8\pi^3} \int d\mathbf{q}$, one obtains

$$g(\mathbf{R}) = \langle |m(0)|^2 \rangle \left[\frac{1}{R} e^{-Rk_0 \left(\frac{T-T_c}{T} \right)^{1/2}} \right] = \langle |m(0)|^2 \rangle \left[\frac{1}{R} e^{-R/\xi(T)} \right] \tag{12}$$

showing that $\xi(T) = \xi_0 \epsilon^{-1/2}$ is indeed a measure of the correlation.

For $T \gg T_c$ one has a fast decay of the correlations while at T_c, where $\xi(T) \rightarrow \infty$, $g(\mathbf{R}) \propto R^{-1}$ (note that $\gamma > 0$).

Thus we have deduced the enhancement and the onset of correlated fluctuations for T approaching T_c from above. It is noted that for $T < T_c$, where the fluctuations occur around $\overline{m}$ and $\alpha = 2a_0(T_c - T)$ analogous divergences are obtained for $\langle |m_{\mathbf{q}=0}|^2 \rangle$, $\xi(T)$ and for the inverse correlation length $k = k_0 \epsilon^\nu$, with the same critical exponents as above T_c.

1.3 The time dependence of the fluctuations

So far we did not have to take into account the time-dependence of the fluctuations. However the life time of the fluctuations (or, in other words, the microscopic "anomalous" dynamics of the local critical variables, such spin-components $\mathbf{s}(t)$ or local density departure $\delta\rho(\mathbf{r}, t) = \rho(\mathbf{r}, t) - \rho_0$ for fluids, to which the macroscopic order parameter can be connected) is of crucial importance in phase transitions. One can remark that NMR-NQR relaxation, with neutron or light scattering, is among the reliable tools to study the time-dependence of the fluctuations, as it will be emphasized by illustrative examples given later on.

In the framework of a general, Landau-type statistical theory the time-dependence of $m(\mathbf{r}, t)$ around the equilibrium value can be discussed by starting from an equation based on the extension of the thermodynamics of the irreversible processes.

For the collective, Fourier component $m_{\mathbf{q}}$, for $\mathbf{q}$ around the critical wave vector $\mathbf{q} = 0$, we write

$$M\frac{d^2 m_{\mathbf{q}}}{dt^2} + \Omega_{\mathbf{q}}\frac{dm_{\mathbf{q}}}{dt} + k_{\mathbf{q}}m_{\mathbf{q}} = 0 \tag{13}$$

namely the equation for a damped harmonic oscillator.[8] The effective elastic constant $k_{\mathbf{q}}$ must be such that for $\mathbf{q} = 0$ $k_{\mathbf{q}=0} \propto (T - T_c)$, since it is related to the curvature of the free energy as a function of m. The relaxation

---

[8] From an extension of the thermodynamics of the irreversible processes the following phenomenological equation for the order parameter has been written

term in Eq.(13) involves a coefficient $\Omega_{\mathbf{q}}$ that can be assumed temperature independent. With respect to the usual relaxation equation

$$\frac{\partial m}{\partial t} = -\frac{1}{c}\frac{\delta f}{\delta m}$$

(with c non-critical kinetic coefficient) and leading to an exponential decay for the fluctuation[8], in Eq.(13) one finds an inertial term, with a generalized mass M. This term is required, in particular, to describe structural phase transitions, while it can be neglected for magnetic PT where the local critical variable can be identified with a spin operator.

From the equation of motion there are two methods to obtain the correlation and the slowing down of the fluctuations. One way is to look for the response of the system to an external, wave-dependent and oscillating field $h(\mathbf{q},\omega) = h_0 e^{-i\mathbf{q}\cdot\mathbf{r}}e^{-i\omega t}$, thus deriving the response function, namely the generalized susceptibility $\chi(\mathbf{q},\omega) = \chi_{\mathbf{q}}'(\omega) + i\chi_{\mathbf{q}}''(\omega)$. Then, by means of the fluctuation-dissipation theorem, the power spectrum of the correlation function $\langle m_{\mathbf{q}}(0)m_{-\mathbf{q}}(t)\rangle_0$ is obtained.

The $\mathbf{q}$ and ω transform of the correlation function $g(\mathbf{r},t) = \langle m(0,0)m(\mathbf{r},t)\rangle_0$ is known as *dynamical structure factor* (DSF):

$$S(\mathbf{q},\omega) = \int e^{-i(\omega t - \mathbf{q}\cdot\mathbf{r})}g(\mathbf{r},t)d\mathbf{r}dt \equiv \int e^{-i\omega t}\langle m_{\mathbf{q}}(0)m_{-\mathbf{q}}(t)\rangle_0 dt \qquad (16)$$

Another method is to construct the correlation function $g_{\mathbf{q}} = \langle m_{\mathbf{q}}(0)m_{\mathbf{q}}(t)\rangle_0$ from Eq.(13). For underdamped normal oscillators, i.e. $\Omega_{\mathbf{q}}^2 \ll 4Mk_{\mathbf{q}}$ the solution of Eq.(13) is

$$m_{\mathbf{q}}(t) = e^{-\frac{\Omega_{\mathbf{q}}}{2M}t}\left\{m_{\mathbf{q}}(0)\cos\omega_{\mathbf{q}}t + \left[\dot{m}_{\mathbf{q}}(0) + \frac{\Omega_{\mathbf{q}}}{2M}m_{\mathbf{q}}(0)\right]\frac{\sin\omega_{\mathbf{q}}}{\omega_{\mathbf{q}}}\right\} \qquad (17)$$

with

_____

$$\frac{dm}{dt} = -c\frac{\partial f}{\partial m} - c\frac{d(\partial f/\partial\dot{m})}{dt} \quad . \qquad (14)$$

Without the last term the equation has the well-known meaning: the speed to approach the equilibrium (after a variation induced by the fluctuation) is proportional to the restoring force. In this case from $f = a(T - T_c)m^2$, expressing m as $m_0 + \delta m$ and linearizing with respect to δm, the regression equation is

$$\delta m(t) = \delta m(0)exp[-t/\tau] \qquad (15)$$

The time constant $\tau \propto (T - T_c)^{-1}$ going to zero at the transition, describes the classical Landau-Kalatnikov slowing down of the fluctuations.

The meaning of the second term in Eq.(14) is better understood when, after the expansion of the energy in terms of the deviation from the equilibrium value and Fourier transform to the collective variable, Eq.(14) is rewritten in the form of Eq.(13) (strictly valid only in the limit $\mathbf{q} \to \mathbf{0}$).

$$\omega_{\mathbf{q}} = \left[\frac{k_{\mathbf{q}}}{M} - \frac{\Omega_{\mathbf{q}}^2}{4M^2} \right]^{\frac{1}{2}}$$

yielding the correlation function[9]

$$g_{\mathbf{q}}(t) = \langle |m_{\mathbf{q}}(0)|^2 \rangle e^{-\frac{\Omega}{2M}t} \left[\cos \omega_{\mathbf{q}} t + \frac{\sin \omega_{\mathbf{q}} t}{\omega_{\mathbf{q}}} \cdot \frac{\Omega_{\mathbf{q}}}{2M} \right] \qquad (18)$$

The DSF turns out

$$S(\mathbf{q}, \omega) \propto \frac{4\Omega_{\mathbf{q}} \omega_{\mathbf{q}}^2 / M}{(\omega_{\mathbf{q}}^2 - \omega^2)^2 + 4\Omega_{\mathbf{q}}^2 \omega^2 / M^2} \qquad (19)$$

namely in the form of two resonant peaks centered at $-\omega_{\mathbf{q}}$ and at $+\omega_{\mathbf{q}}$, of width $\Omega_{\mathbf{q}}/M$.

In general, for any extent of the damping, the solution of Eq.(13) can be written in the form

$$m_{\mathbf{q}}(t) = m_{\mathbf{q}}(0) e^{i\tilde{\omega} t} \qquad , \qquad (20)$$

$\tilde{\omega} \equiv \omega_{\pm}$ being two complex frequencies with

$$Im = \Omega_{\mathbf{q}}/2M \qquad (21a)$$

and

$$Re = \frac{\left[4Mk_{\mathbf{q}} - \Omega_{\mathbf{q}}^2 \right]^{1/2}}{2M} \qquad (21b)$$

If T^* is defined as the temperature at which $\Omega_{\mathbf{q}\sim0}^2 = 4Mk_{\mathbf{q}\sim0}$, for $T < T^*$ both poles lie on the imaginary axis and $\omega_- = -ik_{\mathbf{q}\sim0}/\Omega_{\mathbf{q}\sim0}$ moves towards the origin of the complex plane. Accordingly the transition is approached with ω_- going as $\propto (T - T_c)$.

Thus Eqs.(20) and (21) describe the slowing-down of the fluctuations for $T \to T_c^+$. In terms of the DSF it can be said that above a given temperature one has the temperature dependent resonant peaks of constant width while for $T \leq T^*$ the slowing-down is rather described by a diffusive-type central peak, centered at $\omega = 0$, of width that decreases while the transition is approached. Equivalently, an effective correlation time $\tau_{\mathbf{q}\sim0} \propto \omega_{\mathbf{q}}^{-1} \propto (T - T_c)$ goes to infinity for $T \to T_c^+$.

---

[9] Due to the deterministic character of the motion (Eq.(13)) the statistical ensemble average for the correlation function involves the initial ($t = 0$) condition, i.e. the value of $m_{\mathbf{q}}(0)$. Note that in Eq.(17) $\dot{m}_{\mathbf{q}}(0) = 0$, since the recovery of the collective order parameter towards the equilibrium, after a fluctuation, initiates from zero velocity. The mean square value $\langle |m_{\mathbf{q}}(0)|^2 \rangle$ is related to the temperature by the condition

$$\langle |m_{\mathbf{q}}(0)|^2 \rangle_0 = \frac{1}{2A_{\mathbf{q}}} \int_{-A_{\mathbf{q}}}^{+A_{\mathbf{q}}} m_{\mathbf{q}}^2(0) dm_{\mathbf{q}} = \frac{A_{\mathbf{q}}^2}{3}$$

and $k_{\mathbf{q}} A_{\mathbf{q}}^2 = 2k_B T$.

The DSF under favorable circumstances is directly probed by the cross section for inelastic neutron and light scattering. As we shall see by means of illustrative examples, when the time-dependent perturbation hamiltonian driving the NMR-NQR relaxation process can be expressed in terms of $m(\mathbf{r}, t)$ then the relaxation rate is related to the $\mathbf{q}$-integration of $S(\mathbf{q}, \omega = \omega_{res})$. This is the case of the magnetic time-dependence of the relaxation process driven by the hyperfine field at the nuclear site, or by the time-dependent EFG components induced by the critical lattice dynamics accompanying the ferroelectric transitions or other structural phase transitions.

As already mentioned, the fluctuation-dissipation theorem relates the spectrum of the fluctuations to the response to a dynamic perturbation:

$$S(\mathbf{q}, \omega) = \frac{2\hbar \chi''(\mathbf{q}, \omega)}{1 - exp(-\hbar\omega/k_B T)} \simeq \frac{2k_B T}{\omega} \chi''(\mathbf{q}, \omega) \qquad (22)$$

$\chi''(\mathbf{q}, \omega)$ being the dissipative part of the generalized susceptibility. The static structure factor

$$S(\mathbf{q}) = \frac{1}{2\pi} \int S(\mathbf{q}, \omega) d\omega \qquad (23)$$

yields the mean square amplitude of the collective $m_\mathbf{q}(t)$ and then

$$S(\mathbf{q}) = \langle |m_\mathbf{q}|^2 \rangle = k_B T \chi(\mathbf{q}, 0) \qquad (24)$$

$\chi(\mathbf{q}, 0)$ being the static wave-vector dependent susceptibility.

In view of the analysis of the NMR relaxation data some model forms for the $\mathbf{q}$-dependence of the fluctuations are necessary. For the static structure factor $S(\mathbf{q}) = \langle |m_\mathbf{q}|^2 \rangle$, for temperature close to T_c where the fluctuations at the critical wave vector $\mathbf{q}_c$ (here $\mathbf{q} = 0$) are dominant, one can tentatively adopt the form given by Eq.(9) in term of the correlation length.

The $\mathbf{q}$-dependence of the time-dependent quantities $\tilde{\omega}$ in Eq.(21) is a very complicated issue, in essence involving the dynamics of cooperative effects. Several theoretical approaches have been developed along the years, all with limitations or possible criticisms.

For the present lectures we will rely on general arguments that basically are simplified forms of the dynamical scaling theories. The dynamical part $J_\mathbf{q}(\omega) \equiv S(\mathbf{q}, \omega)/S(\mathbf{q})$ of the DSF is written

$$J_\mathbf{q}(\omega) = \frac{2\pi}{\Gamma_\mathbf{q}} f(\omega/\Gamma_\mathbf{q}) \qquad (25)$$

where $\Gamma_\mathbf{q} = \Gamma_\mathbf{q}(T)$ is a characteristic frequency and $f(\omega/\Gamma_\mathbf{q})$ is a well-behaving function depending on the frequency only through the ratio $\omega/\Gamma_\mathbf{q}$, with a width and an area of the order of the unity (to be determined either by experiments or by the theoretical model[10]).

---

[10] One can use an expansion of the time-dependent correlation function in a series in t, the coefficients being the equal-time correlation functions related to the

Then the amplitude of the fluctuations i.e. $S(\mathbf{q}) = \langle|m_\mathbf{q}|^2\rangle$ and the decay rates $\Gamma_\mathbf{q}$ in (20) are scaled in their $\mathbf{q}$-dependence in terms of the correlation length ξ. In the light of Eq.(8) and (9), with a slight generalization we write

$$\langle|m_\mathbf{q}|^2\rangle \propto \xi^{2-\eta} f(q\xi) \tag{26}$$

where f is a homogeneous function of $q\xi$ and η a correction exponent (for dimensionality $D \geq 2$ η is close to zero). The simplest form of f is

$$f(q\xi) = \frac{1}{1 + q^2\xi^2} \tag{27}$$

in accordance to Eq.(9) and corresponding to the Ornstein-Zernike expansion of the correlation function in the theory for the critical opalescence at the gas-liquid transition (the odd-order terms being dropped out because of the isotropy of the system).

For the decay rate $\Gamma_\mathbf{q}$ in Eq.(25) the analogous of Eq.(26) is

$$\Gamma_\mathbf{q}(T) = \Gamma_{\mathbf{q}_c}(T) g(q\xi) \tag{28}$$

(where for the moment $q_c = 0$). g is the homogeneous function and in its simplest form it reads

$$g(q\xi) = 1 + q^2\xi^2 \equiv f^{-1} \tag{29}$$

According to the general arguments given above, $\Gamma_{\mathbf{q}\approx\mathbf{q}c}$, in Eq.(27) must decrease towards zero for $T \to T_c^+$. It can be assumed of the form $\Gamma_{\mathbf{q}\approx\mathbf{q}_c} \propto \xi^{-z} \propto \epsilon^{\nu z}$ where z is the *dynamical critical exponent* (z in the dynamical scaling also controls the $\mathbf{q}$-dependence of $\Gamma_\mathbf{q}$ at T_c i.e. $\Gamma_\mathbf{q} \simeq Aq^z$, in other words the exponent expressing the $\mathbf{q}$-dependence of $\Gamma_\mathbf{q}$ at T_c is the same as the exponent that expresses the ξ dependence of $\Gamma_\mathbf{q}$ above T_c).

It is noted that combining Eqs.(25)-(29) our scaling relationships correspond to scale the generalized susceptibility in Eqs.(21) and (24) in the form

$$\chi(\mathbf{q}, \omega) = \chi_0 \xi^z f(q\xi, \omega/\xi^z) \tag{30}$$

ξ_0 being the single particle response function. In other words $\chi(\mathbf{q}, \omega)$ can be expressed as a function *independent* of ϵ provided that length and frequencies are rescaled by appropriate powers of ϵ.

Illustrative examples of NMR spin-lattice relaxation will better clarify the role of the scaling hypothesis used in order to achieve approximate expressions for the wave-vector dependence of the critical quantities that correspond to the enhancement and to the slowing-down of the fluctuations. It can be remarked that if one assumes $z\nu = \gamma = 1$ then the so-called thermodynamical slowing-down condition is attained: on approaching T_c from above the correlation growths at the same rate of the life-time of the "islands" where the correlation is effective.

---

lowest moments of $J_\mathbf{q}(\omega)$. Then approximants are often used, the procedure being essentially the attempt to guess the long-time behaviour of $g(\mathbf{r}, t)$ from the short-time behaviour. Another theoretical line is based on the Mori continuous fraction approximation or on Monte Carlo computer simulations.

1.4 NMR relaxation at the gas-liquid critical point

As mentioned the possibility to study the onset of the correlation and the critical dynamics accompanying the phase transition from NMR relaxation is related to the fact that the relaxation rate can be written in terms of the spectral densities at the resonance frequency ω_{res} of the appropriate lattice functions $F_L(t)$ which described the time-dependence of the fluctuating magnetic field (or EFG) at the nuclear site. When $F_L(t)$ can be expanded in series of the critical variable that is the microscopic counterpart of the thermodynamical order parameter, the relationship of the relaxation rate to the q-integrated DST is attained.

For the transition from the gas to the liquid phase on cooling along the critical isochore, the possibility to derive insights on the DSF is offered by modulation of the intramolecular dipole-dipole interactions due to local density fluctuations. By referring to the fluctuations of the dipolar field $\mathbf{h}_{dip}(t)$ due to the fluctuations in the number density $n(\mathbf{r}, t)$ one has

$$T_1^{-1} \propto \sum_{j,j'} \int e^{-i\omega_{res}t} \langle n(\mathbf{r}_j, 0) n(\mathbf{r}_{j'}, t) \rangle \qquad . \tag{31}$$

After the introduction of the collective variables and by assuming for simplicity for the Fourier transform of the positional functions entering in $\mathbf{h}_{dip}$ the condition of q-independence, then

$$T_1^{-1} = \frac{\gamma^4 \hbar^2}{5d^3} \sum_{\alpha=1,2} \int \alpha^2 S(\mathbf{q}, \alpha\omega_{res}) q^2 dq \tag{32}$$

where d is the distance of minimum approach between molecules.

In the light of Eqs.(22), (24) and (25) we write

$$S(\mathbf{q}, \omega) = \frac{\chi_T}{\chi_0} \frac{1}{1 + q^2\xi^2} J_q(\omega) \tag{33}$$

where χ_T is the isothermal compressibility which controls the strength of the density fluctuation. The DSF in Eq.(33) corresponds to the light scattering spectrum in a fluid. As it is known, it has a three-components structure: the Brillouin doublet related to pressure fluctuations at constant entropy (correspondent to resonant self-propagating modes) and a central Rayleigh component due to entropy fluctuations at constant pressure (correspondent to diffusive, non propagating thermal modes).

Since ω_{res} in Eq.(32) corresponds to the low-frequency part of the DSF, we can assume for $J_q(\omega)$ in Eq.(33) a Lorentzian function, with width $\Gamma_\mathbf{q}$ corresponding to the width of the Rayleigh component yielding the decay rate of the density fluctuations. According to the hydrodynamical theory one has

$$\Gamma_\mathbf{q} = D_T q^2 \tag{34}$$

where D_T is the thermal diffusivity, for which

$$D_T \propto \epsilon^\alpha \tag{35}$$

From Eq.(34), with $S(\mathbf{q}, \omega)$ according to the above assumption one would derive the critical contribution $(T \simeq T_c)$ to the relaxation rate in the form

$$T_1^{-1} \propto D_T^{-\frac{1}{2}} \omega_{res}^{-\frac{1}{2}} \propto \epsilon^{-0.35} \omega_{res}^{-\frac{1}{2}} \tag{36}$$

namely a divergent behavior for $T \to T_c^+$ and a weak frequency dependence.

The experimental results for the proton relaxation (obtained a long ago) around the critical point in chloroform evidence the breakdown of Eq.(36). On the other hand, some time before, in the framework of the mode-mode coupling theory, Kawasaki had argued that the hydrodynamical theory had to be significantly corrected: a scaling function $K(q\xi)$ had to be included in Eq.(34), to give

$$\Gamma_\mathbf{q} = D_T q^2 K(q\xi) \equiv D_T q^2 \frac{3}{4} \left[1 + \frac{1}{q^2\xi^2} + \left(q\xi - \frac{1}{q^3\xi^3} \right) \arctan q\xi \right] \tag{37}$$

which in the limit of $q \ll \xi^{-1}$ yields

$$\Gamma_\mathbf{q} = D_T q^2 K(q\xi) \equiv D_T q^2 \left[1 + \frac{3}{5} q^2\xi^2 - \frac{1}{7} q^4\xi^4 \right]$$

By using for the Rayleigh peak in the DST in Eq.(32) the frequency width given by Eq.(37) one has the following theoretical results:

i) the temperature behaviour of T_1^{-1} displays a maximum *above* T_c which moves towards higher temperatures on increasing ω_{res}.

ii) for $T \geq T_c$ there is a weakly divergent behaviour of T_1^{-1} and for $T \simeq T_c$ one has

$$T_1^{-1} = \frac{\gamma^4 \hbar^2}{5d^3} \frac{2}{\pi} \frac{\chi_T}{\chi_0} \frac{1}{\xi^2} \left(\frac{1}{D_T\xi} \right)^{\frac{1}{3}} \frac{1}{\omega_{res}^{\frac{3}{2}}} \simeq A\omega_{res}^{-\frac{3}{2}} \tag{38}$$

As shown in Fig.1, in a temperature range of about 3 degrees above T_c the experimental data satisfactorily agree with the theoretical predictions when Eq.(37) is used for $\Gamma_\mathbf{q}$. In particular the frequency dependence at T_c is close to $\omega_{res}^{-\frac{3}{2}}$ and the maxima in the relaxation rate occur above T_c.

Although several improvements are evidently required for a better agreement between theory and the experimental data, from this study (unfortunately no other one has been carried out along almost three decades, in the author's knowledge) it appears that under proper conditions the spin-lattice relaxation can actually probe the critical density fluctuations. While light scattering is almost confined to the $\mathbf{q}$-range where the corrections to the hydrodynamical theory are hard to detect and neutron scattering at the critical point in a fluid is probably a difficult experiment to be carried out, T_1 appears a natural tool to investigate the non-hydrodynamical regime.

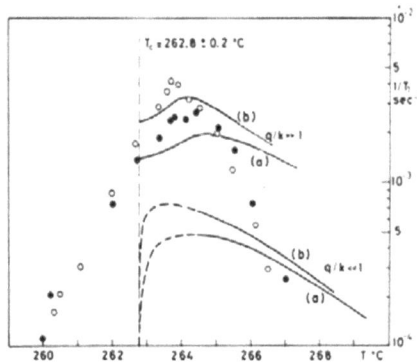

Fig. 1. Critical contribution to the proton spin-lattice relaxation rates in chloroform, at the measuring frequency 20.8 MHz (full dots) and 10 MHz (open dots), as a function of temperature along the critical isochore. The solid lines represent the theoretical behavior according to the theory outlined in the text (see Krynicky et al. and F. Borsa and A. Rigamonti in Further reading).

1.5 NMR-NQR relaxation and spin dynamics for strong electron correlation, with an illustrative example in a 2D pseudo-spin AF

In this Subsection we shall illustrate some basic aspects of the nuclear spin-lattice relaxation driven by the time dependence of the magnetic hyperfine hamiltonian through the spin operator $s(t)$, in systems with strongly correlated electrons. In these cases, most Equations recalled for the correlated fluctuations on approaching the phase transitions can be used, in particular in terms of the correlation length ξ. The illustration for a pseudo-spin 2D-AF structural transition should serve as introduction for the spin dynamics in 2D Heisenberg AF.

The magnetic transition probability W_m is written[11]

$$W_m = \frac{\gamma_h^2}{8} \left[J_+(\omega_0) + J_-(\omega_0) \right] \tag{39}$$

where $J_\pm(\omega)$ are the spectral density for the field $h(t)$ at the nuclear site due to the electron spin $s(t)$. ω_0 is the nuclear Larmor frequency. The spectral densities for $h(t)$ can be transformed to the ones for $s(t)$ by resorting to the time evolution of the spin operator under the total Hamiltonian $\mathcal{H}_T = H_{Zeeman} + H_{exch}$. In the assumption of isotropic exchange Hamiltonian $\mathcal{H}_{exch}$, commuting with Zeeman Hamiltonian, the transformation yields

---

[11] This form of W_m holds for "classical lattice" and weak-collisions approach. In practice this corresponds to assume strong damping in the precessional motions of s, so that the spin components fluctuate in a random way and a bath of low energy spin excitations is present.

$$J_\pm(\omega_0) = 9\gamma_e^2\hbar^2\left[B_1 J_\|(\omega_0) + B_2 J_\perp(\omega_e \pm \omega_0) + A J_\perp(\omega_e \mp \omega_0)\right] \quad, \quad (40)$$

where $J_\|$ is the spectral density for s_z and $J_\perp$ the one for $(s_x + is_y)$. A and $B_{1,2}$ are the well-known "lattice functions" related to the position of $\mathbf{s}$. For NQR the external field is absent and thus in (40) $\omega_e = \gamma_e H_0 = 0$. The same condition can be used in NMR in the case that the precessional motion of $\mathbf{s}$ at ω_e is strongly damped, e.g. in the presence of anisotropy for the total Hamiltonian.

For isotropically-exchange interacting spins, in the high-temperature condition where the exchange frequency is $\omega_{exch} \sim (J/\hbar) \gg \omega_e$ while ω_0 is smaller than the fluctuation frequencies of $\mathbf{h}(t)$, one shows that the correlation function for the spin components is approximately gaussian, with characteristic Heisenberg exchange frequency

$$\omega_E = \left[\frac{8}{3}z\,s(s+1)\right]^{\frac{1}{2}}\frac{J}{\hbar} \qquad (41)$$

(z number of nearest neighbors). Thus for an Heisenberg paramagnet, in the limit $T \to \infty$, by considering for simplicity a powder average, the relaxation rate is written

$$W_m = \sum_i W_m^{(i)} = \sum_i \frac{1}{2}\gamma_h^2 h_i^2 \omega_{exch}^{-1} \qquad (42)$$

where h_i is the amplitude of the fluctuating field at the nuclear site due to $\mathbf{s}_i$[12].

At finite temperature the magnetic correlation length ξ controls the site dependent order parameter and, in terms of the microscopic critical variable $s(t)$, the space-time spin correlation function. The collective spin components $\mathbf{s_q}$ have to be introduced and the transverse local field correlation function $\langle h_+(0)h_-(t)\rangle$ (involved in Eq.(39)) are related to $\langle s_{\mathbf{q}}^\pm(0)s_{-\mathbf{q}}^\pm(t)\rangle$ through the FT of the "lattice functions" A and $B_{1,2}$. Then the spectral densities $J_{\|,\perp}$ lead to the DSF for the spin components and in compact form we write

$$2W = \frac{\gamma_h^2}{2N}\sum_{\mathbf{q}}\left[(h_{\mathbf{q}})^2 S(\mathbf{q},\omega_{res})\right]_\perp = \frac{\gamma_h^2}{2N}k_B T\sum_{\mathbf{q}}\left[(h_{\mathbf{q}})^2\frac{\chi''(\mathbf{q},\omega_{res})}{\omega_{res}}\right]_\perp \quad (43)$$

where $\perp$ means that the product $h_{\mathbf{q}}^2 S(\mathbf{q},\omega)$ related to the transverse components of the local field $\mathbf{h}(t)$ has to be considered. It is noted that $h_{\mathbf{q}}^2$ in Eq.(43) strongly depends on the symmetry properties of the spin fluctuations and it can filter out or enhance the contribution to the relaxation rate arising from the "critical" part of the spin susceptibility. $(h_{\mathbf{q}})^2$ can be expressed in close form once that the magnetic hyperfine Hamiltonian is known. (see Section2).

---

[12] For restricted dimensionality (1D or 2D) long-term persistence of the correlation occurs and Eq.(42) has to be modified.

As we have already mentioned, for the analysis of the experimental data it is useful to modelize the DSF or, in other words, the amplitudes $|S_{\mathbf{q}}|^2$ and the decay rates $\Gamma_{\mathbf{q}}$ of the spin fluctuations. From the equivalent of Eqs.(26)-(29), in the assumption that the spin fluctuations are probed at a time much longer than the microscopic time scales (namely $\omega_e, \omega_0 \ll \Gamma_{\mathbf{q}}$) by expanding $\left(h_{\mathbf{q}}\right)^2$ in Eq.(43) by starting from the value at the critical wave vector, in the assumption that $h_{\mathbf{q}_c}^2 \neq 0$ one arrives at

$$
\begin{aligned}
2W &\approx \left(h_{\mathbf{q}_c}\right)^2 \frac{\gamma_h^2}{2N} \sum_{\mathbf{q}} \left[\xi^{2-\eta} f(q\xi) \frac{g^{-1}(q\xi)}{\xi^{-z}\omega_{exch}} \right] \\
&\simeq \left(h_{\mathbf{q}_c}\right)^2 \frac{\gamma_h^2}{2\omega_{exch}} \xi^{2-\eta+z} \int_0^{q_D\xi} \frac{f(x)}{g(x)} \frac{x^{D-1}}{\xi^{D-1}} \frac{1}{\xi} dx \qquad (44) \\
&\approx \left(h_{\mathbf{q}_c}\right)^2 \frac{\gamma_h^2}{\omega_{exch}} \xi^{z-D+2+\eta}
\end{aligned}
$$

(ξ is in lattice units).[13]

For extreme slowing down of the fluctuation, i.e. $\Gamma_{\mathbf{q}} \ll \omega_e$, condition for which the function $f(\omega/\Gamma_{\mathbf{q}})$ in Eq.(25) can be approximately assumed of the form $\Gamma \approx \Gamma_{\mathbf{q}}^2/\omega^2$, then instead of Eq.(44) one would have $2W \propto \omega_{exch}^{-1} \omega_e^{-2} \xi^{(2-\eta-z)}$.

While this case is hard to occur, a more realistic condition is the one of intermediate range of slowing-down. Then $\omega_{exch} \gg \omega_e$ while $\Gamma_{\mathbf{q}_c} \leq \omega_e$. Here the full dependence of $f(\omega/\Gamma_{\mathbf{q}})$ has to be taken into account, possibly in the form $f \approx \Gamma_{\mathbf{q}}^2/(\Gamma_{\mathbf{q}}^2 + \omega^2)$. A complicate temperature and frequency dependence occurs. The qualitative behavior of W is a flattening for T approaching T_c from above, with an inverse square root dependence from the magnetic field: $W \propto \omega_e^{-1/2} \propto H_0^{-1/2}$.

An illustrative example of studies of critical dynamics in 2D systems and meantime a test for dynamical scaling hypothesis, is offered by the proton T_1 relaxation in the molecular crystal $C_{18}H_{24}$, which at $T_c \simeq 190°K$ undergoes an antiferrodistortive disorder-order transition, twisting the phenylene ring out of the plane formed by the two ends phenil groups. The phenil ring is in a double-well local potential and torsional oscillations bring the proton from one of the well to the other. Thus the decay rates of the fluctuations are described by a pseudo-spin variable $s_z(t) = \pm 1$ controlling the dipole-dipole nuclear Hamiltonian. With the exception of a narrow temperature range close

---

[13] It is noted that $\int_0^\infty \frac{f(x)}{g(x)} x^{D-1} dx$ converges to a number around the unity. As regards the critical exponents, renormalization group theories indicates $z = \frac{D+2-\eta}{2}$ for Heisenberg ferromagnet, $z = \frac{D}{2}$ for Heisenberg AF. Since the critical exponent γ for the static response is $\gamma = (2-\eta)\nu$ and ν is the exponent for ξ, for strong AF or FE correlation one can also write Eq.(44) in the form $W \propto \epsilon^{-\gamma+\nu(D-z)}$. In a planar system, η being small, $W \propto \xi^{-z}$ and the dynamical critical exponent is expected to be $z = 1$.

to T_c the decay rate remains greater than the measuring frequency ω_0. The dominant contribution to the dipolar field is the on-site one and thus only the auto-correlation function $\langle s_z(0)s_z(t)\rangle$ is involved in the relaxation rate and the factor $(h_{\mathbf{q}}^2)$ (Eq.(43)) is $\mathbf{q}$-independent. Thus one can rewrite

$$T_1^{-1} = \omega_{int}^2 \frac{1}{2N} \sum_{\mathbf{q}} \frac{|S_{\mathbf{q}}^{\alpha\alpha}|^2}{\Gamma_{\mathbf{q}}} \tag{45}$$

$\omega_{int} = \gamma_h h$ being the effective strength of the dipolar field at the nuclear site. From the correspondent to Eqs.(26) and (28) and analogously to Eq.(44) (for $D = 2$ and $\eta \approx 0$) one obtains

$$T_1^{-1} = \omega_{int}^2 \Gamma_{\mathbf{q}_c}^{-1} \tag{46}$$

From the experimental data the temperature dependence of the critical decay rate (see Fig.2) $\Gamma_{\mathbf{q}_c}(T)$ is extracted with no adjustable parameter, ω_{int} being derived from the NMR spectra.

The critical frequency is found to go as ξ^{-z}, namely as $\epsilon^{\nu z}$, as expected from the dynamical scaling condition in 2D.

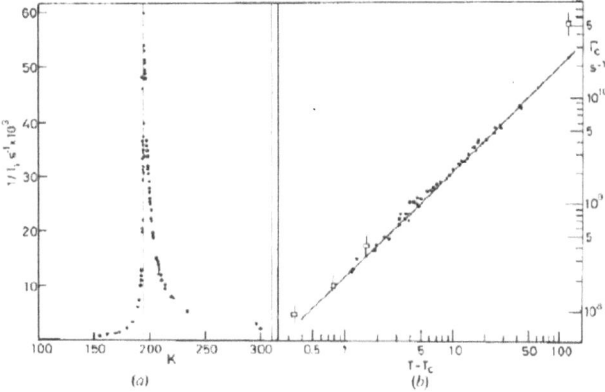

Fig. 2. Spin lattice relaxation rate in p-terphenil on approaching the transition temperature $T_C = 194.5K$ (a). No dependence from the measuring frequency was detected. In part b) the temperature behavior of the critical frequency is reported and compared with the results (open squares) from high resolution neutron back-scattering in the deuterated crystal (see T. Guillon et al. in Further reading.)

Further reading

1. D.L. Goodstein: *States of Matter* (Dover, 1985) Chapter 6
2. L.D. Landau and E.M. Lifshitz: *Statistical Physics* (Pergamon Press, 1959)
3. R.M. White: *Quantum Theory of Magnetism* (McGraw-Hill, 1970) Chapter 1

4. K.A. Muller and A. Rigamonti: *Transizioni di fase nei solidi* Voce quadro dell'Enciclopedia Treccani delle Scienze Fisiche, Vol. **V** pag. 393 (in italian)

5. K.A. Muller, J.C. Fayet, F. Borsa and A. Rigamonti: *Structural Phase Transitions II*, Ed.s K.A. Muller and N. Thomas (Springer Verlag, 1991) Chapter 2

6. F. Borsa and A. Rigamonti in *Magnetic Resonance of Phase Transitions*, Ed.s F.J. Owens, C.P. Poole, H.A. Farach (Academic Press, 1979) Chapter 3

7. A. Rigamonti: *NMR-NQR Studies at Structural Phase Transitions*, Advances in Physics **33**, 115 (1984)

8. P.C. Hohenberg and B.I. Halperin: *Theory of dynamic critical phenomena*, Rev. Mod. Phys. **49**, 435 (1977)

9. K. Krynicki, J. G. Powles and A. Rigamonti, Molecular Physics **34**, 1717 (1977)

10. A Rigamonti: *NMR-NQR Relaxation: Basic Principles and Applications in Paramagnets and in HT_C Superconductors* in *Magnetic Properties of Matter* (Ed.s G. Asti, D. Fiorani and F. Lucari, World Scientific, 1991) pag. 293

11. *Local Properties at Phase Transitions*, Proceedings of the International School "Enrico Fermi", Editors K.A. Muller and A. Rigamonti (North Holland, 1976)

12. T. Guillon, M.S. Conradi and A. Rigamonti, Phys. Rev. B **31**, 4388 (1985)

2 NMR-NQR insights on the phase transitions and spin dynamics in pure and in charge and spin- doped 2D $S = 1/2$ Heisenberg Antiferromagnets

Long abstract of the Chapter
"Correlated spin dynamics and phase transitions in pure and disordered $S = 1/2$ antiferromagnets: insights from NMR-NQR"

by A.Rigamonti, P. Carretta and N. Papinutto in "Novel NMR and EPR techniques", Eds. J. Dolinsek, M. Vilfan and S. Zumer, Springer Series Lecture Notes in Physics, Springer Verlag (2006) pg 351-382

2.1 General Properties of 2DQHAF

La_2CuO_4, the parent of high temperature superconductors, is also among the first experimental realizations of the model two-dimensional (2D), quantum $(S = 1/2)$, Heisenberg antiferromagnet (AF). This compound, beginning with 1987 has triggered strong interest towards low-dimensional quantum magnetism.

In La_2CuO_4 spin dilution is achieved, by $S = 0$ Zn^{2+} or Mg^{2+} for $S = 1/2$ Cu^{2+} substitutions, while charge doping, namely injection of holes in the CuO_2 plane, is obtained by substituting the La^{3+} ions by Sr^{2+}. Attention has to be devoted to a variety of aspects involving static and dynamical properties of the 2D array of Cu^{2+} magnetic ions in super-exchange interaction via Oxygen : the temperature dependence of the in plane magnetic correlation length ξ_{2D}; the critical spin dynamics driving the system towards the long-range ordered state (at $T = 0$ in pure 2DQHAF in the absence of inter-planar interaction); the validity of the dynamical scaling, where ξ_{2D} controls the relaxation rate of the order parameter. One can also analyse the modifications induced by spin dilution and by charge doping, aspects of particular interest in the vicinity of the percolation thresholds, where the AF order is about to be hampered at any finite temperature. This can be considered a situation similar to a *quantum critical point*, where no more the temperature but rather Hamiltonian parameters can drive the transition.

In principle the problems involving the aforementioned aspects could be studied by spectroscopic techniques relying on the electron response. However, electron paramagnetic resonance is often prevented by very broad lines, possibly due to the strong electron-electron correlation. Neutron scattering is a powerful technique but it could suffer of resolution limitation for low-energy excitations or for short correlation length. Thus the nuclei have represented in recent years one of the best tools to investigate the properties of 2DQHAF, by resorting to the hyperfine interactions and to their time-dependence, as explored by NMR-NQR spectra and relaxation.

- D.C. Johnston, in *Handbook of Magnetic Materials* Vol.10, Ed. K.H.J. Buschow (Elsevier 1997) Chapter 1
- S. Chakravarty, B.I. Halperin, D.R. Nelson, Phys.Rev. B **39**, 2344 (1989)
- P. Hasenfratz, F. Niedermayer, Phys. Lett. B **268**, 231 (1991); Z. Phys. B **92**, 91 (1993)
- P.W. Anderson: *Frontiers and Borderlines in Many-particle Physics* Eds. R.A. Broglia and J.R. Schriffer (North Holland 1988)
- S. Sachdev: *Quantum Phase Transitions* (University Press, Cambridge 1999)

2.2 Temperature Dependence of the Correlation Length (in La_2CuO_4 and in CFTD, within Scaling Arguments)

The results obtained in two 2DQHAF (La_2CuO_4 and CFTD) prototypes show how quantitative estimate of the correlation length is derived, concluding that in a wide temperature range the predictions of the renormalized classical (RC) regime are rather well followed.

- A. Rigamonti, F. Borsa, P. Carretta, Rep.Prog. Phys. **61**, 1367 (1998)
- P. Carretta, T. Ciabattoni, A. Cuccoli, A. Rigamonti, V. Tognetti, P. Verrucchi, J. Appl. Magn. Resonance **19**, 391 (2000)

- P. Carretta, A. Rigamonti, R. Sala, Phys. Rev. B **55**, 3734 (1997)
- P. Carretta, T. Ciabattoni, A. Cuccoli, E.R. Mognaschi, A. Rigamonti, V. Tognetti, P. Verrucchi, Phys. Rev. Lett. **84**, 366 (2000)

2.3 Spin and Charge Doped La_2CuO_4

For charge and spin doped 2DQHAF nice insights are derived from ^{63}Cu and ^{139}La NQR in Sr-doped and in Zn (or Mg) -doped La_2CuO_4. It is shown, in particular for spin doping, how the "dilution model", namely the simple correction in the 2D AF Hamiltonian by the probability of presence of the magnetic ion, explains rather well the behaviour of $\xi_{2D}(x,T)$, of the spin stiffness $\rho_s(x)$ and of the ordering temperature $T_N(x)$, when the doping amount x is far from the percolation threshold.

2.4 Near the AF Percolation Thresholds: Spin Stiffness, Correlation Length and Staggered Magnetic Moment

The analysis of the NQR relaxation rates can be extended to the spin doping region where the dilution model would be a crude assumption. In practice only ^{139}La NQR relaxation have been measured in the temperature range of interest and the correlation length has been extracted. The absolute values and the temperature dependences derived in the strong dilution condition are compared with the data for light doping and with the theoretical behaviours, by leaving the spin stiffness as adjustable parameter. One is led to the conclusion that the RC regime holds, also in regards of charge-doped La_2CuO_4.

A quantity of interest for the quantum effects in disordered 2DHQAF is the zero-temperature staggered magnetic moment $\langle \mu_{Cu}(x, T \to 0) \rangle$ along the local quantization axis, namely the dependence of the sublattice magnetization on spin dilution. The staggered magnetic moment is different from the classical $S = 1/2$ value because of the quantum fluctuations, that in turn are expected to increase with spin dilution. The quantity

$$R = \langle \mu_{Cu}(x, 0) \rangle / \langle \mu_{Cu}(0, 0) \rangle$$

has been obtained to a good accuracy from the magnetic perturbation due to the local hyperfine field on ^{139}La NQR spectra or from μSR precessional frequencies and recently evaluated also close to the percolation threshold from neutron diffraction in a single crystal of $Zn - Mg$ doped La_2CuO_4. The x-dependence of R as it results from a combination of NQR and neutron diffraction data can be derived. The classical doping dependence (for $S \to \infty$) as well as the one predicted by the quantum non-linear σ model are not supported by the experimental findings. The data indicate a doping dependence with non-classical critical exponent in substantial agreement with finite-size scaling.

- F.C. Chou et al, Phys. Rev. Lett. **70**, 222 (1993)

- H.H. Klauss et al, Phys. Rev. Lett. **85**, 4590 (2000)
- P. Carretta, F. Tedoldi, A. Rigamonti, F. Galli, F. Borsa, J.H. Cho and D.C. Johnston, Eur. Phys. J. B **10**, 233 (1999)
- B. Keimer et al, Phys. Rev. B **46**, 14034 (1992)
- T. Imai et al, Phys. Rev. Lett. **70**, 1002 (1993)
- M. Acquarone, Physica B **259–261**, 509 (1999)
- P. Carretta, A. Rigamonti, E. Todeschini, L. Malavasi, Acta Physica Polonica B **34** (2003)
- O.P. Vajk, P.K. Mang, M. Greven, P.M. Gehring, J.W. Lynn, Science **295**, 1691 (2002)
- A. Paolone, F. Cordero, R. Cantelli, M. Ferretti, Phys. Rev. B **66**, 094503 (2002)
- A. Paolone, R. Cantelli, F. Cordero, M. Corti, A. Rigamonti, M.Ferretti, Inter. J. Modern. Phys. B **17**, 512 (2003)

2.5 The Cluster Spin-Glass Phase

The main properties of the cluster spin glass phase, occurring when the charge doping is above the percolation threshold in Sr-doped La_2CuO_4, have been widely explored by NMR-NQR spectra and relaxation.

- A. Campana, M. Corti, A. Rigamonti, F. Cordero, R. Cantelli, Eur. J. Phys. B **18**, 49 (2000)
- J.H. Cho, F. Borsa, D.C. Johnston, D.R. Torgeson, Phys. Rev. B **46**, 3179 (1992)
- P.M. Singer, A.W. Hunt, A.F. Cederstrom, T. Imai, cond-mat /0302077 (2003)
- S. Wakimoto, S. Ueki, Y. Endoh, K. Yamada, Phys. Rev. B **62**, 3547 (2000)
- Ch. Niedermayer et al, Phys. Rev. Lett. **80**, 3843 (1998)
- M.H. Julien et al, Phys. Rev. B **63**, 144508 (2001)
- M. Eremin and A. Rigamonti, Phys. Rev. Lett. **88**, 037002 (2002)
- A. Rigamonti, M. Eremin, A. Campana, P. Carretta, M. Corti, A. Lascialfari, P. Tedesco, Intern. J. Modern Physics B **17**, 861 (2003)
- A. Paolone, R. Cantelli, F. Cordero, M. Corti, A. Rigamonti, M.Ferretti, Inter. J. Modern. Phys. B **17**, 512 (2003)
- M. Corti, A. Rigamonti, F. Tabak, P. Carretta, F. Licci, L. Raffo, Phys.Rev.B **52**, 4226 (1995)
- Y.C. Cheng, A.H. Castro Neto, Phys. Rev. B **61**, R3782 (2000)
- A.L. Cheryshev, Y.C. Chen, A.H.Castro Neto, Phys. Rev B **65**, 104407 (2002)
- A.W. Sandvik, Phys. Rev. B **66**, 024418 (2002)
- K. Kato et al, Phys. Rev. Lett. **84**, 4204 (2000)

2.6 The Quantum Critical Point in an Itinerant 2DAF - Effect of Magnetic Field

Finally one should look at the results of a recent NQR-NMR study carried out in a system ($CeCu_{6-x}Au_x$) around the quantum critical point (QCP) ($x = 0.1$) separating Fermi liquid and AF phases, with a field-dependent magnetic generalized susceptibility of 2D character, with anomalous exponent and energy/temperature scaling. The main results obtained by other authors by means of inelastic neutron scattering are confirmed also by NMR-NQR measurements based on low-energy susceptibility over all the Brillouin zone. Novel aspects are pointed out in regards of the role of the magnetic field.

- J.A. Hertz, Phys. Rev. B **14**, 1165 (1976)
- Q. Si et al, Nature **413**, 804 (2001)
- A. Schroder et al, Nature **407**, 351 (2000)
- H. v.Löhneysen et al, Physica B **223-224**, 471 (1996)
- P. Carretta, M. Giovannini, M. Horvatic, N. Papinutto, A. Rigamonti, Phys. Rev. B **68**, 220404 (2003)
- N. Papinutto, M. J. Graf, P. Carretta, M. Giovannini and A. Rigamonti, Physica B **359**, 89 (2005)
- T. Moriya: *Spin Fluctuations in Itinerant Electron Magnetism*, Vol. 56 (Springer, Berlin 1985)
- O. Stockert et al, Phys. Rev. Lett. **80**, 5627 (1998)
- A. Schroder et al, Phys. Rev. Lett. **80**, 5623 (1998)
- A.J. Millis, Phys. Rev. B **48**, 7183 (1993)
- M. Winkelmann, G. Fisher, B. Pilawa, E. Dormann, Eur. Phys. J. B **26**, 199 (2002)
- N. Papinutto, M.J. Graf, P. Carretta, A. Rigamonti, M. Giovannini, K. Sullivan, Physica B 378, 84 (2006)

2.7 Summarizing Remarks

NMR-NQR relaxation can be a valuable tool in order to study the correlated spin dynamics and the phase transitions in systems that in the last decade have called strong interest as models for quantum magnetism and as parents of high temperature superconductors, namely the square planar arrays of $S = 1/2$ magnetic moments in antiferromagnetic interaction. Their rich phase diagram as a function of temperature and of spin dilution or hole injection, can be explored by means of NQR-NMR spin-lattice relaxation measurements. Quantitative estimate of the quantum fluctuations-affected correlation length, spin stiffness and order parameter can be derived by resorting to the integration of the generalized susceptibility, once that the wave vector dependence of the electron-nuclei hyperfine interaction is properly taken into account.

In pure, non-disordered, 2DQHAF the in-plane correlation length $\xi(0, T)$ can be obtained from the relaxation rate. By working in the prototype CFTD, where the exchange constant J is rather small, it has been possible to prove that the classical regime for ξ, with renormalization of the spin stiffness and of the spin wave velocity due to the quantum fluctuations, holds in a wide (T/J) range. No evidence of the crossover to quantum critical regime, expected on the basis of an extension of the non-linear σ model, is observed. Similar conclusion holds also for La_2CuO_4, although only a more limited (T/J) range can be explored in this compound, where $J = 1500K$. On the other hand La_2CuO_4 allows one to perform spin dilution or charge doping, which corresponds to injecting itinerant holes in the CuO_2 plane. The disordered 2DQHAF is thus created and a variety of interesting effects is observed. For moderate spin doping the dilution model is found to lead to a reliable description: the spin stiffness and therefore the correlation length are still the ones pertaining to the renormalized classical regime, once that the probability of a spin vacancy is taken into account in the AF Hamiltonian.

For strong dilution the dilution model is evidently inadequate, the spin stiffness decreasing with increasing the Zn or Mg content x with a x-dependence much stronger when the percolation threshold is approached. Still the spin doped La_2CuO_4 was found to remain in the RC regime, $\xi(x,T)$ displaying a temperature dependence similar to the one for $\xi(0,T)$, once that the spin stiffness is renormalized to the corrected value. The transition to the 3D ordered state occurs at the temperature where $\xi(x,T)$ reaches about the same value as in the pure 2DQHAF, namely about 150 lattice steps.

Similar results have been found also in charge doped La_2CuO_4, for a Sr content $y = 0.016$, not far from the percolation value $y = 0.02$. Again the temperature behaviour of $\xi(y,T)$ appears almost the same as in the pure compound, although with a strongly reduced spin stiffness, and $\xi(y = 0.016)$ at T_N again turns out around 150 lattice steps.

The reduction of the expectation value of the Cu_2^+ magnetic ion as a function of the spin dilution has also been derived. In accordance to neutron diffraction data it has been found that the x-dependence turns out of the form $(x - x_c)^\beta$ with $\beta = 0.45$, close to the one expected from spin wave theories and T matrix description and in agreement with finite-size scaling. The temperature dependence of the staggered magnetic moment seems to follow a rather universal law in terms of the x-dependent Neel temperature, with an abrupt but continuous phase transition and a critical exponent close to 0.2 for small x, possibly increasing to 0.3 for large doping amounts.

A system where a study of the spin dynamical properties at the disorder conditions corresponding to a quantum critical point has been possible, is $CeCu_{5.9}Au_{0.1}$. The NQR-NMR ^{63}Cu relaxation rates have provided enlightening insights on the magnetic response function, particularly when the critical frequency slows down below the resolution limit of neutron scattering and in regards of the role of an external magnetic field. On one side a 2D response, with a critical exponent different from 1 and the energy/temperature scaling have been confirmed from the relaxation measurements involving the **k**-integrated dynamical susceptibility at low energy. On the other hand, it has been shown that below a certain field-dependent temperature the system crosses over to a phase of gapped spin excitations, with quenching for $T \to 0$ of the magnetization fluctuations and unconventional magnetic field scaling.

In short, the nuclei can be used as useful tools in the attempt to unravel the many static and dynamical phenomena occurring in 2DQHAF upon charge and spin doping. The powerfulness of the NQR-NMR measurements, when accompanied by a suitable analysis, is comparable to inelastic neutron scattering. While in some cases fine confirmations of the data found by this technique have been obtained, in other cases the information obtained from NQR-NMR relaxation in regards of low-energy spin excitations have turned out even more subtle and novel aspects have been pointed out, in turn stimulating new scientific work.

3 Short-range correlation in AF spin-1 chains: insights from ^{89}Y NMR in $Y_2BaNi_{1-x}Mg_xO_5$

Long abstract, involving the papers reported as References

As already discussed in previous sections, due to the richness of their phase diagrams, low-dimensional AF spin systems have attracted strong interest. Since the "strength" of quantum effects is enhanced by the reduced dimensionality, exotic behaviors characterize AF spin chains (i.e. linear systems of AF-coupled spins). For instance, in contrast to the classical scenario, where the spins assume a perfect antiparallel configuration at $T = 0$ independently on their value S, the ground state of the quantum Heisenberg chain is never ordered and shows qualitative difference for integer-S and half-integer-S.

- A. Auerbach in *Interacting Electrons and Quantum Magnetism*, Springer-Verlag 1994.
- F. D. M. Haldane, Phys. Lett. **93A**, 464 (1983) and Phys. Rev. Lett. **50**, 1153 (1983).

The correlation properties of low-dimensional AF's are enbedded in the spin-spin correlation length ξ, namely the distance over which the memory of a given spin configuration is preserved. In the paper

- F. Tedoldi, R. Santachiara and M. Horvatić, Phys. Rev. Lett. **83**, 412 (1999).

the authors have pointed out a novel way to measure ξ, that consists in imaging, by Nuclear Magnetic Resonance (NMR), the response of the system to a homogeneous magnetic field, after the introduction of finite size effects.

The compound chosen for the study, $Y_2BaNi_{1-x}Mg_xO_5$, is a prototype one-dimensional, spin-1, Heisenberg AF, where the average size of chain segments $\langle L \rangle$ is easily varied by controlled substitution of x $Ni^{2+}(S = 1)$ by $Mg^{2+}(S = 0)$, to obtain $\langle L \rangle \simeq 1/x$.

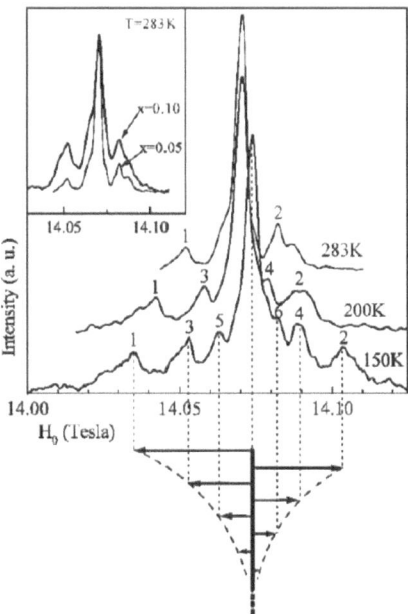

Fig. 3. ^{89}Y NMR spectra in doped Haldane chain Y$_2$BaNi$_{0.95}$Mg$_{0.05}$O$_5$ recorded at fixed frequency of 29.4 MHz by sweeping the magnetic field. Resolved satellite peaks allow one to obtain the magnetization "site by site", as schematically shown at the bottom of the figure, starting from one of the chain edges. Inset shows that the intensity of satellite peaks in Y$_2$BaNi$_{1-x}$Mg$_x$O$_5$ decreases as a function of the average chain length $\langle L \rangle \simeq 1/x$.

In ^{89}Y NMR spectra of Y$_2$BaNi$_{1-x}$Mg$_x$O$_5$, shown in Fig. 3, several satellite peaks are resolved. These peaks can be associated to individual spin sites along the chain, providing a *real-space picture* of the microscopic spin polarization and giving direct access to ξ, as explained below.

In a NMR spectrum acqired at frequency ν_{NMR} in an external field H_0, the signal intensity is proportional to the number of ^{89}Y nuclei probing an internal *local* field h such that the resonance condition $\nu_{NMR} = |\frac{\gamma}{2\pi}(\boldsymbol{H_0} + \boldsymbol{h})|$ is fulfilled ($\gamma/2\pi = 2.0859$ MHz/Tesla is the ^{89}Y gyromagnetic ratio). The local field $\boldsymbol{h}$ reflects, via the hyperfine interaction, the expectation value $\langle S_i^z \rangle$ of the single spin operator at the nearby chain site, $h = A\langle S_i^z \rangle$, where in Y$_2$BaNiO$_5$-based compounds the Ni^{2+}-^{89}Y hyperfine coupling constant was found to be $A = 13$ KGauss. So, each different value $\langle S_i^z \rangle$ will produce a peak in the NMR spectrum at the position given by $H_0 = 2\pi\nu_{NMR}/\gamma - A\langle S_i^z \rangle$. The multi-peak structure shown in Fig. 3 is thus a clear signature of a *non-uniform* magnetization along the chain. Explicit mapping between chain-position i and resolved peaks p_I is suggested by the evolution of the spectra as a function of the chain length. The size of the satellite peaks is reduced on increasing $\langle L \rangle \simeq 1/x$ (inset to Fig. 3) and disappear for $\langle L \rangle \to \infty$, while the central line p_c remains almost the same. This indicates that the peaks p_I should be associated to spins at the boundaries of the chains, while p_c reflects the properties of the "bulk". Then, it looks reasonable correlate the shift of each peak from the central line, $\Delta(p_I) = H_0(p_I) - H_0(p_c)$, to the distance of the ions from the edge of the chain, $i.e.$, to assign the peaks p_I to the positions $i = I$ (and $i = L - I + 1$) following the decreasing magnitude of $\Delta(p_I)$, as illustrated in Fig. 3. Thus the magnetization induced near the edge of the chain is deduced: $\Delta\langle S_I^z \rangle \equiv \langle S_I^z \rangle - \langle S_c^z \rangle = \Delta(p_I)/A$. The alternation of the sign of $\Delta(p_I)$ as a function of I points out the staggered character of the magnetic response, although the applied field is homogeneous. In the light of the correspondence established above, one can study how the local magnetization evolves moving from the boundaries towards the center of the chain. In Fig. 4, left side, the magnitude of $\Delta\langle S_I^z \rangle$ extracted from the shift $\Delta(p_I)$ in $Y_2BaNi_{0.95}Mg_{0.05}O_5$ is reported as a function of I. At all temperatures, $|\Delta\langle S_I^z \rangle|$ is well fitted by an exponential law providing the characteristic decay length $\xi(T)$ of the boundary magnetization. As shown on the right side of Fig. 4, $\xi(T)$ obtained in this way matches the temperature behavior predicted for the $infinite$ chain correlation length reported in

- Y. J. Kim, M. Greven, U. J. Wiese, and R. J. Birgeneau, Eur. Phys. J. B **4**, 291 (1998).

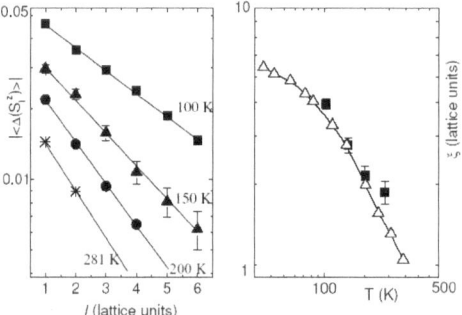

Fig. 4. Left: The magnitude of the site magnetization $\Delta\langle S_I^z \rangle$ (deduced from the shift of the satellites in the spectra) decreases exponentially by moving from the boundaries towards the center of the chain. This defines experimental correlation lengths ξ (full squares) which are compared **(Right)** to the theoretical prediction for the infinite-chain correlation length (open triangles)reported by Kim et al. (see text).

Thus, one concludes that the perturbation developed around spinless impurity (Mg-ion) indeed "reflects" very well, in a broad temperature range, the bulk correlation properties of the ideal unperturbed systems.

This experimental finding has been subsequently confirmed by Monte Carlo simulations of Heisenberg AF spin chains with open boundaries in uniform magnetic field:

- S. Botti, A. Rosso, R. Santachiara and F. Tedoldi, Eur. Phys. Rev. B **63**, 012409 (2001).

These calculations also identify the distribution of chain lengths around the average value $\langle L \rangle$ to be the origin of the progressive increase of the width of the NMR peaks on cooling observed in Fig. 3).

A complete report of the various insight obtained on AF spin chains by NMR investigation of pure and doped Y_2BaNiO_5 can be found in

- F. Tedoldi, Scientifica Acta **XVI**, 1 (2001).

4 An overview on the main results from NMR-NQR studies in HT_c superconductors, with emphasis on the vortex dynamics

Long abstract, involving review articles and papers reported as References

4.1 Generalities

In the NMR-NQR studies of SC cuprates one can envisage three main lines of activity:

i) along the charge and spin doping of precursor antiferromagnets
ii) in the strong charge doped SC regime, above and below T_c
iii) about the flux lines lattice and the thermal excitations of the vortices

Line i) is basically similar to what has been discussed in Section 2. Fundamentals in quantum magnetism have been studied, most in terms of the in-plane correlation length, on the effects of charge and spin doping on the excitations in 2DQHAF in the various regimes of the phase diagram, on the extra magnetic moments associated to doping and their cooperative freezing and on the charge localization and spin texture. Interesting results have been obtained in regards of the low-energy excitations, of the itinerancy vs. localization and on the doping dependence of the gap from itinerant to localized states.

Line ii): NMR-NQR has provided a great deal of nice results about the fundamental hyperfine interactions, on the interplay Fermi-liquid and AF correlations, on the features of the low-energy excitations in the underdoped phase (with the *vexata quaestio* of the spin gap opening and pseudo gap opening above T_c), in regards of the characteristics of the SC phase vis-a vis the results expected from the BCS theory (namely about the coherence peak, the orbital pairing, the anisotropy of the spin fluctuations). We shall briefly overview this line at Subsection (4.2). Furthermore, somewhat unexpected phenomena, like the coexistence of superconductivity and spin freezing in underdoped SC cuprates, with the related striped-phase and wipe-out on cooling of the ^{63}Cu NMR-NQR signals have recently attracted a great deal of interest.

Line iii) regards the studies of the structure of the flux lines lattice, of their thermal excitations and the character of the motions, by means of NMR line narrowing, spin lattice relaxation and echo dephasing. This line shall be discussed in some more detail in Subsection(4.3).

- D. Brinkmann and M. Mali: NMR Basic Principles and Progress **Vol. 31** (Springer, 1996)
- C. Berthier, M. H. Julien, M. Horvatic and Y. Berthier, J. Phys. I France **6**, 2205 (1996)
- K. Asayama, Y. Kitaoka, G-q Zheng and K. Ishida, Progress NMR Spectroscopy **28**, 221 (1996)
- A. Rigamonti, F. Borsa, and P. Carretta, Rep. Prog. Phys. **61**, 1367 (1998)

4.2 The superconducting phase (above and below T_c)

In the normal state the spin-lattice relaxation rates have been measured for a variety of nuclei and interpreted in terms of the generalized magnetic susceptibility $\chi''(\mathbf{q}, \omega)$, in general by referring to one of the two limiting situations: strong correlation, namely $(U/t) \gg 1$ in the Hubbard Hamiltonian, and for $(U/t) \ll 1$ in a scenario of Fermi-like carriers with AF correlation.

In the first case the excitations are basically overdamped spin waves for localized magnetic moments, the interpretation is similar to the one for 2DQHAF, with an effective temperature- and doping- dependent frequency, controlled by the in-plane correlation length.

For the opposite limit $(U/t) \ll 1$, the interpretation of the relaxation rates has been most based on a generalized susceptibility of the form

$$\chi(\mathbf{q}, \omega) = \frac{\chi^0(\mathbf{q}, \omega)}{[1 - \chi^0(\mathbf{q}, \omega)I(\mathbf{q})]}$$

Here the bare susceptibility in a Fermi gas is related to the density of states at E_F, while the Stoner-like enhancement factor $I(\mathbf{q})$ is due to the AF correlation.

The main conclusions, particularly in comparison to the BCS superconductors, can be summarized as follows.

For $T > T_c$ the relaxation rates W are site-dependent, because of the filtering factors $h_\mathbf{q}^2$ (discussed in Section 2) and reflect the different temperature dependence of the excitations in different regions of the Brillouin zone. For ^{17}O the filtering factor is peaked at $\mathbf{q} = \mathbf{0}$, only a moderate increase of W with respect to the BCS value (small enhancement due to the Stoner like factor) is detected and the interpretation is most in terms of the density of states, for the $\mathbf{q}$-integrated spectral density.

For Cu nuclei a strong enhancement of W with respect to the expected W_{BCS} is observed: the filtering factor is peaked at $(\pi/a, \pi/a)$ and a sizeable and temperature dependent AF correlation is active.

From the comparison of the relaxation rate in optimally doped and in underdoped YBCO, a spin and charge pseudo-gap is noticed to open up at T^*, well above T_c, basically corresponding to a transfer of excitations from the low frequency to the high frequency range[14]. A long list of attempts of explanations of the pseudo-gap opening has been produced, involving electronic crossover, charge density waves, instability, AF or SC fluctuations of non Ginzburg-Landau character, polarons and stripes of various nature. For a recent mis-a-point, see Norman et al. in the References.

Below T_c the main conclusions achieved, with a significant role of NMR measurements (Knigth shift, the Gaussian part of the echo dephasing and T_1) are the following. The SC state is a singlet spin state; the zero temperature gap is about $4T_c$, thus indicating strong coupling; the orbital symmetry of the order parameter is d-wave; the relaxation rate goes as $W \propto T^3$ (a feature common to heavy fermions SC) and it is rather well accounted for by the BCS density of states, with the modifications taking into account the 2D character and the d-wave orbital symmetry.

As already mentioned another topic of more recent and strong interest, where NMR studies have provided meaningful insights, is the phenomenon of the microscopic coexistence of spin-freezing and superconductivity in the underdoped phase of cuprates, particularly LSCO. This implies a kind of extension to the superconducting regime of the cluster spin-glass phase described for charge doped 2DQHAF. Related effects are the onset of charge-stripes and the marked slowing-down of spin fluctuations, causing wipe-out of the Cu NQR signals, since the echo dephasing times become very short.

For a comprehensive analysis, see Julien et al.

- M.R. Norman, D. Pines and C. Kallin, Advances in Physics **54**, 715 (2005)
- M. Julien, A. Campana, A. Rigamonti, P. Carretta, F. Borsa et al., Phys. Rev. B **63**, 144508 (2001)
- P. Carretta, A. Lascialfari, A. Rigamonti, P. Tedesco, F. Tedoldi and A. Larionov, Phys. Rev. B **69**, 104512 (2004)
- A. Uldry and P.F. Meier, Phys. Rev. B **72**, 094508 (2005)

---

[14] For the shift in the $\mathbf{q}$-range, NMR and neutron scattering are somewhat in contrast: from T_2 one could deduce for $T > T^*$ a global increase in the excitations at $\mathbf{q} = \mathbf{q}_{AF}$ while neutron scattering does not detects effects on the coherence length.

4.3 The flux lines lattice and their thermal excitation

The cuprates are superconductors strongly of II type. Therefore an external magnetic field penetrates inside the sample in form of flux lines, each carrying the elementary fluxon $\Phi_0 = 2.7 \cdot 10^{-7}$ Gauss cm^2. Typically for YBCO (oriented powder with field along the **c**-axis) in a field of 6 Tesla, the vortices are accommodated on a 2D triangular lattice, at a distance of about 200 Å, a factor at least ten larger than the diameter of the vortex core. The London penetration length λ_L is about 1500 Å and thus by moving along a given x-direction in the CuO_2 plane one finds a kind of 'ripple' of the magnetic field $\mathbf{H}(x)$, with temperature-dependent amplitude of about $50 - 100$ Gauss, going towards zero for $T \to T_c(\mathbf{H})$ from below.

Above a certain temperature (the **irreversibility line**) field- and somewhat sample-dependent, the vortices move, the features of their motion being possibly in form of pieces of flux lines ('pancakes') diffusing in the plane, with or without correlation among adjacent planes.

NMR line width, echo dephasing and T_1 have been used as microscopic tools to study the structure of the flux lines (FL) lattice and their thermal excitations.

Because of the spatial distribution of the magnetic field $\mathbf{H}(x)$ (that can be approximately deduced from the solution of the London equations for δ-like field line) instead of an intrinsic NMR line, in general one detects an asymmetric broad line, with temperature dependent second moment proportional to $[\lambda_L(T)]^{-4}$. Below the irreversibility temperature T_{irr}, in practice no direct evidence of FL motions is detected in line narrowing or T_1, while some evidence of motion has been claimed from the echo dephasing (which is much more sensitive, however involving an effective correlation time poorly defined).

Above T_{irr}, from the extra narrowing (with respect to the one related to $\lambda_L(T)$) of the ^{89}Y NMR line in YBCO family (particularly YBCO124) information on an effective correlation time reaching about 50 μs and further decreasing on approaching T_c have been achieved. Better insights have been obtained from high-resolution ^{17}O NMR spectra in YBCO123, with evidence of coexistence of liquid and solid-like vertex behaviour in a certain temperature range.

More quantitative studies have been possible from the field and temperature dependence of the relaxation rate related to the time dependence of the local field components induced by the motions of the FL, most in YBCO oriented powers and ^{89}Y relaxation. The starting point is the expansion of the local field components $h_{\rho,z}$ (in cylindrical coordinates and for the regions outside the vortices) in terms of the displacements of the FL from the equilibrium position. Then the correlation function for the transverse field components is written in terms of the mean square displacements from the equilibrium position, with an ensemble average, over the range from the FL core up to the intra-vortices distance l_e.

The final result is

$$T_1^{-1} = \frac{1}{2}\gamma^2 h_e^2(T)j(\omega_L)$$

where $j(\omega)$ is the spectral density of that correlation function and the effective field driving the relaxation process reads

$$h_e^2 = \frac{8}{3}\left[\frac{d}{(\xi\lambda_L)^2}\right]k_B T\left[2\ln\left(\frac{\lambda_L}{l_e}\right)\right] \qquad (T > T_{irr})$$

d being the inter-plane distance.

The interplay between h_e (which decreases on increasing temperature) and the field and temperature dependencies of the maxima of the spectral density (expected when the effective frequency $\langle\omega_q\rangle = Dq_e^2$, with D the pseudo-diffusion constant of the vortices and $q_e = l_e^{-1}$, is around ω_L) leads to a nice structure of the relaxation rates $W(T, \mathbf{H})$. Maxima in W are detected at different temperatures for different fields, while W_{max} do not increase with decreasing field as one would expect from the popular argument $W_{max} \propto \omega_L^{-1}$. Information on the barrier for the vortex depinning and its field dependence can be derived from the analysis of the relaxation rates.

From measurements of the spin echo dephasing some further information can be obtained in regards of the degree of correlation in the pancake motions along the $\mathbf{c}$-axis.

An additional contribution to the nuclear relaxation comes from the excitations within the vortex cores (vortex core relaxation), where the nuclei relax as in the normal non-SC phase and possibly with other pseudo-particles excitations. The contribution to the relaxation from the vortex cores goes approximately as $T_1^{-1} \propto HT$ and can be studied in detail in non-cuprate SC, where at moderate magnetic field (e.g. 5-10 Tesla) the vortex cores are a sizeable part of the whole sample.

- M. Corti, B.J. Suh, F. Tabak, A. Rigamonti, F. Borsa et al., Phys. Rev. B **54**, 9469 (1996)
- A. Lascialfari, A. Rigamonti and I. Zucca, Phys. Rev. B **71**, 214510 (2005)

5 Superconducting fluctuations: an overview of the studies based on the spin susceptibility (Knight shift and T_1) and on diamagnetic susceptibility

5.1 Introductory aspects

The superconducting (SC) transition is a second-order one and for $T \to T_c$ from above, the enhancement and slowing down of the fluctuations described in Section 1 occur. As already mentioned, for the SC transition the fluctuations take particular aspects, the order parameter being complex:

$$\psi(\mathbf{r}) = \sqrt{n_c}\, e^{i\theta(\mathbf{r})}. \qquad (47)$$

n_c is the density of Cooper pairs, while the phase $\theta(\mathbf{r})$, through its gradient, is part of the velocity of the pairs. While above T_c the average of $\psi(\mathbf{r})$ is zero, (no long range order) because of the fluctuations the average of the modulus $[\langle|\psi|^2\rangle]^{\frac{1}{2}}$ is different from zero. This corresponds to the generation of local pairs, without long-range spatial coherence, which decay slower and slower for $T \to T_c^+$.

In conventional (BCS) superconductors the mean field theories apply quite well, basically because the coherence volume ξ^3 includes a large numbers of fluctuating pairs. At variance, in layered SC cuprates the coherence length ξ is very small, the carriers density is reduced, one has anisotropy and high T_c, and thus a marked enhancement of the fluctuations occurs.

As recalled in Section 1, a suitable description of the superconducting fluctuations (SF) can be given in terms of the Ginzburg-Landau (GL) approach[15] based on the expansion of the free energy density $f(T, \psi)$

$$f(T,\psi) = f_{normal} + \alpha|\psi|^2 + \frac{\beta}{2}|\psi|^4 + \frac{1}{2m^*}\left|\left[\frac{\hbar}{i}\nabla - \frac{e^*}{c}\mathbf{A}\right]\psi\right|^2 \qquad (48)$$

Eq.(48) include the spatial variation and the magnetic field, through the vector potential $\mathbf{A}$. In the Gaussian approximation the term in β is neglected. The derivation of the collective amplitude and of the decay times of the fluctuations for $\beta = 0$ and in the absence of the magnetic field can be given in a way similar to the one shown in Section 1. The order parameter is expanded in free-particle eigenfunctions

$$\psi(\mathbf{r}) = \sum_{\mathbf{k}} \psi_{\mathbf{k}} e^{i\mathbf{k}\cdot\mathbf{r}} \qquad (49)$$

Then $f \equiv f(T,\psi) - f_{norm} = \sum_{\mathbf{k}}\left(\alpha + \frac{\hbar^2}{2m^*}k^2\right)|\psi_{\mathbf{k}}|$ and from the thermodynamical average over all the possible values of the order parameter

$$\langle|\psi_{\mathbf{k}}|^2\rangle \equiv \frac{\int |\psi_{\mathbf{k}}|^2 e^{-\frac{f}{k_B T}} d^2\psi}{\int e^{-\frac{f}{k_B T}} d^2\psi} \qquad (50)$$

one obtains

$$\langle|\psi_{\mathbf{k}}|^2\rangle = \frac{k_B T}{\alpha(1 + k^2\xi^2)} = \frac{\langle|\psi_{\mathbf{k}=0}|^2\rangle}{1 + k^2\xi^2} \qquad (51)$$

where

$$\xi(T) = \left[\frac{\hbar^2}{2m^*\alpha}\right]^{1/2} = \xi(0)\left(\frac{T_c}{T - T_c}\right)^{1/2} \propto \epsilon^{-1/2}$$

is the correlation length involved in the correlation function $g(0, \mathbf{R}) = \langle\psi^*(0)\psi(\mathbf{R})\rangle_0$. One can envisage the SF as generating metastable SC droplets of size $\xi(T)$.

---

[15] As shown by Gor'kov, the GL theory can be derived from the BCS theory.

For the time-dependence of the SF, required to describe the transport properties and NMR T_1, one can start from the basic equation for the deviation out of equilibrium

$$\frac{\partial \psi}{\partial t} = -\frac{1}{\gamma}\frac{\delta f}{\delta \psi^*} \tag{52}$$

where γ here is the kinetic coefficient. From Eq.(48), with $\mathbf{A} = 0$ and neglecting non linear terms,

$$-\hbar\gamma\frac{\partial \psi}{\partial t} = \alpha\psi - \frac{\hbar^2}{2m^*}\nabla^2\psi \tag{53}$$

which for the uniform mode at $k = 0$ yields an exponential relaxation of the order parameter with the GL relaxation time $\tau_{GL} = (\hbar\gamma/\alpha) \propto (T_c/t - T_c) \propto \epsilon^{-1}$.

For $k \neq 0$, from Eq.(53) the linearized time-dependent GL equation becomes

$$-\tau_{GL}\frac{\partial \psi}{\partial t} = \left(1 - \xi^2\nabla^2\right)\psi.$$

From Eq.(49) one sees that each mode decays exponentially in time, i.e. $\psi_{\mathbf{k}}(t) = \psi_{\mathbf{k}}(0)exp(-t/\tau_{\mathbf{k}})$, with

$$\tau_{\mathbf{k}} = \frac{\tau_{GL}}{1 + k^2\xi^2} \tag{54}$$

namely the equivalent of the slowing down of the fluctuations seen in Section 1.

From the SC current density $\mathbf{j} = 2e|\psi|^2\frac{\hbar}{m}\nabla\theta$ and Eq.(53) for the imaginary part, the SF contribution to the conductivity turns out

$$\sigma_{SF} = \frac{(e^*)^2}{2m}\sum_{\mathbf{k}}\langle|\psi_{\mathbf{k}}|^2\rangle\,\tau_{\mathbf{k}}. \tag{55}$$

By comparing this equation with the Drude conductivity $\sigma_0 = (e^2/m)n\tau$ [with n number of electrons per cm^3 and τ average collision time], one realizes that $\langle|\psi_{\mathbf{k}}|^2\rangle$ is indeed the collective number density of Cooper pairs while $\tau_{\mathbf{k}}$ is a life-time of the collective fluctuations.

5.2 The SF contributions to NMR shift and T_1

One can derive the SF contribution to conductivity and to other transport properties such as the spin-lattice relaxation within the Fermi liquid scenario, without specifying the nature of the interactions. It should be noted that it is not required that the system is a Fermi liquid but just that the charge and spin excitations are of fermionic character.

The term with a direct correspondence to the description of fluctuating Cooper pairs is the one given in Eq.(55) (called *Aslamazov-Larkin*, (AL)), with the most marked singularity at T_c. The AL term has no effect on T_1. A schematic justification of the lack of contribution to the nuclear relaxation from the AL term can be obtained by recalling that the nuclear transition requires a spin-flip of the scattered electron, through the transverse spin operators $S_\pm$ in the hyperfine hamiltonian. Therefore the spin-flip would imply the crossover of the pair to the triplet state.

A term that in principle is effective for the nuclear spin-lattice relaxation is the one introduced by *Maki-Thompson* (MT). Being of purely quantum nature the MT term cannot be expressed in the GL scenario. It is a kind of pairing of an electron with itself along intersecting trajectories and on physical grounds it is expected to be very sensitive to pair breaking mechanism, such as the external magnetic field. The MT contribution to T_1 can be expressed in terms of a decay rate $\Gamma_{\mathbf{k}}$ of diffusive character that goes to zero for $\mathbf{k} = 0$ when $T \to T_c^+$, so that a weak divergence in the relaxation rate $(T_1^{-1})_{MT}$ is predicted.

SF imply the subtraction of electrons to create the fluctuating pairs. Thus one has a negative contribution for $T \to T_c^+$ from the reduction in the density of states $\rho(E)$ and then of the single-particle contribution (DOS contribution). The DOS term causes a decrease in the static susceptibility and therefore in principle it could imply an effect on the Knight Shift K_{SF}:

$$K_{SF}^0 = K^0 \left[1 - \left(2\frac{T_c}{T_F} \right) \ln \epsilon^{-1} \right]$$

(K^0 Knight shift in the absence of SF, with $K^0 \propto \rho(E_F)$).

In classical BCS superconductors, some evidences the effects of SF on T_1 and K_{SF} have been obtained in nanoparticles (see Mac Laughlin, in "Further reading"). In cuprates high-temperature SC, from high resolution ^{17}O NMR spectra in external magnetic field ranging from 2 up to 24 Tesla, a smooth crossover in K_{SF} close to $T_c(H)$, with suppression in high field of the SF and a dimensionality crossover has been evidenced (see Bachman *et al.*, in "Further reading").

As regards T_1 a possible way to evidence the SF is when the relaxation rates W_{DOS} and W_{MT} have different sensitivity to the magnetic field. Unfortunately, some theoretical estimates (see Mosconi *et al.*) seem to indicate that DOS and MT terms are affected to comparable extent by the field. Thus it turns out that the SF are actually hard to evidence from NMR relaxation measurements.

From a collection of data in optimally doped YBCO it seems that a small decrease of $W(H, T \to T_c^+)$ is observable, due to the W_{DOS} term. However one is left with the problem of the concurrent effect of the spin-gap opening (see Section 4) that might be present even around $T \approx T_c^{max}$. Furthermore it seems that no effect of SF is detectable for $^{63,65}Cu$ while field-dependent ^{17}O T_1 has been claimed (see Mitrovic *et al.*)(For a through discussion see Chapter 11, in the book by Larkin and Varlamov reported in "Further reading").

5.3 Superconducting fluctuations and fluctuating diamagnetism

More convincing studies of the SF can be carried out from the related diamagnetic effects and particularly from the isothermal magnetization curves $M(T \approx T_c^+, H)$.

A qualitative argument to understand the generation of a diamagnetic susceptibility above T_c is the following. To each evanescent pair, of size ξ_{GL} one can attribute the diamagnetic susceptibility $\chi_{dia}^c \approx -e^2 \xi_{GL}^2 / mc^2$, a direct extension from atomic physics. From $n_c = \sum_{\mathbf{k}} \langle |\psi_{\mathbf{k}}|^2 \rangle$ and Eq.(51), by integrating over the Brillouin zone the susceptibility turns out

$$\chi_{dia} \propto -\epsilon^{\frac{D}{2}-2} \tag{56}$$

Almost the same result is obtained from the GL free energy density (Eq.(48)), in the Gaussian approximation (i.e. $\beta = 0$), with $\mathbf{A} = \frac{1}{2}\mathbf{H} \times \mathbf{r}$. By expanding the order parameter in terms of the eigenfunctions for Landau levels, from $\langle f \rangle = -k_B T \ln Z$, evaluating the partition function Z to the second order in the field, and from

$$M_{dia} = -\frac{d\langle f \rangle}{dH} \tag{57}$$

one derives a diamagnetic magnetization M_{dia} linear in the field, defining the susceptibility in a form close to the one in Eq.(56). In this way the effect of the magnetic field in modifying the transition temperature and possibly affecting the fluctuating pairs is *disregarded*.

A first step towards a more suitable description goes back to Prange, who provided the exact solution for the partition function Z. Then the divergence in χ_{dia} for $T \to T_c^+$ was correctly predicted to occur at $T_c(H)$ and for $T \approx T_c(H)$ the magnetization was found to go as $M_{dia} \propto -H^{1/2}$.

Since the early measurements of the fluctuating diamagnetism in BCS metals by Tinkham and coworkers, a dramatic breakdown of any theory neglecting the effect of the field in suppressing the fluctuating pairs was evidenced. A microscopic theory to integrate the GL approach and avoiding the breakdown for finite field was given by Gor'kov, by taking into account short wave-length fluctuations and non locality effects.

A simple interpretative model that allows one to grasp the basic effects of the field on the Cooper's pairs is to apply the zero-dimensional (0-D) condition to the evanescent SC droplets of size $\xi(T)$[16]. In this model in the free energy functional $f[\psi(\mathbf{r})]$ (involved in the partition function $Z[\psi(\mathbf{r})]$ to be integrated for Z and then derived for M_{dia}) $\psi(\mathbf{r})$ is assumed site independent, the gradient of $\psi(\mathbf{r})$ goes to zero and the potential vector is $\langle \mathbf{A} \rangle = H^2 d^2/10$, as for a sphere of diameter d. Then in the Gaussian approximation, from Eq.(48), by identifying d with $\xi(T)$ one obtains for the single droplet free energy

---

[16] Although this assumption is evidently crude, still leads to the field dependence of M_{dia} qualitatively correct, likely because are the droplets of size $\xi(T)$ to give the maximum diamagnetic screening.

$$f_{0D} = -k_B T \ln \frac{\pi}{\alpha\left(\epsilon + \frac{4\pi\xi^2}{\Phi_0^2}\langle A^2\rangle\right)} \tag{58}$$

and then M_{dia} is obtained:

$$M_{dia}(\epsilon, H) = -k_B T \frac{\dfrac{2\pi^2\xi_0^4\epsilon^{-1}}{5\Phi_0^2}H}{\epsilon + \dfrac{\pi^2\xi_0^4\epsilon^{-1}H^2}{5\Phi_0^2}} = C\frac{H}{H_{up}^2 + H^2} \tag{59}$$

with $\chi_{dia}(H \to 0) = C/H_{up}^2$.

It is noted that Eq.(59) predicts an upturn in the field-dependence (where $|M_{dia}|$ reverses its field-dependence and starts to decrease on increasing H) at

$$H_{up} = \frac{\sqrt{5}\Phi_0^2}{\pi\xi_0^2}\epsilon$$

This 0D model is grossly approximated, since the order parameter does change over a length of the order of ξ[17]. Still the magnetization curve $M_{dia}(T = const, H)$ are rather well justified, as it can be see for the BCS SC MgB_2 (see Lascialfari *et al.* 2002, in "Further reading"). When the zero-temperature coherence length is in the range of ~ 100 , the upturn field is found in the range $100 \div 1000$ Oe, depending on temperature.

In SC cuprates since $\xi(0) \approx 3 - 5$ lattice steps the upturn in M_{dia} vs H should be observable only at very high field, say in the range $10 - 30$ Tesla. In fact in optimally doped YBCO the GL theory, with no-field quenching of the Cooper pairs, works pretty well. As already mentioned the magnetization curves are linear in H for $T \geq T_c^+$ while very close to T_c they display the expected $H^{\frac{1}{2}}$ dependence. Good agreement with the 3D anisotropic scaling relationship for $M_{dia}(T_c)/H^{1/2}$ is also found for optimally doped SC cuprates.

At variance in underdoped cuprates of the YBCO and the LSCO family a quite different behavior of M_{dia} vs H is detected (see Lascialfari *et al.* 2002 and 2003 in "Further reading"). A dramatic enhancement of χ_{dia} for evanescent field is found and the magnetization curves display an upturn field in the range below 1000 Oe. This upturn behavior in non-optimally SC cuprates cannot be ascribed to the breakdown of the GL theory in finite field as in MgB_2, since ξ_0 is a few . Instead it has been shown (see Lascialfari *et al.* 2002 and Roman 2003 in "Further reading"), that enhancement of χ_{dia} and field effect are related to a different type of fluctuating diamagnetism, typical of systems with charge inhomogeneities. A kind of precritical regime is generated, with $|\psi| \neq 0$ in local, non-percolating regions, lacking of long range

---

[17] At variance, the model works quite well in nanoparticles, where the magnetization curves have been obtained even in the critical region, namely for $\beta \neq 0$ in close form and compared with experimental findings, see Bernardi *et al.* in "Further reading".

order and experimenting marked fluctuation in the phase. The fluctuations involve vortices and antivortices in the plane and 3D vortex loops. A theory based on the phase fluctuations in an anisotropic XY liquid of vortices rather well accounts for a variety of interesting effects observed in the fluctuating diamagnetism of underdoped cuprates.

Further reading

1. M. Thinkam: *Introduction to Superconductivity* (McGraw-Hill, 1996) Chapter 8
2. A. Larkin and A.A. Varlamov: *Theory of Superconducting Fluctuations* Oxford U.P. (2005) Chapters 2 and 4
3. E. Bernardi, A. Lascialfari, A. Rigamonti, L. Roman, et al., Phys. Rev. B **74**, 134509 (2006)
4. A. Lascialfari, T. Mishonov, A. Rigamonti, P. Tedesco and A. Varlamov, Phys. Rev. B **65**, 180501 (2002)
5. A. Lascialfari, A. Rigamonti, L. Roman, P. Tedesco, A. Varlamov and D. Embriaco, Phys. Rev. B **65**, 144523 (2002)
6. A. Lascialfari, A. Rigamonti, L. Roman, A. Varlamov and I. Zucca, Phys. Rev. B **68**, 100505 (2003)
7. A. Sewer and H. Beck, Phys. Rev. B **64**, 014510 (2001)
8. L. Roman, Int. J. Mod. Phys. **17**, 423 (2003)
9. L. Cabo, F. Soto, M. Ruibal, J. Mosqueira and F. Vidal, Phys. Rev. B **73**, 184520 (2006)

Muon Spin Rotation

Introduction to μSR.

Roberto De Renzi

Department of Physics and CNISM, Viale G.P. Usberti, 7a, I-43100 Italy

We briefly introduce the principles of muon spin spectroscopy, considering in particular what happens with molecules and magnets.

1 Introduction

The following notes are meant to introduce the principles of Muon Spin Spectroscopy. They are divided as follows: The first section briefly recalls the historical origins and the principles upon which the technique is based; Section II describes the main types of experimental setup; Section III distinguishes what may happen to the muon at implantation, before the measurement starts; The detection of paramagnetic muon molecules is outlined in section IV; a brief account of what happens in magnetic material is given in section V; finally specific muon relaxation functions are described in section VI.

Besides the books and articles cited explicitly in these Sections, it is worth referring the reader to the main papers, books and web sites available on the subject [1, 2, 4, 3]. Among the more topical reviews it is worth quoting the recent special issue of the Journal of Physics, Condensed Matter [5], and recalling that a rather detailed account of μSR research is contained in the Proceedings of the ten past International Conferences on Muon Spin Rotation, Relaxation and Resonance (the first seven editions on *Hyperfine Interactions*, the three most recent on *Physica B*).

2 The rules of the game

2.1 A bit of history.

The experimental technique of μSR applied to condensed matter originates from the discovery of the violation of parity in decays regulated by weak interactions. The Nobel Prize was awarded to Lee and Young (1957) [6] "for their penetrating investigation of the so-called parity laws" (from the motivation of the prize). These investigation were confirmed by madame Wu and collaborators [7] in their experiment on the β-decay of ^{60}Co, and, almost at the same time, in the parallel experiment of Richard L. Garwin, Leon M. Lederman, and Marcel Weinrich [8] on the muon decay, which illustrates parity violation more directly.

LETTERS TO THE EDITOR   1415

current for a given time delay. A typical run is shown in Fig. 2. As an example of a systematic check, we have

Observations of the Failure of Conservation of Parity and Charge Conjugation in Meson Decays : the Magnetic Moment of the Free Muon*

RICHARD L. GARWIN,† LEON M. LEDERMAN, AND MARCEL WEINRICH

Physics Department, Nevis Cyclotron Laboratories, Columbia University, Irvington-on-Hudson, New York, New York

(Received January 15, 1957)

LEE and Yang[1-3] have proposed that the long held space-time principles of invariance under charge conjugation, time reversal, and space reflection (parity) are violated by the "weak" interactions responsible for decay of nuclei, mesons, and strange particles. Their hypothesis, born out of the τ-θ puzzle,[4] was accompanied by the suggestion that confirmation should be sought (among other places) in the study of the successive reactions

$$\pi^+ \to \mu^+ + \nu, \qquad (1)$$
$$\mu^+ \to e^+ + 2\nu \qquad (2)$$

They have pointed out that parity nonconservation implies a polarization of the spin of the muon emitted from stopped pions in (1) along the direction of motion and that furthermore, the angular distribution of electrons in (2) should serve as an analyzer for the muon

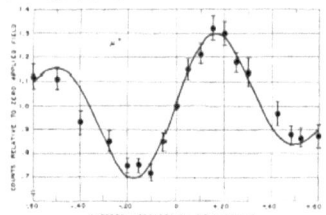

FIG. 2. Variation of gated 3-4 counting rate with magnetizing current. The solid curve is computed from an assumed electron angular distribution $1 - \frac{1}{3}\cos\theta$, with counter and gate-width resolution folded in.

Various other materials were investigated for μ^+ mesons. Nuclear emulsion as a target was found to have a significantly weaker asymmetry (peak-to-valley ratio of 1.40 ± 0.07) and it is interesting to note that this did not increase with reduced delay and gate width. Neither was there any evidence for an altered moment. It seems possible that polarized positive and negative muons will become a powerful tool for exploring magnetic fields in nuclei (even in Pb, 2% of the μ^- decay into electrons[5]), atoms, and interatomic regions.

Fig. 1. The first column and the main picture from ref. [8] are shown on the left. Due to parity violation the positron counts from the decays of a muon population are made to oscillate in time, as the spins of the muons are coherently precessing around an external magnetic field. This implies that the muons preferentially emit their decay positron along the direction of their spin, instead of uniformly distributed in all directions, as it would happen for parity conserving events.

Fig. 1 shows an excerpt from the muon decay paper, and the bottom line in the right column indicates that the authors had already the intuition of the potential use of the newly discovered effect as a solid state technique.

2.2 The muon

Thanks to parity violation the muon may be produced with a large spin polarization, nearly 100%, starting from an accelerated particle beam, most often protons. It is then implanted as a probe in solids, liquids and gasses. It generally ends up inside the sample in few equivalent, most energetically-convenient sites, retaining its initial spin polarization. Parity violation again allows the detection of the muon spin coherent behaviour from the direction of emission of its decay-positron (electron). This is accomplished by averaging over an ensemble of muons which are implanted one by one, or in bunches, at subsequent times.

Charge	$\pm$		
Mass	$m_\mu = 0.1126\,m_p = 206.7684(6)\,m_e = 105.65849(4)\,\mathrm{MeV/c^2}$		
Spin	$I_\mu = \frac{1}{2}$		
Magnetogyric ratio	$\frac{\gamma_\mu}{2\pi} = 135.53875(6)\,\mathrm{MHz/T}$		
Mean lifetime	$\tau_{\mu^+} = 2.19703(4)\,\mu s$		

Table 1. Muon properties. Values of other quoted constants: electron mass $m_e = 9.1093826(16)$, $10^{-31}\,\mathrm{kg} = 0.51099907(15)$ MeV, speed of light in vacuum $c = 299792458$ m/s, electron charge $e = 1.60217653(14)\,10^{-19}$ C, Planck's constant $h = 6.6260693(11)\,10^{-34}$ Js. The muon magnetogyric ratio is given by $\gamma_\mu = \frac{e}{m_\mu}(1 + \frac{g_\mu - 2}{2})$, where the measured quantity is $1 + \frac{g_\mu - 2}{2} = 1.001165923(8)$.

The muon is a lepton[9] and its properties are listed in Tab. I. In order to produce intense muon beams suitable for implantation the decay of charged pions is exploited. Since negative muons in matter are strongly attracted by nuclei their fate often tell more about the nuclear interactions in the sample than about its solid state properties. For this reason from now on we specialize to positive muons. In the rest frame of the pion, which decays according to:

$$\pi^+ \rightarrow \mu^+ + \nu \tag{1}$$

the outgoing muon and neutrino have opposite linear momentum, equal to 29.79 MeV/c. This value is obtained by imposing the conservation of the relativistic energy:

$$\sqrt{m_\pi^2 c^4 + p_\pi^2 c^2} = \sqrt{m_\mu^2 c^4 + p_\mu^2 c^2} + \sqrt{m_\nu^2 c^4 + p_\nu^2 c^2} \tag{2}$$

with $p_\pi = 0$, $p_\nu = -\pi_\mu$, $m_\nu = 0$ (effectively) and $m_\pi = 139.57\,\mathrm{MeV/c^2}$. The muon kinetic energy is then

$$E_k = \sqrt{m_\mu^2 c^4 + p_\mu^2 c^2} - m_\mu c^2 = 4.119\,\mathrm{MeV/c^2} \tag{3}$$

Comparing the terms in Eq. 2 is is easy to see that, whereas the neutrino is entirely relativistic (if its mass were zero, it would travel at the speed of light), the muon is far from the fully relativistic limit, having $\beta = 0.272$. Equation 3 sets the energy that a muon originating from a pion at rest has to spend to get through thin vacuum windows along the beamline and to implant inside the sample.

2.3 Muon production

The decay of charged pions, Eq. 1, is more precisely:

$$\pi^+ \rightarrow \overline{\mu}^+ + \nu_\mu$$
$$\pi^- \rightarrow \mu^- + \overline{\nu}_\mu,$$

the bar indicating a leptonic antiparticle (the decay conserves the number of leptons, zero on both side of the arrow).

In order to obtain the pions, for a start, a proton beam of sufficient energy is aimed at a solid target and a transport channel for charged particles is tuned to focus one open surface of the target, at one end, onto the sample, at the opposite end.

The energy of the primary proton beam must be above the threshold for pion production. To determine this threshold, consider that the dominant channel for positive pions is

$$p + (p, n) \rightarrow n + (p, n) + \pi \tag{4}$$

where the second particle on the left (proton or neutron) is at rest.

Assuming the proton and neutron mass both equal to $M = 938 \text{MeV}/c^2$ and studying the system in the center of mass, where the two particles have equal and opposite momentum q we suppose that the threshold momentum for pion production (mass $m = 139.57 \text{MeV}/c^2$) is for the head-on collision, where the proton turns into a neutron and bounces off with momentum p in the reverse direction, whereas the pion momentum x is opposite to p and so is the final momentum of the original particle at rest $(-p + x)$. This situation is sketched in Fig. 2.

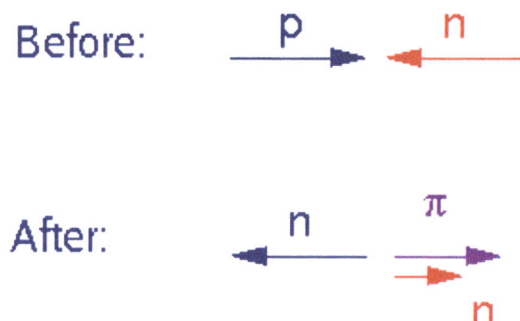

Fig. 2. Pion production: momentum is conserved before and after the head-on collision.

Energy conservation imposes that:

$$2\sqrt{M^2 + q^2} = \sqrt{M^2 + p^2} + \sqrt{M^2 + (p - x)^2} + \sqrt{m^2 + x^2} \tag{5}$$

One can see that the minimum q is that for which $p=x=0$, i.e. $q \geq \sqrt{Mm + m^2/4} = 361.93$ MeV/c which means $q' \geq 790.6$ MeV/c in the laboratory frame ($q' = 2q\sqrt{M^2 + q^2}/M$). This implies that the threshold kinetic energy of the proton beam in the laboratory frame is $\sqrt{M^2 + q'^2} - M = 281.5$ MeV.

In practice the cross section increases [10] rapidly up to $q' = 1.1$ GeV/c, corresponding to a beam energy of 500 MeV, representing an optimum in some sense.

Pions of a well defined linear momentum are selected with a *transport channel* from the primary target on which the proton beam is impinging. The target is often pyrolitic graphite, with special care for cooling, since typical proton beams at muon facilities are from hundreds to thousands of μA.

A *transport channel* for charged particles is roughly equivalent to an optical beam and it is composed of magnetic lenses and magnetic prisms. The function of a (chromatic) lens is provided by two or three quadrupole magnets at short distance from each other. Quadrupoles have two opposed north poles and two opposed south poles each. The function of the prism is produced by a dipole magnet, a *bender*, which acts as a momentum selector. Muons, pions, electrons and any other particle of the same momentum and charge are transported down such a beam.

An additional important element is the *electrostatic separator*, with crossed electric and magnetic fields. This device acts on a straight line as a velocity selector, since the two fields can be tuned to compensate each other for a specific mass. Hence it removes the other particles from the muon beam.

At best, the cross section of the beam at one of its subsequent foci is equal to cross section of the pion source on the primary target, which, in turns, is determined by the proton beam cross section and, possibly, by the target geometry (i.e. its thickness, for muon beams extracted at right angles to the protons) .

2.4 Muon spin polarization

The role of the violation of parity in yielding spin polarized muon beams can be directly identified in the rest frame of the pion, again.

We already considered the kinematics of this decay. We can now concentrate on angular momentum conservation, recalling that the π^+ is a spinless particle. Since, instead, the neutrino has spin one half, like the muon, the angular momenta of the two particles must come out opposite to each other. Furthermore the neutrino is highly relativistic: assuming that it is massless, its spin would be aligned with its linear momentum, i.e. its helicity[11] is equivalent to chirality[12] (handedness) multiplied by $\hbar/2$.

The violation of parity in this event is equivalent to stating that only neutrinos of negative helicity (spin antiparallel to linear momentum) exist. Positive helicity is conversely the only possibility for antineutrinos, as shown in Fig. 3. This is often simplified by saying that, although in the macroscopic world the mirror image of any possible event is also a possible event, this symmetry is broken in the microscopic realm of elementary particles

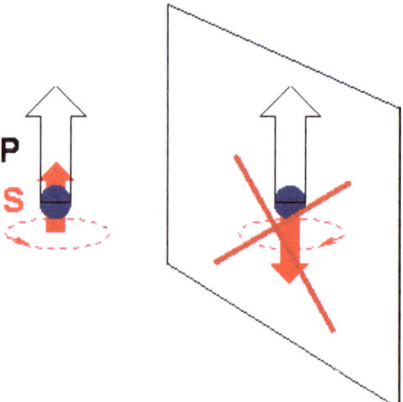

Fig. 3. The figure shows an antineutrino, with spin parallel to momentum; its mirror image should have spin antiparallel to momentum (negative helicity) because angular momentum is an axial vector that reverses in the mirror image, but this state does not exist in nature. For neutrinos the opposite holds: only those with negative helicity exist.

This statement and the conservation of angular momentum impose that also the muon from the decay must have negative helicity. This is illustrated in Fig. 3 and Fig. 4.

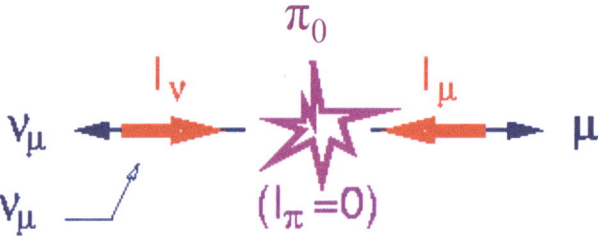

Fig. 4. Pion decay in its rest frame: the muon must have negative helicity like the neutrino.

Therefore if we select by transport a very small solid angle of directions diverging from the primary target, and a momentum of exactly 29.79 MeV/c, the beam will be predominantly of muons from the decays of pions at rest on the surface of the target, and very nearly[13] 100% spin-polarized backwards with respect to linear momentum.

By the same argument *forward* negative muons would be produced in a similar geometry. This simple description, performed in the rest frame of the pion, is valid in practice if we consider pions decaying on the surface of the production target, yielding the so-called *surface muon* beams. The initial muon direction is preserved by all the magnetic elements of conventional beam transport "optics", i.e. by the focusing and bending actions of magnetic fields on charged particle trajectories, to the sample. Bending elements can finely tune the linear momentum of transported particles. At the low momenta of surface muon beams the high field versions of separators are called *spin rotators*. Besides selecting velocity (mass), the magnetic field in these devices makes the muon spin precess by a large angle and the electric field may be tuned to compensate the corresponding turn in the geometrical trajectory by the same angle, yielding therefore a straight trajectory and a predetermined initial spin direction.

The surface beam is only one of the possible schemes for the production of muon beams, the most convenient for the majority of μSR experiments, since it provides fully polarized muons of relatively low energies, which stop in condensed matter within few hundreds micron from the surface (roughly 80 μm for Pb and 1 mm for water). In this way muons are probes of the bulk and require moderate amounts of material.

However higher energy beams can be obtained, simply by selecting the corresponding linear momentum in the transport channel. In this case the transported particles are pions and sufficient length of the transport path must be allowed for most pions to decay (with mean lifetime $\tau_\pi = 26.033(5)$ ns). This is accomplished by superconducting solenoids, on dedicated beamlines, where the helical path is much longer than the physical length of the device. The use of higher energy muons (typically 80 to 100 MeV/c) is mandatory for pressure experiments, where the muons must penetrate a massive pressure cell.

2.5 Parity violation in the muon decay

Parity violation is also the key to the detection of the muon spin evolution in time. As in the case of the muon production it is implemented in angular momentum conservation with the constraint that the positron and the neutrinos:

$$\overline{\mu}^+ \rightarrow \overline{e}^+ + \nu_e + \overline{\nu}_\mu \tag{6}$$

distinguish between left and right. They must have well defined helicities (negative for the particle ν_e and positive for the antiparticles $\overline{e}^+$ and $\overline{\nu}_\mu$).

Since Eq. 6 represents a three body decay, there is a continuum of possible geometries, as shown in the sketch on the left) and, correspondingly, a range of energies for the emitted positron ($E \leq 52.83$MeV). The correlation imposed on the directions of the muon spin and of the positron momentum is however simple at maximum positron energy, which corresponds to the two neutrino being both emitted opposite to the positron. In this particular collinear case, which is close to that shown in Fig. 5:

- the two neutrinos do not contribute to angular momentum, since they are collinear and of opposite helicity
- the positron must have positive helicity, hence it must come out in the direction of the muon spin in order to conserve angular momentum.

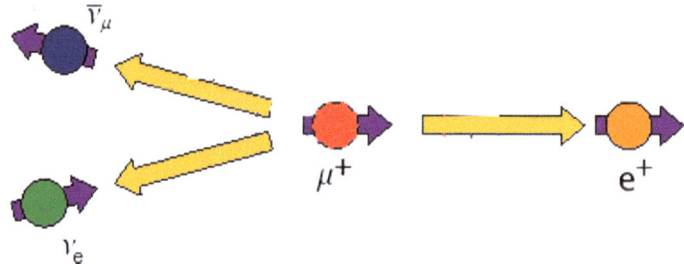

Fig. 5. Muon decay, conservation of linear (long arrows) and angular momentum (short arrows), together with parity violation, produces a correlation between the muon spin direction and the positron linear momentum.

This correlation allows the experimental determination of the muon spin direction from the identification of the positron momentum direction. The correlation is reduced for lower energies. More precisely, the probability distribution function for the positron emission is correlated to the instantaneous direction of the muon spin by:

$$P(\theta) \propto 1 + A(E)\cos\theta \qquad (7)$$

where $A(E)$ is an asymmetry factor that depends on the energy of the emitted positron, equal to 1 for $E = E_{max}$.

The distribution vs. both angle and energy is given in terms of $x = E/E_{max}$ as:

$$W(x,\theta) = \frac{E(x)}{4\pi}[1 + A(x)\cos\theta] \qquad (8)$$

with $E(x) = 2x^2(3-2x)$ and $A(x) = (2x-1)/(3-2x)$. These two functions are shown in Fig. 6. The third shown function is the weighted asymmetry spectrum, $E(x)a(x)/2$. The polar plot in Fig. 6, inset a, shows the probability distribution lobe for the maximum asymmetry, $A(E_{max}) = 1$: the probability

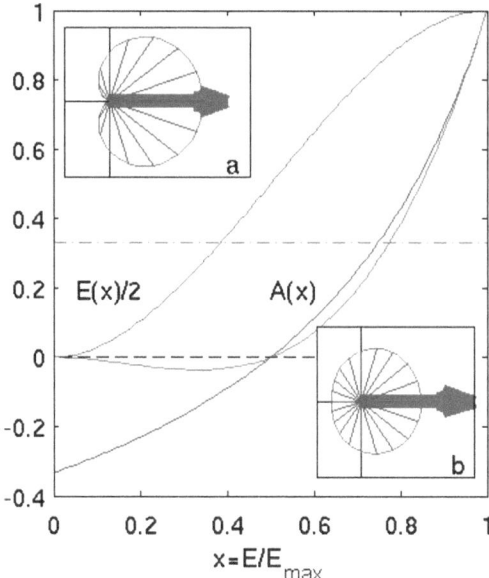

Fig. 6. Energy spectrum, E, and asymmetry spectrum, A, vs. reduced energy, $x = E/E_{max}$. The third solid curve is the weighted asymmetry, $E(x)a(x)/2$. The top inset (a) shows the polar plot of the emission probability at the maximum asymmetry $A = 1$, as one would observe if only maximum energy positron were detected. The bottom inset (b) shows the polar plot of the emission probability at the average asymmetry $A = \frac{1}{3}$, that is ideally observed if all positrons are detected.

in each direction is proportional to the length of the blue segment along that same direction. The average over all energy of the function $A(E)$ is equal to $\frac{1}{3}$, which is the ideal experimental asymmetry value when no positron electron discrimination is performed. The corresponding polar plot of the probability is shown in inset b.

2.6 Longitudinal field

Let us build the simplest possible experimental setup: we place two positron detectors in front of the sample, one in the muon spin direction (F for forward) and one in the opposite direction (B, for backward), as shown in Fig. 7.

With spin polarized muons the F detector will count more and the B detector less than with unpolarized muons. If we simply count events vs. the lifetimes of the individual muons, we shall record rates that follow an exponential decay with mean lifetime $\tau_\mu = 2.2\,\mu S$.

$$dN(t) = N_0\, e^{-t/\tau_\mu}\,(1 + A cos\theta) dt\, d\theta \tag{9}$$

The unpolarized muon rate, $N_0 e^{-t/\tau_\mu}$ corresponds to the central black curve, the polarized rate in the F and B detectors are correspondingly labeled. Note that the polar plot on Fig. 7 (right) is for the maximum asymmetry, $A(52.8\text{MeV}) = 1$, for clarity, whereas the count rates in Fig. 7 (left) are shown for the more realistic average asymmetry $\overline{A} = \frac{1}{3}$.

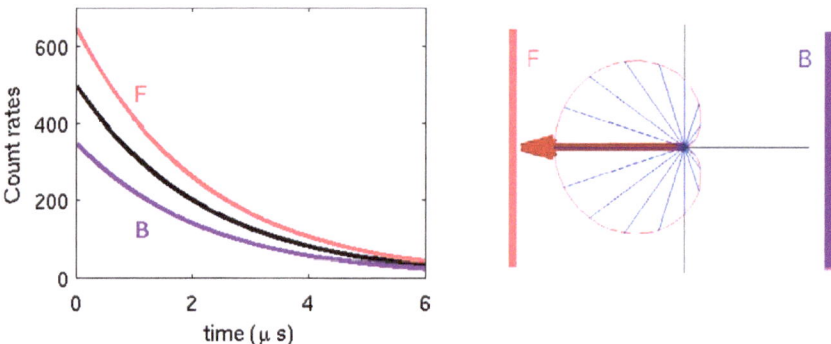

Fig. 7. Left: Count rates in the same hypothetical experiment, with unpolarized muons (middle curve), with muons polarized along the direction towards which the detector is facing (Forward, top curve), and with muons polarized along the opposite direction (Backward, bottom curve); average asymmetry $\overline{A} = 0.33$. Right: the geometry of the Forward and Backward detectors, together with the probability lobe for maximum asymmetry $A = 1$.

In order to obtain the count rates of a real experiment Eq. 9 must be integrated over a finite time bin and over the solid angle covered by the detector. Both integrations lead to an average, i.e a reduction of the observed asymmetry from the theoretical $\frac{1}{3}$ value.

2.7 Transverse field

Another very simple experimental setup is when an external magnetic field $\boldsymbol{B}$ is applied perpendicular to the incoming muon spin direction. This is the setup of the original experiment of Fig. 1.

After implantation the muon spin precesses in a plane perpendicular to $\boldsymbol{B}$, describing an angle $\theta = \gamma B t$ with its own initial direction, and the probability distribution lobe for the emission of the positrons precesses with it. Averaging over an ensemble of muons will then modulate harmonically the decay rate of the ensemble in time, obtained from the positron count rates in a single detector, as shown in Fig. 8.

$$dN(t) = N_0\, e^{-t/\tau}(1 + A\,cos\gamma Bt)dt\, d\theta \qquad (10)$$

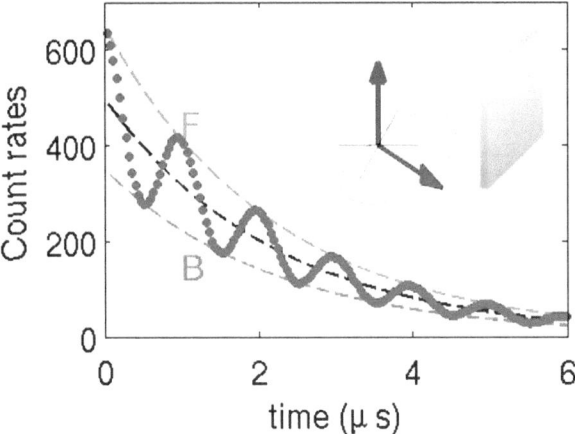

Fig. 8. Count rates in the transverse field case. The inset shows that, while the muon spin precesses around the vertical magnetic field, the probability lobe sweeps past the detector modulating the count rates between the limits appropriate for a forward detector (dashed F curve) and for a backward detector (dashed B curve).

Muons are implanted either one by one, at continuous sources, or in bunches of hundreds, at pulsed sources. Individual lifetimes are thus recorded for millions of muon events, each contributing with one count in one time bin. All these events are accumulated independently, over long times, minutes to hours. Coherence is guaranteed by the fact that all muons at implantation (time zero in the plot) have the same spin polarization.

2.8 Muon interactions

At each interstitial site the muon spin $I = \frac{1}{2}$ is subject to interactions with the other spins, those of neighbouring electrons and nuclei. A classification of these interactions similar to that of NMR may be performed, with the simplifying condition that the muon itself, being spin one-half, does not possess an electric quadrupole moment. Hence the muon spin does not couple to the electric field gradient, although it definitely may produce one.

Assuming that a uniform external magnetic field B will be aligned along z, we may write:

$$\frac{\mathcal{H}}{h} = \frac{1}{2\pi} \left\{ \left[-\gamma_\mu I_z + \gamma_e \sum_{i=1}^{n_e} s_{iz} - \sum_{k=1}^{n_n} \gamma_k I_{kz} \right] B \right. \tag{11}$$

$$+ \gamma_\mu \boldsymbol{I} \cdot \sum_{i=1}^{n_e} \hbar \gamma_e \left[\frac{\boldsymbol{l}_i}{r_i^3} + \frac{-\boldsymbol{s}_i + 3\hat{r}_i(\boldsymbol{s}_i \cdot \hat{r}_i)}{r_i^3} + \frac{8\pi}{3} \boldsymbol{s}_i \delta(\boldsymbol{r}_i) \right] \tag{12}$$

$$\left. + \gamma_\mu \boldsymbol{I} \cdot \sum_{k=1}^{n_n} \hbar \gamma_k \frac{-\boldsymbol{I}_k + 3\hat{r}_k(\boldsymbol{I}_k \cdot \hat{r}_k)}{r_k^3} \right\} \tag{13}$$

where the first row (11) contains the Zeeman interactions of all the involved spins, s_i for the i-th electron and I_k for the k-th nucleus, the second (12) row contains the magnetic interactions between the muon and the electrons and third (13) row those between the muon and the nuclei.

The sum over k in line 13, where the vector $\boldsymbol{r}_k = r_k \hat{r}_k$ connects the muon and the k-th nucleus, may be recognized as the classical dipolar field, $\boldsymbol{B}_d$, from nuclear point dipoles, each contributing with

$$\boldsymbol{B}_{kd} = \hbar \gamma_k \frac{-\boldsymbol{I}_k + 3\hat{r}_k(\boldsymbol{I}_k \cdot \hat{r}_k)}{r_k^3} \tag{14}$$

where the nuclear coordinates may be treated classically.

In contrast, in line 12 the coordinates should be considered as electron operators. However, since the electron dynamics is much faster than the muon spin precessions, we shall always average these operators over the electron wave functions, recovering a spin-only Hamiltonian. The first term in line 12 represents the interaction with the orbital momentum $\boldsymbol{l}_i$ of each electron $\hbar \gamma_e \frac{\boldsymbol{l}_i}{r_i^3}$, and it is often neglected because of the quenching of orbital momentum by crystalline fields. The second term of line 12 typically accounts for the distant magnetic ions, yielding a purely dipolar field, like that of Eq. 14, by

$$\boldsymbol{B}_{id} = \hbar \gamma_e \frac{-\boldsymbol{s}_i + 3\hat{r}_i(\boldsymbol{s}_i \cdot \hat{r}_i)}{r_i^3}, \tag{15}$$

For a and the last term which corresponds to the dipolar contribution when the muon and the i-th electron overlap. For electrons involved in the chemical bond with the muon, hence whose wave functions overlap, the same second term, together with the third one, contribute to the so-called hyperfine coupling. This may happen directly, if the $i - th$ unpaired electron of spin $\boldsymbol{s}_i$ is involved in the bond. Then the second term gives rise to the pseudo dipolar hyperfine (tensorial) and the third term gives rise to the Fermi contact hyperfine (scalar) couplings:

$$\delta \tilde{\nu} + \nu_0 = \frac{\mu_0}{4\pi} \frac{\gamma_\mu}{2\pi} \boldsymbol{I} \cdot \hbar \gamma_e \left[\langle \Psi | \frac{-\boldsymbol{s}_i + 3\hat{r}_i(\boldsymbol{s}_i \cdot \hat{r}_i)}{r_i^3} | \Psi \rangle + \boldsymbol{s} \frac{8\pi}{3} |\Psi(0)|^2 \right] \tag{16}$$

respectively the first and second contributions in parenthesis.

In magnetic materials the muon is often bound to an anion (oxygen, fluorine, etc..), which is non magnetic, i.e. the bond is formed, in first approximation, only by paired electrons. In this case a second nearest neighbour magnetic cation, M, may spin polarize the Mu bond, say, M-O-Mu, giving rise to a net spin density at the muon. This will yield the so-called *super-hyperfine* coupling. Averaging over the intervening electrons, the second and third terms act as a tensorial coupling:

$$\delta\tilde{\nu} + \nu_0 = \frac{\gamma_\mu}{2\pi}\boldsymbol{I} \cdot \left[\delta\tilde{A} + A_0\right] \cdot \boldsymbol{S} \tag{17}$$

formally equivalent to those of Eq. 16, although here $\boldsymbol{S}$ may be a composite spin, like that of the ground state of a magnetic cation.

3 The muon fate at implantation

Muons come to rest inside the sample and, in most cases, this happens without loss of spin polarization. The initial energy of the muon, of the order of 4 MeV, is huge compared to the typical energies of electrons (from few eV, for valence shells to hundreds of keV, for inner shells in heavy elements) with whom collisions take place. Therefore the initial collisions lead to ionization (Coulomb scattering) and in this process the muons leave behind a wake of ions and electrons.

When below the ionization threshold the muons generally travel quite a long way away from this region and the final thermalization depends on solid state or chemical details of the environment. In this context the positive muon may be regarded as a light isotope of hydrogen, whose complex epithermal chemistry reactions in the host material determine its successive fate.

Muons seldom replace an atom of the host compound: this may be the case when cation vacancies are abundant or when a MU-H exchange takes place, e.g. in organic compounds. More often the muon ends up at "'interstitial"' sites, in crystals, or equivalently, at addition sites, in molecules. The spin dynamics of the thermalized muon depends dramatically on whether it ends up in a "coherent" or in an "incoherent" spin state with its environment. "Coherent" spin states are formed:

- when the muon is attached to a *paramagnetic* molecule (in the chemical sense), that is when a MO state with an unpaired electron, also called *free radical*, is formed; this instance is characteristic of unsaturated organic molecules and the prototype of such a free radical is notably muonium, a hydrogen atom with the proton replaced by the muon.
- when the muon is bound to a *diamagnetic* molecule; a molecular orbital (MO) type of state is formed in which two electrons of opposite spin occupy a bond between an atom and this light hydrogen isotope; the only influence of the host on the muon is a chemical Larmor frequency shift, i.e. a tiny

diamagnetic effect due to the polarization of the bond currents, which is however too feeble to be measured within a few muon lifetimes.

- when the muon is localized interstitially in a *magnetically ordered* environment where it experiences hyperfine and dipolar couplings to the ordered magnetic moments of the electron.
- when the muon is in close dipolar interaction with few isolated nuclei having non vanishing magnetic moment. This is typically the case of the F-Mu-F center [14] formed in several fluorides.

The spin environment may be "incoherent" because of fast temporal fluctuations, e.g. when the magnetic moments of nearby electrons fluctuate rapidly on the time-scale of the muon spin dynamics. This is the "paramagnetic" cases in the condensed matter sense, i.e.:

- in metals, where free electrons (the Fermi liquid) are characterized by Pauli paramagnetism; in this case electron charge screening would generally prevent the formation of a bound state, and spin dependent scattering with the electron liquid would prevent the observation of a bound state, if it did form. Knight shift may be measured for metals, although high fields are required to observe it within a few muon lifetimes.
- in the disordered state above the magnetic ordering temperature of a magnetic material.

Another incoherent spin environment, frequently encountered in condensed matter, is the nearly static, spatially inhomogeneous case of the unpolarized (or very weakly polarized) nuclear spins, when their number is large enough to allow a classical statistical treatment, rather than the exact quantum computation of the F-Mu-F center [14].

3.1 Muon sites

Quite often the fate of the muon is not unique. A number of distinct final states may be obtained with different probabilities. These states may simply correspond to distinct interstitial sites, due to alternative chemical bonds. Sometimes different sites correspond also to very different electronic configurations, as it is the case for intrinsic semiconductors, where often a metastable antibonding position can be occupied, e.g. in Si at low temperatures, together with a more stable bond center position.

In this respect the situation is more complex than it would be for most chemical species at thermal equilibrium, both because of the epithermal energy that the muon initially possesses, and because of the short timescale over which it is observed, which allows the detection of the metastable states.

Note that muon locations may be further differentiated in a magnet if the direction of the ordered electron magnetic moments produces different dipolar fields at otherwise equivalent sites. The multiple sites are effectively accounted for by considering that the total asymmetry A of the full implanted muon ensemble is divided into the partial asymmetries of the n distinct muon sites, according to:

$$A(t) = \sum_{j=1}^{N} A_j(t) \tag{18}$$

where the initial partial asymmetry, $A_j(0) = Ap_j$, may be written in terms of the stopping probability p_j at the j-th site (with $\sum_j p_j = 1$).

Sometimes one refers to a muon spin polarization function P_j at each site:

$$A_j(t) = A_j(0)P_j(t) \tag{19}$$

4 Paramagnetic chemical species

When the muon binds to a molecule with at least one spin-unpaired electron the interactions with the electron spin become dominant.

We shall describe in some more details this species, the muonic adduct radical, quite typical of muons implanted in molecular compounds containing unsaturated organic molecules, i.e. double or triple bonds. In these cases the muon actually modifies greatly its surrounding by forming a unique molecule, very different from the others. The state thus formed turns out to be a very sensitive probe in itself, yielding a wealth of valuable information via its specific spectroscopic signatures.

Muonium and radical states are directly observable: muonium may be found in vacuum (evaporated from suitable hot spongy surfaces) and inside matter in a few cases (e.g. in quartz, inside the fullerene cage, in a slightly distorted form in silicon). Radicals are observed in many unsaturated compounds.

We shall start describing the simplest of these states, the muonium atom. Fig. 9 shows the Breit-Rabi diagram of the energy levels of muonium. In zero field the upper degenerate state is the spin triplet of electron and muon spins, whereas the lower state is the singlet. Precessions can be detected corresponding to the nearly degenerate triplet transitions at low transverse field, shown in the figure by the double arrows on the left.

The spin states at high field are shown by the labels ($\pm$ stand for spin up and down, for electron, first, and muon). The transitions that become detectable are the two shown by the double arrows at high fields, which represent the spectroscopic signature of a radical. This description is valid also for more complex radicals since in the high field condition the additional interactions with other nuclei are decoupled. We shall now proceed to discuss this point in some more details.

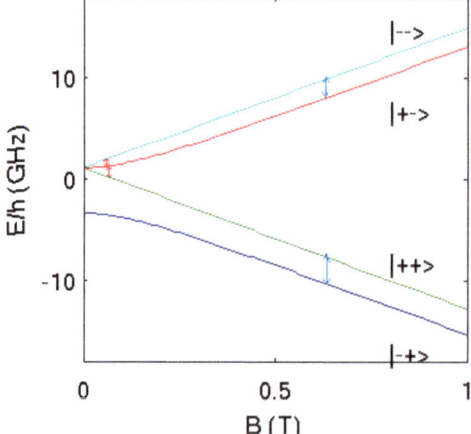

Fig. 9. Energy levels of muonium (in frequency units) vs. external magnetic field. Low field triplet transitions and high field, muon-spin-flip transitions are shown by double arrows.

4.1 Muonium precessions

Muonium is the muon equivalent of the hydrogen atom and the simplest radical state. When free muonium is formed in its 1s ground state, the muon spin interacts with that of the bound electron, governed by the hyperfine Hamiltonian $\mathcal{H}$:

$$\frac{\mathcal{H}}{h} = -[\nu_\mu I_z - \nu_e s_z + \nu_0 \mathbf{I} \cdot \mathbf{s}] \qquad (20)$$

where $2\pi\nu_\mu = \gamma_\mu B$, $2\pi\nu_e = \gamma_e B$ are the muon and electron Larmor frequencies, due to the Zeeman interaction with the magnetic field, and $\nu_0 = \frac{\mu_0}{3\pi}\gamma_\mu\gamma_e\hbar|\psi(0)|^2 = 4.4$ GHz is the hyperfine frequency, proportional to the square modulus of the electron wave function ψ at the muon position. Note that the Hamiltonian of Eq. 20 is invariant under rotation, thanks to the spherical symmetry of the 1s ground state.

The energy levels of this spin Hamiltonian, a 4x4 matrix, are very easy to recognize for $B = 0$, since the hyperfine term is invariant under rotations and the addition of the two angular momenta $I, s = \frac{1}{2}$ yields two eigenstates: the triply degenerate $J = 1$ state (triplet) and the non-degenerate $J = 0$ state (singlet).

The spin matrix for finite fields and its diagonalization are easy to calculate with matlab or octave. The field dependence of the energy levels is plotted in Fig. **??** in frequency units ($\gamma_e/2\pi = 27.992$ GHz/T). For $B > \frac{2\pi\nu_0}{\gamma_e}$ the electron and muon states are decoupled as the labels in the plot imply.

It is easy to see that in high fields the selection rule is $\Delta m = 1$ for the muon only, in other words in a high transverse field the the electron is in an eigenstate, whereas the muon spin is not. Hence the two allowed muon transitions are between states of equal electron spin polarization. The field dependence of these high field transitions is shown in Fig. 9.

4.2 The adduct radical

The free adduct radical differs from free muonium in that the unpaired electron wave function is delocalized over many more nuclei, several of which may possess a spin. A marked difference with the 1s ground state of muonium is that the electron wave function is not spherically symmetric around each of these nuclei, including the muon. This brings in two modifications to the Hamiltonian of Eq. 20: the hyperfine coupling becomes a tensor (its trace is the isotropic hyperfine coupling, whereas the traceless tensor is the pseudodipolar contribution), and additional terms appear:

$$\frac{\mathcal{H}_n}{h} = -\nu_n I_{nz} - \mathbf{I}_n \cdot \nu_{0n} \cdot \mathbf{s} \tag{21}$$

for each nucleus in the molecule, with spin I_n. The dimensions of this Hamiltonian quickly grow with the number m of nuclei involved, as $2^2 \cdot N = 4 \prod_{n=1}^{m}(2I_n + 1)$. However it is still rather simple to obtain the muon transitions in the high field condition, the so-called Paschen-Bach regime where $B > \max\{\omega_0, \omega_{0n}\}/\gamma_e$. In this regime all the spins are decoupled and the eigenvalues of I_z, s_z, I_{nz} are good quantum numbers. Therefore the measurement of I_x can induce only muon spin transitions, whereas all other quantum numbers must remain constant. In practice one has N replicae of the corresponding "muonium-like" energy levels, determined only by the muon and electron spin.

Fig. 10 shows the energy levels obtained with one additional proton and hyperfine frequencies $\nu_0 = 500$ MHz and $\nu_{op} = 150$ MHz, as appropriate for the cyclohexadienyl radical obtained by Mu addition to benzene (although that radical actually has six hydrogens).

Allowing for a hyperfine traceless anisotropy tensor $\delta\tilde{\nu}_0$ one has an effective Hamiltonian:

$$\frac{\mathcal{H}}{h} = -\nu_\mu I_z + \nu_e s_z - \mathbf{I} \cdot (\nu_0 + \delta\tilde{\nu}_0) \cdot \mathbf{s} \tag{22}$$

valid only at high fields. The tensor $\delta\tilde{\nu}_0$ can be written as:

$$\delta\tilde{\nu}_0 = \delta\nu_0 \begin{pmatrix} -\frac{1}{2} & 0 & 0 \\ 0 & -\frac{1}{2} & 0 \\ 0 & 0 & 1 \end{pmatrix} \tag{23}$$

in the reference frame of its principal axes, ξ, η.

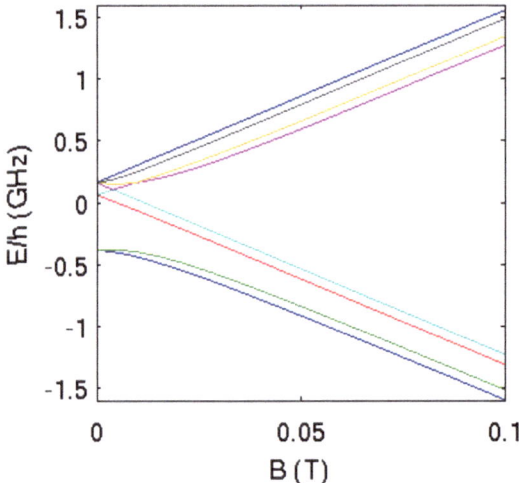

Fig. 10. Three-spin-radical energy levels (in frequency units, see text) vs. external magnetic field.

The observation of narrow muon transitions requires either a single crystal, to select the tensor orientation, or a liquid, to average out the effects of the anisotropy by fast reorientation. Isotropic reorientation averages to zero the traceless tensor $\delta\tilde{\nu}_0$ and reduces Eq. 22 to the same form of Eq. 20 (isotropic muonium), the only difference being a reduced value of the isotropic hyperfine frequency ν_0, due to the delocalization of the unpaired electron over the whole molecule.

A radical in a liquid at high fields then shows two precession frequencies:

$$\nu_{12} \approx \nu_0/2 + \frac{\gamma_\mu}{2} B \tag{24}$$

$$\nu_{34} \approx -\nu_0/2 + \frac{\gamma_\mu}{2} B \tag{25}$$

such that their difference is equal to ν_0 and their sum approximately equal to $\gamma_\mu B$.

5 Magnetically ordered materials

Let us assume for simplicity that the ordered moments $\hbar\gamma_e \mathbf{S}$ are directed along a known, single crystallographic direction. This can be the case of a uniaxial ferromagnet in zero applied field. A uniaxial antiferromagnet may yield a similar situation.

The local field at the muon site is due to a spin Hamiltonian of the type described in Eq. 15 and 17 as

$$\frac{\mathcal{H}}{h} = \frac{\gamma_\mu}{2\pi}\mathbf{I} \cdot \left[\frac{\mu_0}{4\pi}\hbar\gamma_e \sum_i \frac{3\hat{r}_i(\mathbf{S}_i \cdot \hat{r}_i) - \mathbf{S}_i}{r_i^3} + (\delta\tilde{A} + A_0) \cdot \mathbf{S}\right] \qquad (26)$$

where, in fact, the electronic term in square brackets acts as an effective magnetic field, $\boldsymbol{B}_\mu$, on the muon. Note that $\boldsymbol{B}_\mu$ is not necessarily parallel to $\boldsymbol{S}$, due to the tensorial character of the dipolar and pseudo-dipolar interactions, which justifies, as we already noted, that crystallographically *equivalent* interstitial muon sites in the unit cell might experience different local fields, which has important consequences both for the amplitudes of muon precessions in an oriented single crystal sample and for the vector composition of the local field with externally applied magnetic fields. We shall briefly review the two topics.

5.1 Magnetic single crystal

We shall treat this example a bit pedantically, as a useful exercise. Let us treat formally the general case, specializing for the time being to the geometry of a typical instrument, the General Purpose Spectrometer (GPS) of PSI, whose basic geometry is sketched in Fig. 11.

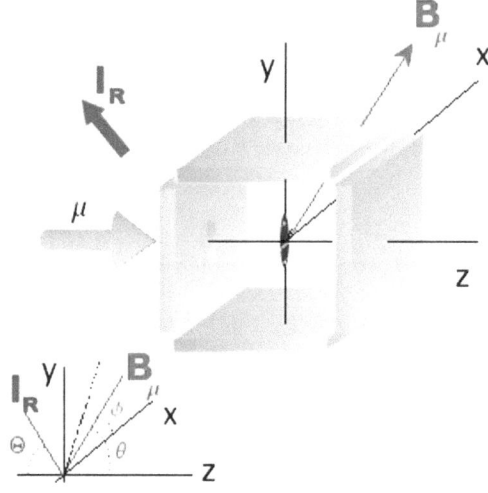

Fig. 11. Typical setup, e.g. from GPS. Only four out of five GPS positron detectors are shown for clarity. The μ arrow indicates the muon momentum, $\boldsymbol{I}_R$ its spin, and $\boldsymbol{B}_\mu$ the local field.

We need to specify the geometry in detail. The direction $\hat{\zeta}$ is defined by the polar angles θ, ϕ in the GPS reference frame $\hat{x}, \hat{y}, \hat{z}$, shown in the figure.

Likewise the initial muon spin $\boldsymbol{I}_R$ lies along $\hat{Z}$, in the $\hat{y}, \hat{z}$ plane, forming an angle Θ with the $-\hat{z}$ direction. The spin rotator (subsection 2.4) can select $\Theta \approx 0$ or 50 degrees.

If we call $\hat{\zeta}$ the direction of the local field $\boldsymbol{B}_\mu$ we need to define two additional axes that form an orthogonal frame with $\hat{\zeta}$, say, $\hat{\xi}, \hat{\eta}$. We arbitrarily chose them such that: $\hat{\xi}$ lies in the x, y plane, and is obtained as $\hat{\xi} = \mathcal{R}_z(\phi)\hat{y}$; $\hat{\eta}$ lies in the plane defined by $\hat{z}$ and $\boldsymbol{B}_\mu$, and is obtained as $\hat{\eta} = \mathcal{R}_\xi(\frac{\pi}{2} - \theta)\hat{z}$.

In the absence of an applied field the spin Hamiltonian is simply

$$\frac{\mathcal{H}}{h} = \frac{\gamma_\mu}{2\pi} B_\mu I_\zeta \tag{27}$$

With this notation we can write the time dependent muon polarization observed in the forward-backward detectors, along z, in terms of the density matrix ρ as:

$$P_z(t) = Tr[\rho\sigma_z(t)] \tag{28}$$

Since initially the muon is in the spin state with eigenvector of σ_Z equal to 1 it is easy to write the density matrix [15] in this basis, $|m\rangle = |\pm\rangle$, as

$$\rho = \begin{pmatrix} 1 & 0 \\ 0 & 0 \end{pmatrix} \tag{29}$$

hence we can simply write $\rho = \frac{1}{2}(1 + \sigma_Z)$. It is however more convenient to express all matrices in the basis of σ_ζ, such that the Hamiltonian of Eq. 27 is diagonal. The required transformations are:

$$\sigma_Z = \sigma_\xi \sin\Theta \cos\phi - \sigma_\eta(\cos\Theta \sin\theta \tag{30}$$
$$+ \sin\Theta \cos\theta \sin\phi) + \sigma_\zeta(\sin\Theta \sin\theta \sin\phi - \cos\Theta \cos\theta) \tag{31}$$
$$\sigma_z = \sigma_\eta \sin\theta + \sigma_\zeta \cos\theta \tag{32}$$
$$\sigma_y = \sigma_\xi \cos\phi - \sigma_\eta \cos\theta \sin\phi + \sigma_\zeta \sin\theta \sin\phi \tag{33}$$

and, since $Tr(1+\sigma_Z)\sigma_z(t) = Tr\sigma_Z\sigma_z(t)$ (Pauli matrices are traceless), the forward-backward polarization turns out to be:

$$P_z(t) = \frac{1}{2} \left[\langle+|\sigma_Z|+\rangle\langle+|\sigma_z|+\rangle + \langle-|\sigma_Z|-\rangle\langle-|\sigma_z|-\rangle \right. \tag{34}$$
$$\left. + \langle+|\sigma_Z|-\rangle\langle-|\sigma_z|+\rangle e^{-i\gamma_\mu B_\mu t} + \langle-|\sigma_Z|+\rangle\langle+|\sigma_z|-\rangle e^{i\gamma_\mu B_\mu t} \right]. \tag{35}$$

The first two terms represent the constant, *longitudinal* polarization, due to the field component parallel to the initial muon spin direction, while the last two terms correspond to the *transverse* polarization, which precesses around the local field.

An alternative, perhaps more intuitive interpretation of this expression is to consider the first matrix element in each term of the sum as the projection of the initial muon state $|\psi\rangle$ over the eigenstates of the Hamiltonian, which is the weight of the corresponding expectation value of the observed quantity, σ_z. We could then simply write $P_z(t) = \langle\psi(t)|\sigma_z|\psi(t)\rangle$, with $|\psi(t)\rangle = \sum_m c_m e^{i\omega_m t}|m\rangle$, $m = \pm$. Thus the density matrix [15], $\rho_{mn} = c_m^* c_n$, reappears in $P_z(t)$, leading to the same expression obtained above.

After a little algebra, substituting the relevant σ operators, Eqs. 30, into Eqs. 34 and recalling that σ_ζ is diagonal, while $\langle+|\sigma_\xi|-\rangle = 1$, $\langle+|\sigma_\eta|-\rangle = -i$, we get

$$P_z(t) = \cos\theta(\sin\Theta\sin\theta\sin\phi - \cos\Theta\cos\theta) +$$
$$\sin\theta\left[\sin\Theta\cos\phi\sin\gamma_\mu B_\mu t - (\cos\Theta\sin\theta + \sin\Theta\cos\theta\sin\phi)\cos\gamma_\mu B_\mu t\right] \quad (36)$$

The same calculation yields

$$P_y(t) = Tr[\rho\sigma_y(t)]$$
$$= \sin\phi\sin\theta(\sin\Theta\sin\theta\sin\phi - \cos\Theta\cos\theta) +$$
$$\left[\sin\Theta(\cos^2\phi + \cos^2\theta\sin^2\phi) + \cos\Theta\sin\theta\cos\theta\sin\phi\right]\cos\gamma_\mu B_\mu t$$
$$+ \cos\Theta\sin\theta\cos\phi\sin\gamma_\mu B_\mu t. \quad (37)$$

Table 2 summarizes the most used configurations, distinguishing for each observation direction $\alpha = y, z$ the constant, longitudinal component, from the transverse, precessing component. We incidentally note that starting from the muonium Hamiltonian of Eq. 20, one can follow exactly the same procedure outlined above to obtain the muonium transitions and their amplitudes, discussed in Sec. 4.1.

5.2 Powder average

Let us assume the same local field as above in a polycrystalline sample. Since the spectrum of precession frequencies is unaffected by the orientation of the individual crystallite in which each muon has stopped, in the absence of an external field the amplitudes of the longitudinal and transverse precession signals are simply obtained by taking the powder average of Eqs. 36 and 37, $\overline{P}_{z,y}(t) = \frac{1}{4\pi}\int_0^\pi \sin\theta d\theta \int_0^{2\pi} d\phi P_{z,y}(t;\Theta,\theta,\phi)$ All terms linear in $\sin\phi$ and $\cos\phi$ average to zero and it is straightforward to check that

$$\overline{P}_z(t) = -\frac{1}{3}\cos\Theta\left[1 + 2\cos\gamma_\mu B_\mu\right]$$
$$\overline{P}_y(t) = -\frac{1}{3}\sin\Theta\left[1 + 2\cos\gamma_\mu B_\mu\right] \quad (38)$$

	SPIN ROTATOR OFF ($\Theta = 0$)			
	P_y		P_z	
	Longitudinal	Transverse	Longitudinal	Transverse
$\begin{pmatrix} B_\mu \parallel \hat{x} \\ \theta = \frac{\pi}{2}, \phi = 0 \end{pmatrix}$	0	$\sin\gamma_\mu B_\mu t$	0	$-\cos\gamma_\mu B_\mu t$
$\begin{pmatrix} B_\mu \parallel \hat{y} \\ \theta = \frac{\pi}{2}, \phi = \frac{\pi}{2} \end{pmatrix}$	0	0	0	$-\cos\gamma_\mu B_\mu t$
$\begin{pmatrix} B_\mu \parallel \hat{z} \\ \theta = 0 \end{pmatrix}$	0	0	1	0
	SPIN ROTATOR ON ($\Theta \approx \frac{\pi}{3}$)			
	P_y		P_z	
	Longitudinal	Transverse	Longitudinal	Transverse
$\begin{pmatrix} B_\mu \parallel \hat{x} \\ (\theta = \frac{\pi}{2}, \phi = 0) \end{pmatrix}$	0	$\sin(\gamma_\mu B_\mu t + \Theta)$	0	$-\cos(\gamma_\mu B_\mu t + \Theta)$
$\begin{pmatrix} B_\mu \parallel \hat{y} \\ (\theta = \frac{\pi}{2}, \phi = \frac{\pi}{2}) \end{pmatrix}$	$\sin\Theta$	0	0	$-\cos\Theta\cos\gamma_\mu B_\mu t$
$\begin{pmatrix} B_\mu \parallel \hat{z} \\ (\theta = 0) \end{pmatrix}$	0	$\sin\Theta$	$-\cos\Theta$	0

Table 2. Longitudinal and transverse muon polarization for different geometrical conditions, from Eqs. 36, 37

Intuitively, the one-to-two proportion between the averaged longitudinal and the transverse terms displayed by Eqs. 38 is due to the fact that precessions take place around field components orthogonal to the initial muon spin direction, i.e. two possible Cartesian components, while the constant term arises from field components parallel to the initial muon spin direction, and there is only one such component.

It is simpler to recognize the result of Eqs. 38 without spin rotation ($\Theta = 0$). In this case the U-D counters measure zero polarization and the F-B counters detect a fraction $\frac{1}{3}$ of the polarization as longitudinal and the remaining fraction, $\frac{2}{3}$, as transverse, in the proportions shown in Fig. 12. The example in the figure also shows that, quite typically, both the transverse precessing component and the longitudinal one relax in time, and that, furthermore, the two relaxation rates are generally not the same.

Longitudinal relaxation in this context can only be due to time dependent interactions and it is a T_1 process, whereas transverse relaxation will be probably dominated by inhomogeneity of the local field (a "so-called" T_2^* process). For this distinction we refer to the NMR chapters, which applies equally well to muons.

6 Relaxation functions

We shall treat here only one aspect that is specific of μSR, referring the reader to the NMR chapters for the basic treatment of spin relaxation.

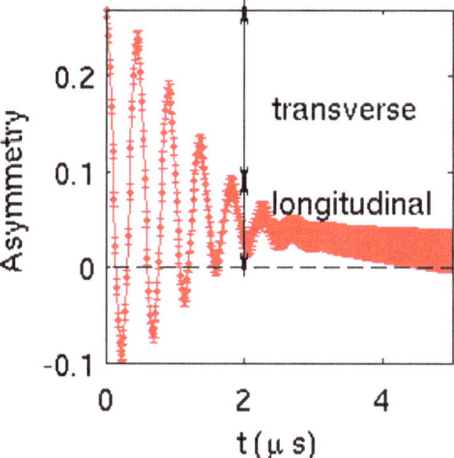

Fig. 12. Powder average of a the signal from a single muon site in a magnetically ordered material., The total asymmetry, $A = 0.27$), is subdivided into two thirds precessing and decaying in time, plus one third decaying at different (generally slower) rate.

We recall in brief that we distinguish two types of relaxation processes: *transverse* relaxation, due to dephasing of precessions, which occur without energy dissipation (also called T_2, or *spin-spin* processes); *longitudinal* relaxation, due to recovery of thermodynamic equilibrium, which requires contact with a thermal reservoir (e.g. the *lattice*) to dissipate the excess spin energy stored in the *out-of-equilibrium* initial muon polarization (also called T_1, or *spin-lattice* processes).

The experimental condition that allows to measure separately the two terms is the geometry of the experiment: whether the initial muon spin polarization, $\boldsymbol{P}(0) \parallel \hat{z}$ is *longitudinal* or *transverse* to the static component of the local field $\boldsymbol{B}_0$. In order to observe a pure *longitudinal* relaxation we need to have $\boldsymbol{B}_0 \parallel \boldsymbol{P}_0$ and the maximum amplitude will be recorded in detectors along $\hat{z}$. In order to observe a pure *transverse* relaxation we need to have $\boldsymbol{B}_0 \perp \boldsymbol{P}_0$, say along $\hat{x}$ and the maximum amplitude will be recorded in the yz plane. Examples of these geometries, including intermediate cases, have been described in section 5.1.

The models for these relaxation functions are always derived from more general NMR descriptions (nuclei may have spin larger than $\frac{1}{2}$ and experience more complicated interactions; the subcase of spin $\frac{1}{2}$ applies to muons).

It is a general feature that *longitudinal* relaxation can only be due to *time dependent* interactions, such as those introduced by time dependent external fields (i.e. light, or radio frequency), thermally populated excitations (phonons, magnons, etc.). For *transverse* relaxation one generally distinguishes the T_1-*like* component, due to time dependent (secular) interactions, from the static component, due to an inhomogeneous distribution of time independent interactions. Since the simple μSR experiment (without spin-echoes) cannot separate the two, *transverse* relaxation is often dominated by the larger inhomogeneous static contribution.

Muons may easily diffuse already at moderate temperatures. For instance they diffuse in pure metals, down to very low temperatures, and in oxides, already below room temperature. This, which is a condition much more rarely encountered with nuclei, gives rise to a specific time dependence of the local fields.

One very specific ability of muons is to measure spin relaxation in the absence of an applied magnetic field. Zero field relaxation may be measured also with Nuclear Quadrupole Resonance, given a nucleus with spin larger than $\frac{1}{2}$ and with a strong enough quadrupolar interaction. Also NMR may allow this measurement, in special conditions and with a *field cycling* apparatus. In this case the nuclei are polarized in a large magnetic field, which is then rapidly reduced to measure spin dynamics in low or zero field. How rapidly a large field may be varied imposes restrictions both on the sample nature and on the accessible relaxation rates.

With muons the experiment is particularly straightforward, and has virtually no restrictions, since the polarization is provided by parity violation. A peculiar μSR experimental condition, therefore, is when only relatively weak, random magnetic fields are present, like, for instance, in the case of nuclear dipolar fields. The relaxation functions appropriate for this case were obtained by Kubo (Ryogo Kubo, 1920 - 1995) and coworkers and it is illustrated in the next section.

6.1 Kubo-Toyabe functions

The relaxation function that one observes e.g. with nuclear dipolar fields is the static *Kubo-Toyabe* function, from the name of the two authors who calculated it first. Its Gaussian form is appropriate for fields whose distribution is a Gaussian of standard deviation $\frac{\Delta}{2\pi\gamma}$, centered at zero, $p_G(B) = \frac{\sqrt{2\pi}}{\Delta}e^{-\frac{1}{2}(2\pi\gamma_m uB/\Delta)^2}$. This is roughly [16] the case of a nuclear species dense on a lattice (abundant isotopes).

The relaxation function is given by

$$G_G(t) = \frac{1}{3}\left[1 + 2(1 - \Delta^2 t^2)e^{-\frac{\Delta^2 t^2}{2}}\right] \tag{39}$$

and it may be worked out in details [17, 18] by means of a simple stochastic model.

In the opposite limit of randomly diluted nuclei on a lattice the Lorentzian field distribution $p_L(B) = \frac{\gamma_\mu}{\Delta}\frac{1}{1+(2\pi\gamma_\mu B/\Delta)^2}$ is more appropriate [19] and the corresponding relaxation function is

$$G_L(t) = \frac{1}{3}\left[1 + 2(1 - \Delta t)e^{-\Delta t}\right] \qquad (40)$$

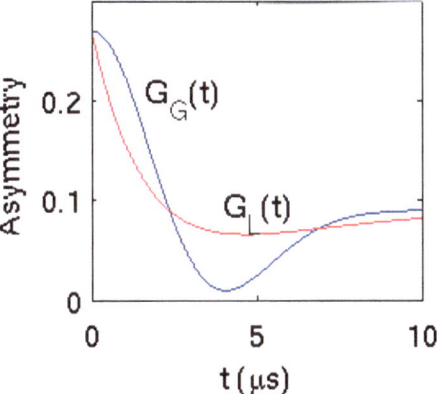

Fig. 13. Kubo Toyabe Gaussian (G_G) and Lorentzian (G_L) relaxation functions, displayed with a (typical) asymmetry of 0.27.

Fig. 13 shows that both functions recover one third of the initial asymmetry at long times, where the one third fraction corresponds to the average longitudinal component of the local fields over the distribution. However the Gaussian relaxation shows a more marked dip before that recovery, which is the reminder of the overdamped precession around the transverse component of the local field.

The cited papers [17, 18] describe also a stochastic treatment of the relaxation function that includes the effects of the dynamics, represented by a characteristic time τ of the time correlations of the local field, both in vanishing and non vanishing external field B_0.

Dynamics is taken into account by means of the so-called Markov chain, or *strong collision* process, i.e. collisional events take place such that within the characteristic time τ all memory is lost of the previous evolution (the more familiar case where this approach is employed is perhaps that of the relaxation time τ in the classical Drude model of metallic conduction [20]).

In *strong external field B_0* this model obtains two results for the *transverse geometry* case, which are valid quite more generally:

- that in the limit of *slow modulation* of a stochastic interaction from a Gaussian distribution $p_G(B)$, the so called *static limit* where $\omega_0 = 2\pi\gamma_\mu B_0 > \Delta \gg \frac{1}{\tau}$, the relaxation function turns out to be

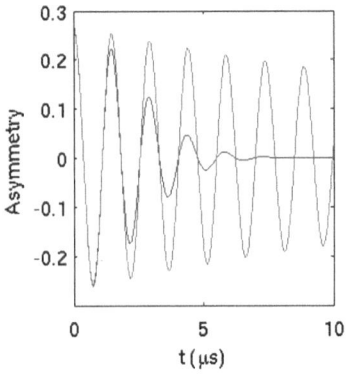
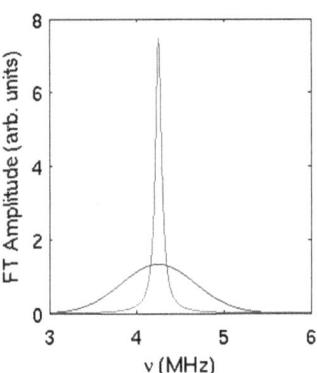

Fig. 14. Left: transverse precession with static (damped oscillations) and motionally narrowed relaxation (much less damped); right: Fourier transform of the two signals, the broader is relative to the static relaxation function.

$$G(t) = \cos(\omega_0 t + \phi)e^{-\frac{\Delta^2 t^2}{2}} \qquad (41)$$

where the relaxing term $e^{-\frac{\Delta^2 t^2}{2}}$ is proportional to the Fourier transform of the distribution $p_G(B)$, i.e. the lineshape itself reflects the distribution (see the static functions in Fig. 14).

- that in the limit of "fast modulation", $\frac{1}{\tau} \gg \Delta$, the same Gaussian distribution produces the so-called "motional narrowing" of the line: the lineshape narrows and the relaxation becomes slower

$$G(t) = \cos(\omega_0 t + \phi)e^{-\Delta^2 \tau t} \qquad (42)$$

where the Fourier transform linewidth is reduced to $\frac{1}{T_2} = \Delta^2 \tau$. This case exemplified by the less damped signal of Fig. 14, is typically obtained with nuclear dipolar fields when fast muon diffusion sets in, and also in the presence of a local instantaneous hyperfine field with an electron moment undergoing paramagnetic fluctuations.

The static limit in the *longitudinal geometry* yields a relaxation function which requires numerical methods to be calculated (e.g. in a fitting routine), The expression derived by Kubo for the Gaussian field distribution is:

$$G(\omega_0, t) = 1 - \left[\frac{\Delta^2}{\omega_0^2} \left(1 + \cos \omega_0 t \, e^{-\frac{\Delta^2 t^2}{2}} \right) - \frac{\Delta^4}{\omega_0^3} \int_0^t dx \, \sin \omega_0 x \, e^{-\frac{\Delta^2 x^2}{2}} \right]. \quad (43)$$

although a more useful equivalent expression for the numerical calculation is perhaps

$$G(\omega_0,t) = 1 + 2\Delta^2 \int_0^t x\,dx\,e^{-\frac{\Delta^2 x^2}{2}} \left[\left(\frac{\Delta^2}{\omega_0^2} - 1 \right) \frac{\sin\omega_0 x}{\omega_0 x} \right.$$
$$\left. - \frac{\Delta^2}{\omega_0^2}\cos\omega_0 x \right] \quad (44)$$

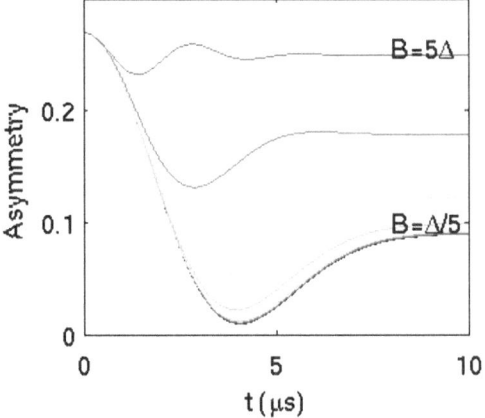

Fig. 15. The effect of increasing longitudinal fields on the Gaussian Kubo-Toyabe function. The plotted functions represent asymmetries, $AG(t)$.

The function of Eq. 44 is shown in Fig. 15. We quote here the equivalent expression for the Lorentzian field distribution

$$G(\omega_0,t) = 1 + 2\frac{\Delta}{\omega_0^2} \int_0^t dx\, \frac{e^{-\Delta x}}{x^2} \left[(1 + \Delta x - \omega_0^2 x^2)\frac{\sin\omega_0 x}{\omega_0 x} - (1 + \Delta x)\cos\omega_0 x \right]$$
$$(45)$$

Finally, the longitudinal field relaxation can be calculated in the presence of a dynamical process producing on average one jump in each time interval τ. The jump produces a change to a different value extracted according to a probability drawn from the same distribution (e.g. Gaussian, $p_G(B)$). This is the so-called strong collision limit of a Markoffian process (i.e. with no memory), appropriate for instance for muon diffusion in zero external field.

The relaxation function corresponding to this model may be computed [17, 18] via a simple expansion, making use of its Laplace transform (which requires some caution to be treated numerically [21]), and it is shown in Fig. 16.

Its behaviour may be understood as follows:

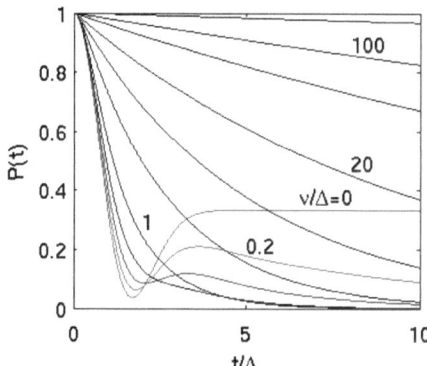

Fig. 16. Time dependence of the muon polarization; time is scaled by the parameter Δ of the static distribution, and so is the jump frequency $\nu = 1/\tau$

- for slow jump rates $\nu < 1$ only the long time longitudinal tail is affected (see e.g. the curve labeled $\nu/\Delta = 0.2$ in the figure);
- for fast jump rates $\nu > 1$ there is "motional narrowing" of the relaxation process, i.e. it becomes a simple exponential decay, and the slower, the faster the jump rate (see the curves labeled $\nu/\Delta = 20 - 100$ in the figure).

A useful analytical approximation for this expression, valid also in the longitudinal field case, may be found in ref. [22].

References

1. A. Schenck, *Muon Spin Rotation: Principles and Applications in Solid State Physics* (Adam Hilger, Bristol, 1985).
2. S. Blundell, Contemporary Physics 40 **40**, 175 (1999).
3. A. Schenck and F. Gygax, in *Handbook of magnetic Materials*, edited by K. Buschow (Elsevier, Amsterdam, 1995), vol. 9, p. 57.
4. S. L. Lee and R. Cywinski, eds., *Muons in Physics, Chemistry and Materials, Proceedings of the Fifty First Scottish Universities Summer School in Physics*, Handbook of magnetic Materials (Institute of Physics, Reading, Berkshire, 1999).
5. E. R. Heffner, J. Phys. Cond. Mat. **16**, S4403 (1986).
6. Look: http://nobelprize.org/nobel_prizes/physics/laureates/1957.
7. C. Wu and al., Phys. Rev. **105**, 1413 (1957).
8. R. L. Garwin, L. M. Lederman, and M. Weinrich, Phys. Rev. **105**, 1415 (1957).
9. Look: http://pdg.lbl.gov/2005/listings/lxxx.html.
10. Look: http://pdg.lbl.gov/2005/hadronic-xsections/hadronicrpp_page11.pdf.
11. Look: http://en.wikipedia.org/wiki/Helicity_%28particle_physics%29.
12. Look: http://en.wikipedia.org/wiki/Chirality_%28physics%29.

13. The lower limit on the mass of the muon neutrino (presently $m_\nu < 2.2\,\text{Mev}/\text{c}^2$) is set also by this argument. A finite mass implies that, notwithstanding parity violation, the spin of the neutrino has a component transverse to momentum, and so does the muon, resulting in a reduction of the spin polarization. Any larger mass than the quoted limit would result in a reduction of the experimental muon spin polarization in this kind of beams, although maybe not easily appreciable in actual experiments.

14. J. Brewer, S. Kreitzman, D. Noakes, E. Ansaldo, D. Harshman, and R. Keitel, Phys. Rev. B **33**, 7813 (1986).

15. We recall that the density matrix of a state $|\psi\rangle = \sum_j c_j |a_j\rangle$, where $|a_j\rangle$ is a complete basis, has matrix elements $\rho_{jk} = c_j^* c_k$.

16. M. Celio and P. Meier, Phys. Rev. B **27**, 1908 (1983).

17. R. Kubo, Hyperfine Interactions **8**, 731 (1981).

18. R. S. Hayano, Y. Uemura, J. Imazato, N. Nishida, T. Yamazaki, and R. Kubo, Phys. Rev. B **20**, 850 (1979).

19. R. E. Walstedt and L. R. Walker, Phys. Rev. B **9**, 4857 (1974).

20. N. W. Ashcroft and N. D. Mermin, *Solid state physics* (Holt-Saunders, Philadelphia, 1976).

21. For an explanation of the numerical pitfalls to be avoided look e.g.: http://www.fis.unipr.it/ derenzi/dispense/pmwiki.php?n=MuSR.KTDyn.

22. A. Keren, Phys. Rev. B **50**, 10039 (1994).

Muon spin rotation of organic compounds

Stephen J Blundell[1]

Oxford University Department of Physics, Clarendon Laboratory, Parks Road,
Oxford OX1 3PU, United Kingdom s.blundell@physics.ox.ac.uk

1 Introduction

Solid-state physics has been traditionally concerned almost exclusively with the study of *inorganic* elements, alloys and simple compounds. However, in the last few years there has been increasing interest in various new organic materials, the building blocks of which are not atoms but molecules. The shape of these molecules and their electronic structure play a crucial rôle in determining the resulting crystallographic structure, and hence the observed electrical and magnetic properties. The technique of muon-spin rotation (μSR) can be applied to many different types of organic compound, including molecular magnets, molecular conductors, conducting polymers and molecular superconductors. In this chapter, I review various recent developments in this field.

2 Molecular magnetism

The field of molecular magnetism has seen rapid change in the last decade or two, driven by advances in chemical synthesis [1, 2]. Newly prepared materials are typically characterized initially using magnetometry; however, measurements of magnetic susceptibility χ extract

$$\chi = \lim_{\delta H \to 0} \frac{\delta M_{\mathrm{av}}}{\delta H},\tag{1}$$

where M_{av} is given by $M_{\mathrm{av}} = \frac{1}{V} \int_V M(\mathbf{r}) \, \mathrm{d}^3 r$, a volume averaged magnetization. In contrast, from μ^+SR data one can extract the staggered magnetization distribution $\rho(M)$ in zero applied field; thus if there are N crystallographically independent muon sites (in most molecular magnetic materials that we have studied, we find that N is 1, 2 or 3), such that a fraction f_i of the muons implant at the ith site, then one can assume that the measured muon polarization function $A(t)$ (neglecting weakly relaxing terms due to longitudinal relaxation) follows

$$A(t) \propto \sum_{i=1}^{N} f_i \int \rho(M) \mathrm{e}^{-\lambda_i t} \cos(\alpha_i M t) \, \mathrm{d}M, \tag{2}$$

where α_i is a constant which depends on the dipolar coupling between the local magnetization M and the muon at site i, and λ_i is a relaxation rate. If the sample has uniform staggered magnetization M_0, so that $\rho(M) = \delta(M - M_0)$, then

$$A(t) \propto \sum_{i=1}^{N} f_i \mathrm{e}^{-\lambda_i t} \cos(\alpha_i M_0 t). \tag{3}$$

Muons can hence allow the temperature dependence of M_0 to be determined and have the useful advantage that they can demonstrate rather easily that M_0 is a characteristic of the entirety of the sample, and not of a minority impurity phase.

2.1 Purely organic magnets

An early candidate organic ferromagnet was tanol suberate, which is a bi-radical with formula $(C_{13}H_{23}O_2NO)_2$ (for molecular structure, see Fig. 1(c)). The spin density is found to be located on the NO group and almost equally shared between the oxygen and the nitrogen atoms [3]. The magnetic suscep-tibility measured down to liquid helium temperatures follows a Curie-Weiss law with a positive Curie temperature $(+0.7\,\mathrm{K})$ and it therefore looked like a good candidate for an organic ferromagnet. The specific heat exhibits a λ anomaly [4] at $0.38\,\mathrm{K}$, but tanol suberate was actually found to be an an-tiferromagnet with a metamagnetic transition in a field of 6 mT, resulting in ferromagnetic spin alignment [5, 6]. In tanol suberate the molecules are arranged in sheet-like layers in which the magnetic moments are almost local-ized [7]. Muon-spin rotation (μSR) experiments [8] yield clear spin precession oscillations (see Fig. 1(b)) corresponding to a uniform staggered magnetiza-tion in the material. The temperature dependence of the precession frequency (Fig. 1(c)) fits to $\nu_\mu(T) = \nu_\mu(0)(1 - T/T_C)^\beta$ where $\beta = 0.22$, consistent with a two-dimensional XY magnet [9] and also with the temperature dependence of the magnetic susceptibility [10], suggesting that the ordered state is domi-nated by two-dimensional interactions.

Following these early attempts, organic ferromagnetism was first achieved using a member of a family of organic radicals called nitronyl nitroxides. The unpaired electron in nitronyl nitroxides is mainly distributed over the two NO moieties, although some unpaired spin density is also distributed over the rest of the molecule. The central carbon atom of the O–N–C–N–O moiety is a node of the SOMO. Nitronyl nitroxides are chemically stable but the vast majority of them do not show long range ferromagnetic order. Therefore, the discovery of long-range ferromagnetism in one of the crystal phases (the β phase) of $para$-nitrophenyl nitronyl nitroxide ($C_{13}H_{16}N_3O_4$, abbreviated to p-NPNN) was particularly exciting, even though the transition temperature

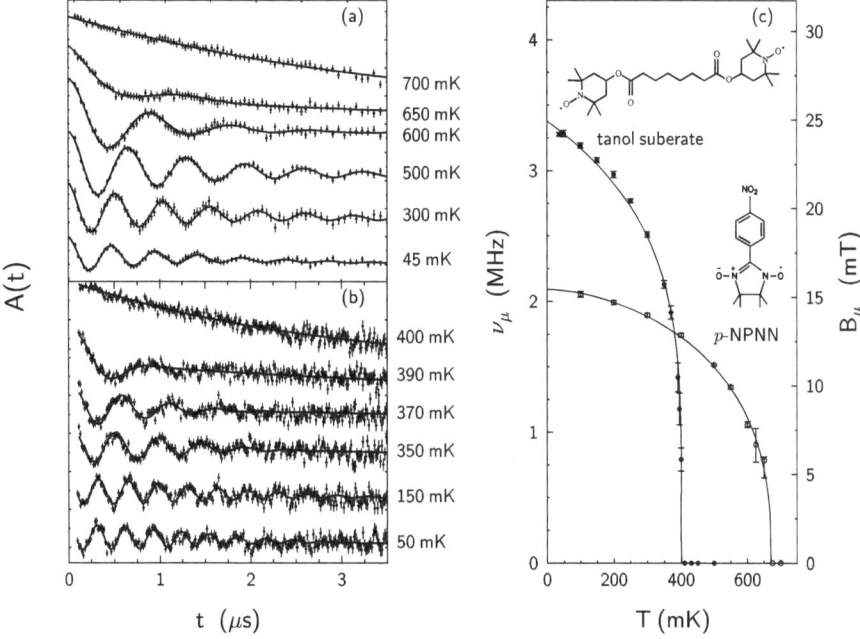

Fig. 1. Zero-field muon spin rotation frequency in (a) the organic ferromagnet p-NPNN (after [13]) and (b) the organic antiferromagnet tanol suberate (after [8]). In both cases the data for different temperatures are offset vertically for clarity. (c) Temperature dependence of the zero-field muon spin rotation frequency in p-NPNN and tanol suberate.

was a disappointingly low 0.65 K [11] and only present in one of its crystal phases.

The transition to ferromagnetic order was indicated by a λ-type peak in the heat capacity and a divergence in the AC susceptibility at the critical temperature [11, 12]. The magnetization below T_C saturates in a very small field, demonstrating that p-NPNN is a very soft ferromagnet. μSR experiments on p-NPNN (Fig. 1(a)) show the development of coherent spin precession oscillations below T_C [13, 14]. The temperature dependence of the precession frequency and the corresponding local field at the muon-site is shown in in Fig. 1(c). This is fitted to a functional form $\nu_\mu(T) = (1 - (T/T_C)^\alpha)^\beta$ yielding $\alpha = 1.7 \pm 0.4$ and $\beta = 0.36 \pm 0.05$. In this expression, the parameter β controls the behaviour near T_C, while the parameter α determines the power law at $T \ll T_C$. This is consistent with three-dimensional long range magnetic order [13, 14]. Near T_C the critical exponent is as expected for a three-dimensional Heisenberg model (one expects $\beta \approx 0.36$ in this case). At low temperatures the reduction in local field with increasing temperature is consistent with a Bloch-$T^{3/2}$ law ($\alpha = 1.5$), indicative of three dimensional spin waves.

The overlaps in all these materials which favour ferromagnetism appear to agree with the McConnell mechanism [15]: as a result of spin polarization effects, positive and negative spin-density may exist on different parts of each molecule; intermolecular exchange interactions tend to be antiferromagnetic, so if the dominant overlaps are between positive (majority) spin-density on one molecule and negative (minority) spin density on another molecule, the overall intermolecular interaction may be ferromagnetic. Though the mechanism for ferromagnetism is electronic, the low values of T_C imply that the magnetic dipolar interactions will play a rôle in contributing to the precise value of T_C. Dipolar interactions are also particularly important in determining the easy magnetization axis [13, 16]. This too depends on the crystal structure, which in turn depends on the molecular shape.

2.2 Chemically complex molecular magnets

The application of μSR to a more complex system can be illustrated with an example. I consider the family of layered molecule-based magnets which has general formula $[Z^{III}Cp_2^*][M^{II}M^{III}(ox)_3]$ and and which consists of an eclipsed stacking of bimetallic oxalate-based extended layers, separated by layers of organometallic cations [17]. Here M^{II}, M^{III} and Z^{III} are paramagnetic transition metal cations, ox is the oxalate anion, and Cp^* is pentamethylcyclopenta-dienyl. These compounds are some of the most chemically complex magnetic materials yet studied with μSR. Fig. 2 shows μSR data for a powder sample of one member of the family, $[FeCp_2^*][MnCr(ox)_3]$. For temperatures above T_c the relaxation of the spin ensemble (not shown) is well described by an exponential function, corresponding to dynamic fluctuations in the local magnetic field at the muon stopping site. Below the transition temperature, we observe clear oscillations, characteristic of a quasistatic field at the muon site. It is possible to identify three separate spin precession frequencies, suggesting three distinct muon sites in the material. The precession frequencies are proportional to the order parameter, and can be extracted as a function of temperature (as shown in Fig. 2) in order to determine the critical exponents describing the phase transition [18, 19].

2.3 Low-dimensional molecular magnets

Low-dimensional magnetic systems, in particular one-dimensional magnets, have attracted much attention because they are, in principle, easier to treat theoretically than three-dimensional (3D) systems [20]. If the interchain interaction J' can be ignored, a spin-$\frac{1}{2}$ Heisenberg antiferromagnetic chain (with Hamiltonian $H = J\sum_i \mathbf{S}_i \cdot \mathbf{S}_{i+1}$, where J is the intrachain exchange) will show magnetic susceptibility χ as given by Bonner and Fisher [21], with a maximum value at $T_{max} = 1.282J/k_B$. However, at very low temperatures

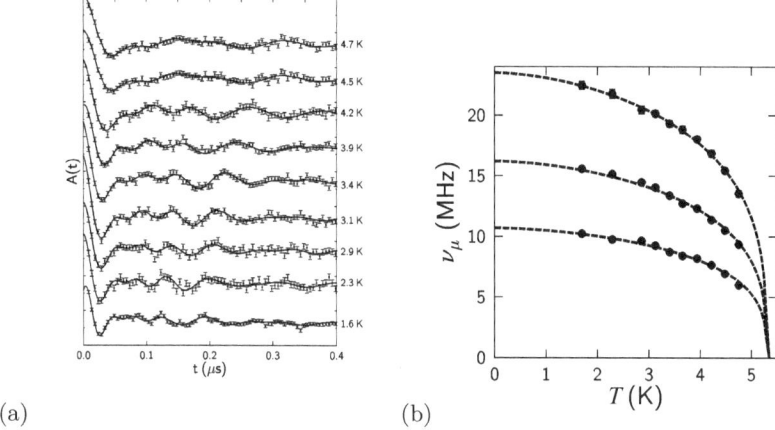

(a)   (b)

Fig. 2. (a) Positron decay asymmetry spectra for [FeCp$_2^*$][MnCr(ox)$_3$] measured below the ordering temperature. (b) The temperature dependence of the precession frequencies, which allows the extraction of critical parameters relating to the phase transition. After [18, 19].

there may be expected to be a departure from Bonner-Fisher behaviour with a singularity in dχ/dT as the temperature approaches zero [22]. In molecular systems both the intrachain and interchain exchange interactions can be controlled, to some extent, using chemical modifications.

An isolated one-dimensional antiferromagnetic spin-chain will never show magnetic long range order (LRO), but correlated regions of spatial extent $\xi(T)$ will develop, where $\xi(T)$ can be estimated using $\xi(T) \approx 2|J|S(S+1)/k_BT$. Clearly $\xi \to 0$ only at $T = 0$. Real crystal systems contain many chains and have non-zero interchain interactions, and they will generally show 3D magnetic LRO, albeit at very low temperature. For example, Sr$_2$CuO$_3$ has a Néel temperature (T_N) of 5 K, despite having an intrachain exchange interaction $J \sim 1300$ K [23, 24]. The strong quantum fluctuations due to the one-dimensionality renormalize the moment on the Cu cation down to $\sim 0.06\mu_B$ [24]. In isostructural Ca$_2$CuO$_3$ the chain separation is slightly smaller and hence the interchain interaction is larger, so that both T_N and the ordered moment are somewhat larger (11 K and $\sim 0.1\mu_B$ respectively [24]). It is the presence of even a very weak interchain exchange J' that can drive the system into 3D LRO below a non-zero T_N. A simple estimate of the transition temperature can be made by matching k_BT_N to the exchange energy of correlated segments in neighbouring chains. Thus $k_BT_N \approx \xi(T_N)|J'|S(S+1)$ and hence we have that

$$k_BT_N/|J| \approx |J'/J|^{1/2}. \tag{4}$$

Recently, substantially more accurate expressions have been obtained for T_N as a function of $|J'/J|$ for quasi-one-dimensional (Q1D) and quasi-two-dimensional (Q2D) antiferromagnetic Heisenberg models on a cubic lattice using Monte Carlo simulations [25]. These are

$$\left| \frac{J'}{J} \right| = \frac{\ln(\lambda x) + \frac{1}{2} \ln \ln(\lambda x)}{4cx} \tag{5}$$

with $x = |J|/k_B T_N$, $\lambda = 2.6$ and $c = 0.233$ for Q1D systems, and

$$\left| \frac{J'}{J} \right| = \exp \left(b - \frac{4\pi \rho_s}{k_B T_N} \right), \tag{6}$$

with $\rho_s = 0.183J$ and $b = 2.43$ for Q2D systems. By using Eqns 5 and 6 a measurement of T_N and J can allow even a small $|J'/J|$ to be extracted.

Muons have a particular advantage in the case of low-dimensional magnets. Because the correlation length ξ in an antiferromagnetic chain grows on cooling, the heat capacity exhibits a rather broad maximum as the entropy of the spins consequently decreases with the increasing correlation. Thus when 3D ordering sets in at T_N, the transition is associated only with a rather small change in entropy and thus gives rise to a tiny peak in the heat capacity, the size of which decreases as $|J'/J|$ decreases. This effect is shown in recent Monte Carlo simulations for Q2D systems [26], and means that identifying 3D ordering in very anisotropic magnets using heat capacity can be challenging. In contrast, the transition from a non-long-range ordered state, even one with dynamic correlations of large spatial extent, to a 3D long-range ordered state is rather straightforward using μ^+SR.

Copper pyrazine dinitrate [Cu(C$_4$H$_4$N$_2$)(NO$_3$)] is a $S = \frac{1}{2}$ chain in which Cu^{2+} ions are linked via pyrazine molecules with an intrachain exchange $J \approx 0.91$ meV (so that $J/k_B \approx 10.3$ K) [27], see Fig. 1. This exchange is considerably weaker than that in Sr$_2$CuO$_3$ ($J = 2200$ K [28]) so that the critical field $\mu_0 H_c = 2J/g\mu_B$ is more easily accessible to experiment in CuPzN [27]. This has allowed inelastic neutron scattering to probe the two-spinon continuum in CuPzN and the multiple overlapping continua with incommensurate soft modes that can be induced by an applied field [29]. However, 3D LRO was not found by heat capacity measurements performed down to 70 mK [30], seemingly implying that $|J'/J| < 10^{-4}$.

Example ZF μ^+SR spectra are shown in Fig. 4 [31]. In spectra measured for $T < 0.11$ K oscillations in the asymmetry are observed (Fig.4(a)). These oscillations are characteristic of a quasi-static local magnetic field at the muon stopping site, which causes a coherent precession of the spins of those muons with a component of their spin polarization perpendicular to this local field (expected to be 2/3 of the total polarization). There are two independent muon sites, so $N = 2$ in Eqn 3. The frequency of the oscillations is given by $\nu_i = \alpha_i M_0/2\pi$. Any fluctuation in magnitude of these local fields will result in a relaxation of the oscillating signal, described by relaxation rates

Fig. 3. Molecular structure of CuPzN, showing the chains of Cu^{2+} ions bridged by pyrazine ($C_4H_2N_2$) and each one additionally coordinated by two NO_3 groups.

λ_i (whose temperature dependence is shown in Fig. 4(e)). The presence of oscillations at low temperatures in CuPzN provides convincing evidence that this material is magnetically ordered below ≈ 0.11 K. From fits of ν_i (see Fig. 4(d)) to the form $\nu_i(T) = \nu_i(0)(1 - T/T_N)^\beta$ close to T_N, one can estimate $T_N = 107(1)$ mK, leading to a value of the ratio $|k_B T_N/J| = 0.0103(1)$ [31]. Substituting this value of T_N into Eqn 5 yields $|J'|/k_B \simeq 0.046$ K and a ratio $|J'/J| = 4.4 \times 10^{-3}$. The failure of earlier heat capacity measurements to detect an anomaly at T_N [30] can be attributed to the large anisotropy (small $|J'/J|$) in this material which renders the entropy change at T_N to be so small that it makes a negligible contribution to the measured heat capacity.

Similar studies (see e.g. [32, 33]) have been performed on a number of low-dimensional molecular magnets containing copper, and one can use the measured T_N from μ^+SR and J from susceptibility to obtain an estimate for J' using Eqns 5 and 6. The results of various low-dimensional copper compounds are plotted in Fig. 5(a) and Fig. 5(b). Note that anisotropy has much less of an effect in reducing T_N/J for Q2D systems (Fig. 5(b)) than for Q1D systems (Fig. 5(a)). For the organic radical-ion salt DEOCC-TCNQF$_4$ [34], which contains linear chains of stacked molecules with a much stronger intrachain exchange ($J = 110$ K) than CuPzN, but very weak interchain interactions, experiments have confirmed the absence of magnetic LRO down to 20 mK, showing that $|J'/J|$ is significantly less than 10^{-4}. This implies

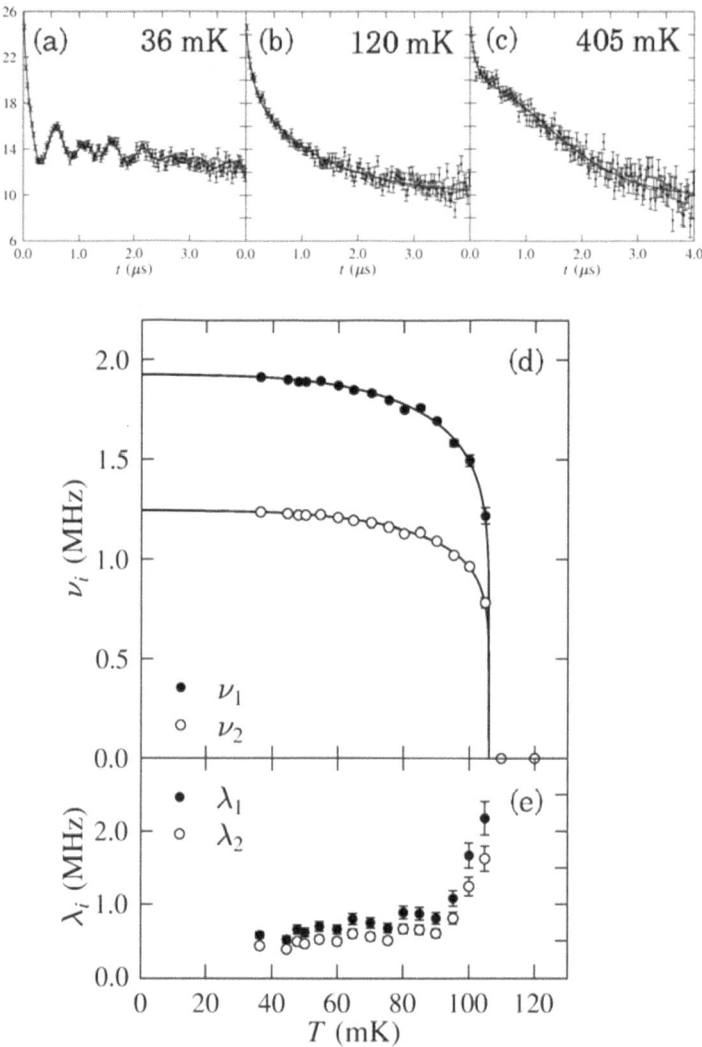

Fig. 4. ZF μ^+SR spectra for CuPzN at three temperatures [31]: (a) $T = 36$ mK: Oscillations are observed in the asymmetry, characteristic of magnetic LRO. A fit is shown to Eqn 3. (b) $T = 120$ mK: the spectra are described by two components. A fast relaxing signal, probably due to the existence of a radical state, and a larger component, seen to be exponential in form. (c) $T = 405$ mK: The fast relaxing component is still evident, although the larger component now shows a Gaussian form. (d) Muon precession frequencies ν_i with fits shown to the expression $\nu_i(T) = \nu_i(0)(1 - (T/T_N)^\alpha)^\beta$. (e) Transverse relaxation rates λ_i are seen to increase as the magnetic transition is approached from below.

that DEOCC-TCNQF$_4$ is one one of the most ideal examples of a $S = \frac{1}{2}$ Heisenberg antiferromagnetic chain yet discovered.

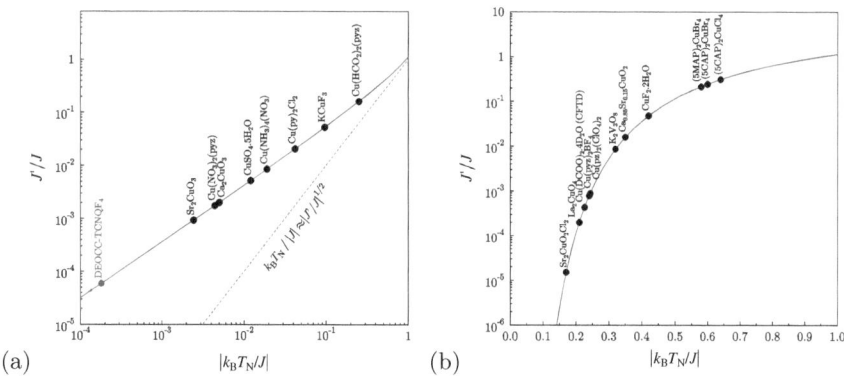

(a)   (b)

Fig. 5. (a) The anisotropy of Q1D antiferromagnetic spin-half chains as a function of $k_B T_N / J$ according to Eqn 5 (the dashed line shows Eqn 4). The Néel temperature and anisotropy in DEOCC-TCNQF$_4$ has not been determined, but it is known that $T_N < 0.002$ K and $|J'/J| < 10^{-4}$. (b) The anisotropy of Q2D antiferromagnetic spin-half systems as a function of $k_B T_N / J$ according to Eqn 6.

3 Muonium states

To understand the behaviour of muons in organic materials, it is helpful to consider the nature of muonium (abbreviated Mu= $\mu^+ e^-$) states (see also [35]).

3.1 Muonium

In muonium the electronic spin and the muon-spin are coupled by a hyperfine interaction which we will initially assume is isotropic. This leads to two energy levels, a lower triplet state and a higher singlet state. In a magnetic field the triplet levels split and the energy levels move. Muonium can be considered as an isotope of hydrogen, with a Bohr radius and Bohr energy close to the value of hydrogen (see Table 1). The Hamiltonian of a free, isotropic muonium state, in a magnetic field **B** parallel to the z-axis is

$$H/\hbar = -\gamma_\mu \mathbf{I} \cdot \mathbf{B} + \gamma_e \mathbf{S} \cdot \mathbf{B} + A\mathbf{S} \cdot \mathbf{I} = -\gamma_\mu I_z B + \gamma_e S_z B + A\mathbf{S} \cdot \mathbf{I}, \quad (7)$$

where **I** and **S** are the muon and electron spin operators and A is the (isotropic) hyperfine coupling constant (expressed as an angular frequency, so that $\hbar A$ has the units of energy). In this equation, all terms are in angular frequency units. The eigenvalues and eigenvectors of this Hamiltonian can be calculated, and the results are shown in Table 2. As a basis of the spin Hilbert-space we use the product of the muon and electron spin-up and spin-down vectors, $|\Psi\rangle = |\chi_\mu\rangle \otimes |\chi_e\rangle = |\chi_\mu\chi_e\rangle$. The eigenstates at low fields are one triplet state and one singlet state.

Property	Muonium=μ^+e^-	Hydrogen=p^+e^-
Mass	$0.1131m_H$	m_H
	$207.77m_e$	$1837.2m_e$
Reduced mass	$0.9956\mu_H$	μ_H
Bohr radius	$1.0044a_0$	$a_0 = 0.5292\text{Å}$
Ionization energy	$0.9956R_\infty$	$R_\infty = 13.6058$ eV
Hyperfine coupling constant	$3.1423A_H$	A_H
	$2\pi\times4463.3$ MHz	$2\pi\times1420.4$ MHz

Table 1. Properties of atomic muonium and atomic hydrogen (protium).

Suppose we implant muons in a sample and apply a magnetic field in the direction of the spin polarization. It then can be shown that the muon spin polarization

$$P_z(t) = \frac{1 + 2(B/B_0)^2}{2\left\{1 + (B/B_0)^2\right\}} + \frac{1}{2\left\{1 + (B/B_0)^2\right\}} \cos(\omega_{24}t) \tag{8}$$

eigenvector	eigenvalue
$\|\Psi_1\rangle = \quad \| \uparrow_\mu\uparrow_e\rangle$	$\frac{E_1}{\hbar} = \frac{A}{4}(1 + 2mB/B_0)$
$\|\Psi_2\rangle = \alpha\| \uparrow_\mu\downarrow_e\rangle + \beta\| \downarrow_\mu\uparrow_e\rangle$	$\frac{E_2}{\hbar} = -\frac{A}{4}\left(1 - 2\sqrt{1 + (B/B_0)^2}\right)$
$\|\Psi_3\rangle = \quad \| \downarrow_\mu\downarrow_e\rangle$	$\frac{E_3}{\hbar} = \frac{A}{4}(1 - 2mB/B_0)$
$\|\Psi_4\rangle = \beta\| \uparrow_\mu\downarrow_e\rangle - \alpha\| \downarrow_\mu\uparrow_e\rangle$	$\frac{E_4}{\hbar} = -\frac{A}{4}\left(1 + 2\sqrt{1 + (B/B_0)^2}\right)$

$$B_0 = A/\left\{\gamma_\mu + \gamma_e\right\} \ , \ m = (\gamma_e - \gamma_\mu)/(\gamma_e + \gamma_\mu) \approx 0.9904,$$

$$\alpha = \frac{1}{\sqrt{2}}\sqrt{1 - \frac{B/B_0}{\sqrt{1+(B/B_0)^2}}} \ , \ \beta = \frac{1}{\sqrt{2}}\sqrt{1 + \frac{B/B_0}{\sqrt{1+(B/B_0)^2}}}$$

Table 2. Eigenstates and eigenvalues of Muonium. For free muonium, $B_0 = 0.1585$ T.

where $\omega_{24} = A\sqrt{1 + (B/B_0)^2}$ and for free muonium, $B_0 = 0.1585$ T. The incoming muons are spin polarized, while the electrons which are picked up to form the muonium are not spin-polarized. Therefore the initial states will be a combination of $|\uparrow_\mu \uparrow_e\rangle$ and $|\uparrow_\mu \downarrow_e\rangle$, see Table 2.

In all cases the muonium can be studied by measuring precession signals in an applied magnetic field, or by using a technique known as repolarization. In this latter method a longitudinal magnetic field is applied to the sample, along the initial muon-spin direction, and as the strength of the magnetic field increases, the muon and electron spins are progressively decoupled from the hyperfine field. For isotropic muonium, half of the initial polarization of implanted muons is lost because of the hyperfine coupling, but this is recovered in a sufficiently large applied field, allowing an estimate of the strength of the hyperfine field. This occurs because on increasing the magnetic field B, the Zeeman energy dominates, and the electron and muon spins become decoupled, (i.e. $\alpha \to 0$ and $\beta \to 1$ in Table 2). This "repolarization" is given by

$$P_z(B) = \frac{1 + 2(B/B_0)^2}{2\{1 + (B/B_0)^2\}}, \qquad (9)$$

see Equation 8. For anisotropic muonium, the repolarization is slightly more complicated, see for example [36].

In metallic samples the muon's positive charge is screened by conduction electrons which form a cloud around the muon, of size given by a Bohr radius. Thus μ^+, rather than muonium, is the appropriate particle to consider in a metal. (The endohedral muonium found in alkali fulleride superconductors is the only known example of a muonium state in a metal.) In insulators and semiconductors screening cannot take place so that the muon is often observed in these systems either as muonium or is found to be chemically bound to one of the constituents, particularly to oxygen if it is present. Isotropic muonium states are found in many semiconducting and insulating systems. The value of the hyperfine coupling strength is close to that for vacuum (free) muonium if the band gap is large. For materials with smaller band gap the hyperfine coupling is lower reflecting the greater delocalization of the electron spin density on to neighbouring atoms. This occurs because the greater the disparity between the levels of the states to be mixed, the smaller is the degree of mixing (i.e. the more the muonium retains its atomic character [37]). In Si, a substantial fraction of neutral muonium is also found in a most unexpected place, wedged into the centre of a stretched Si–Si bond for 'bond-centre', see [38] for a review of muonium states in semiconductors). This state is extremely immobile, and surprisingly turns out to be the thermodynamically more stable site. Its hyperfine coupling is much lower than that of the tetrahedral state, typically less than 10% of the vacuum value. Furthermore the coupling is very anisotropic, with axial symmetry about the $\langle 111 \rangle$ crystal axis (i.e. along the Si–Si bonds) [39, 40, 41].

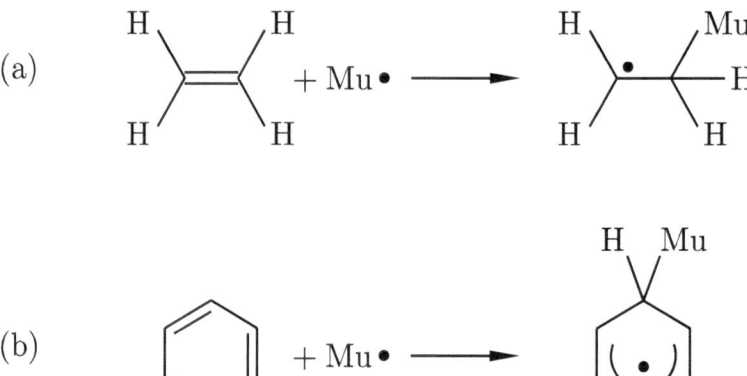

Fig. 6. Muonium addition to (a) ethene and (b) benzene.

3.2 Muonium in organic molecules

Muonium states can also be formed in many organic systems and allow a unique form of radical spectroscopy [42]. Muonium adds to unsaturated bonds to form muoniated free radicals. Addition of muonium at the carbon–carbon double bond in ethene (see Figure 6(a)) produces the muoniated ethyl radical. The radical is electrically neutral, but has an electronic doublet ground state due to the unpaired electron remaining on the unlabelled carbon atom. As described below, it is the hyperfine coupling between the muon and the unpaired electron which can be the source of information on the molecular dynamics (showing, for example, that $-CH_2Mu$ reorients less readily than $-CH_3$). As another example, addition to benzene (C_6H_6) leads to the muoniated cyclo-hexadienyl radical (C_6H_6Mu), as shown in Figure 6(b). The advantage of the muon technique is that one can work with concentrations down to just one muoniated radical at a time in an entire macroscopic sample. In contrast ESR detection needs $\sim 10^{12}$ radicals in a cavity, forbidding measurements at high temperatures where the radicals become mobile and terminate by combination. The technique has been applied to radicals in various environments [42], including those absorbed on surfaces [43], and also to liquid crystals [44].

3.3 Muonium in non-conducting polymers

Figure 7 shows transverse-field μSR data for polybutadiene which is a non-conducting polymer. The diamagnetic precession frequency ν_μ is clearly visible, together with artifacts (labelled c) from the cyclotron at 50, 100 and 200 MHz. The two additional frequencies, labelled ν_1 and ν_2, are due to the

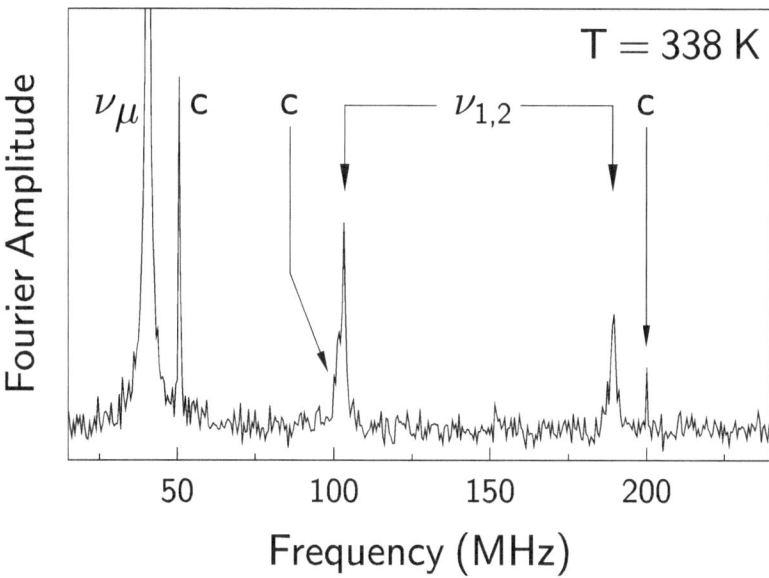

Fig. 7. FFT of transverse-field μSR data for polybutadiene with a field of 0.3 T. (After [45].)

hyperfine coupling with the radical electron and the correlation spectrum shown in Figure 8(a) allows an extraction of the temperature dependence of the hyperfine coupling. At high temperatures, a strong narrow signal is observed, due to efficient dynamical orientational averaging of the hyperfine anisotropy [46]. On cooling, the lines become broader as the polymer dynamics slow down.

Similar behaviour is seen for the $\Delta M = 0$ transition in the ALC spectrum (see Figure 8(b)), but the $\Delta M = 1$ transition shows the opposite behaviour, since it requires some residual anisotropy for oscillator strength. For $T \leq 225$ K, all the radical signals become too broad to extract reliably from the background [46, 45]. The dynamics close to the glass transition at 165 K can be followed using longitudinal-field μSR [46]. Similar effects are found for polystyrene [47]. The main point to be made about polybutadiene is that the muoniated radical is highly localised (see Figure 9), because the double bonds are separated from one another by polyethylene segments. Any modulation of the hyperfine coupling arises therefore from polymer motion. This is not the case for conducting polymers, as I shall discuss in the following section.

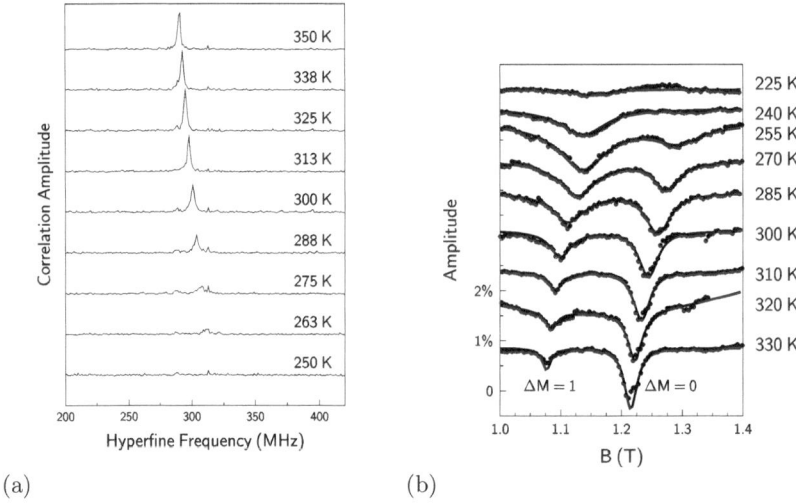

Fig. 8. (a) The associated correlation spectrum for polybutadiene. (b) ALC data for polybutadiene (After [45].)

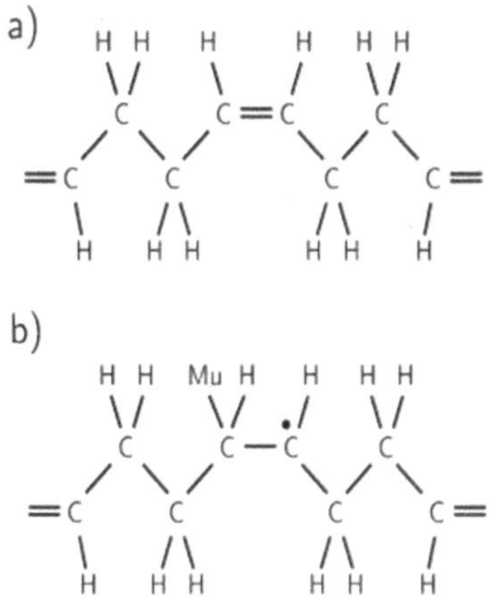

Fig. 9. (a) Raw polybutadiene and (b) the muoniated radical state.

4 Conducting polymers

Conducting polymers have attracted interest from a fundamental point of view because of the different type of mobile defects which can be found in them, including solitons [48] and polarons [49, 50]. The reaction between muonium and trans-polyacetylene [51, 52, 53, 54, 55] produces a diamagnetic, neutral muon defect and a highly mobile unpaired spin (a soliton, see Figure 10). Because the unpaired spin is mobile, the hyperfine coupling between it and the muon defect is intermittent [56]. In other polymers, such as polyaniline (PANI) or polyphenylvinylene (PPV), the muon-induced defect is a negatively charged polaron [57, 58]. In each case, the excitation diffuses up and down the chain but cannot cross the point at which the muon is bonded to the chain which therefore acts as a barrier. Every time the excitation briefly revisits the muon, the muon-electron hyperfine coupling is turned on and then off, so that successive visits progressively relax the muon polarization. Measurement of the magnetic field dependence of this relaxation yields the spectral density function associated with the excitation random walk and can be used to infer the dimensionality of the diffusion [51, 59]. This occurs because the relaxation rate is connected with the noise power (or spectral density), $J(\omega_\mu)$, in the fluctuating magnetic field at the muon Larmor frequency, $\omega_\mu = \gamma_\mu B$, associated with that particular magnetic field B, where γ_μ is the muon gyromagnetic ratio. Sweeping the magnetic field allows one to extract the frequency distribution of the fluctuations.

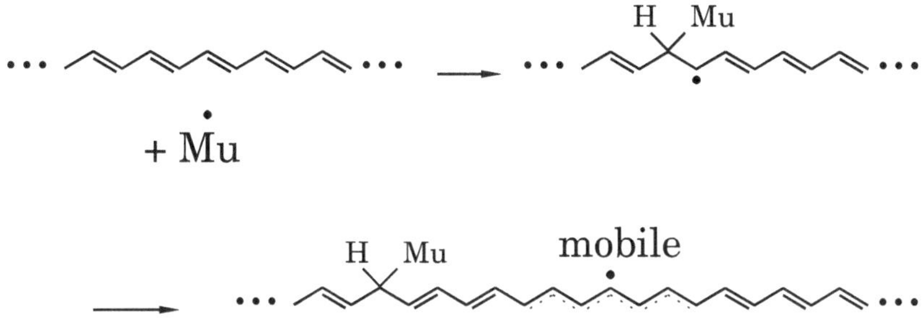

Fig. 10. Muonium interaction with trans-polyacetylene to produce a diamagnetic radical and a mobile neutral soliton.

Polaronic motion in *doped* conducting polymers can be measured using NMR and ESR [60]; for these techniques the motional linewidth contributions are proportional to the carrier density and so measurements on conducting polymers are restricted to doped materials (see e.g. [61]). Muons are uniquely sensitive to polaron transport in *undoped* materials (in which there is no significant carrier density to provide an NMR or ESR signal). In contrast to transport experiments, which provide results which are inevitably dominated by the slowest component of the transport process, muon measurements can provide information on the intrinsic transport processes governing the mobility of an electronic excitation along a chain [62].

The measured muon relaxation data were fitted using the theory of Risch and Kehr [63] for a muon interacting through hyperfine coupling ω_0 to the spin density on the chain site to which it is bonded (this is illustrated in Figure 11).

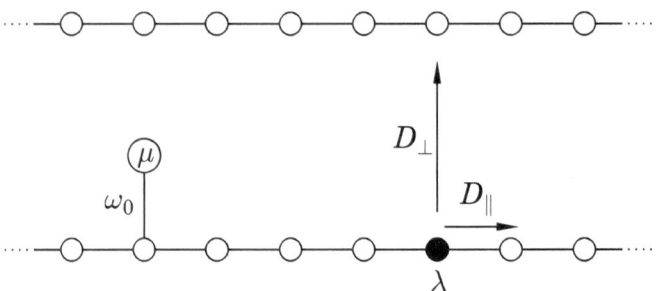

Fig. 11. Schematic diagram of the polaron diffusion model used in the text which is based on that of Risch and Kehr but including interchain diffusion (after [46]). The muon interacts through hyperfine coupling ω_0 to the spin density on the chain site to which it is bonded. The electronic spin defect (black filled circle) rapidly diffuses up and down the chain with intrachain and interchain diffusion rate $D_\parallel$ and $D_\perp$ respectively. λ is the electron spin flip rate, assumed faster than the reciprocal of the experimental time scale.

In this model, when the hyperfine coupling is switched on, the muon electron system evolves according to a Hamiltonian given by equation 7 with $A = \omega_0$. When the hyperfine coupling is switched off, the electron and muon spins are decoupled and separately evolve according to the Zeeman terms in the Hamiltonian. In addition, the electron spin is subject to random spin-flips at a rate λ. The fluctuating spin density induced by an electronic spin defect rapidly diffusing up and down the 1D chain leads to a relaxation function of the form $G(t) = \exp(\Gamma t)\mathrm{erfc}(\sqrt{\Gamma t})$ for $\lambda t_{\max} \gg 1$, with erfc signifying the complementary error function, $t_{\max}$ the experimental timescale and Γ a relaxation parameter. For $D_\parallel > \omega_0 > \lambda$, Γ has an inverse magnetic field dependence given by

$$\Gamma = \frac{\omega_0^4}{2\omega_e D_\parallel^2}. \tag{10}$$

The delocalization of the electronic defect in the neighbourhood of the attached muonium is not known, but a neutral spin-$\frac{1}{2}$ defect is believed to be delocalised over six lattice constants in finite-trans polyacetylene on the basis of calculations [64]. In the Risch-Kehr model, the Brownian motion of an extended defect is replaced by an effective hopping process of a localized defect [63].

Data for polyaniline [62] and various derivates of PPV [65] fit to the Risch-Kehr model very well, and the measured temperature dependence of Γ as a function of field allow the extraction $D_\parallel$ using equation 10. The extracted $D_\parallel$ are then fitted well by a simple model for the transport in which $D_\parallel \propto (\Sigma_0 + \Sigma_r)^{-1}$ where Σ_0 is a temperature independent scattering term and

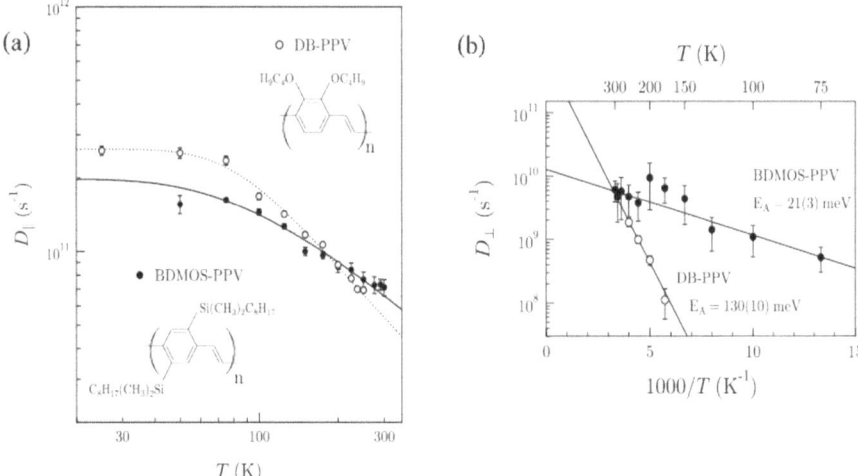

Fig. 12. (a) The temperature dependence of the intrachain polaron diffusion rates in DB-PPV and BDMOS-PPV derived from the muon-spin relaxation. The main figure shows $D_\parallel$, and the solid line is a fit to a simple model of the transport, with a phonon activation energy of 22 meV (DB-PPV) or 11 meV (BDMOS-PPV). (b) An activated plot of the interchain diffusion rate $D_\perp$ for the two polymers. The fitted activation energies are 130(10) meV and 21(3) meV [47, 65].

Σ_r is a phonon scattering term proportional to the number of modes excited, so that it is proportional to $(\exp(E_r/k_B T) - 1)^{-1}$. A scattering contribution proportional to phonon number is expected in theories of phonon-limited polaron transport [66, 67]. In each case, the energy of the phonon mode is found to be in agreement with a mode associated with whole chain librations or mixed chain torsions and ring librations which has been observed in inelastic neutron scattering [62, 65]. These data are shown in Figure 12 for two derivatives of PPV. The field B_c, at which crossover to a 3D diffusion regime is observed, can be used to estimate the interchain diffusion $D_\perp \approx \gamma_e B_c$. The interchain transport is assisted by thermal motion, rather than hindered by it, so that $D_\perp$ increases with increasing temperature (usually with an activated dependence), while $D_\parallel$ decreases.

μSR has been very successful at studying a variety of different types of polymer [68] and these studies have recently been extended to polymer thin films using very slow muons [69].

5 Molecular superconductors

To understand the usefulness of muons, recall that the two important length-scales in superconductors are the penetration depth, λ, which controls the ability of the superconductor to screen magnetic fields, and the coherence length, ξ, which controls the lengthscale over which the order parameter can vary without undue energy cost. If the former is sufficiently greater than the latter (the condition is that $\lambda > \xi/\sqrt{2}$) the material is a type II superconductor which if cooled through its transition temperature, T_c, in an applied magnetic field remains superconducting everywhere except in the cores of the superconducting vortices which usually are arranged in a triangular lattice. Each vortex is associated with a magnetic flux equal to one flux quantum $\Phi_0 = h/2e$. The distance between vortices, d, is such that the number of vortices per unit area $2/(\sqrt{3}d^2)$ equals the number of flux quanta per unit area B/Φ_0. Thus $d \propto B^{-1/2}$. In general the vortex lattice will be incommensurate with the crystal lattice and, except at very high magnetic field, the vortex cores will be separated by a much larger distance than the unit-cell dimensions. Implanted muons will sit at certain crystallographic sites and thus will randomly sample the field distribution of the vortex lattice.

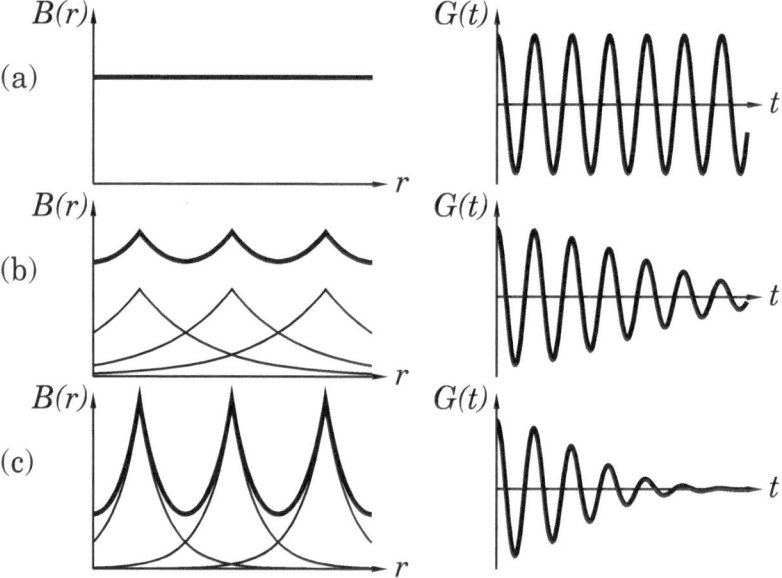

Fig. 13. The field distribution inside a superconductor as a function of position and the corresponding muon-spin relaxation function for three cases: (a) the normal state, (b) the superconducting state, (c) as (b) but with a shorter penetration depth.

In the normal state $(T > T_c)$ with a transverse field B, all muons precess with frequency $\omega = \gamma_\mu B$ (Figure 13(a)). In the superconducting state however the muons implanted close to the vortex cores experience a larger magnetic field than those implanted between vortices. Consequently there is a spread in precession frequency, resulting in a progressive dephasing of the observed precession signal (Figure 13(b)). The larger the penetration depth, the smaller the magnetic field variation and the less pronounced the dephasing (compare Figure 13(b) and (c)). In fact, the relaxation rate σ of the observed precession signal is related to the penetration depth using

$$\sigma = \gamma_\mu \langle B(\mathbf{r}) - \langle B(\mathbf{r}) \rangle_{\mathbf{r}} \rangle_{\mathbf{r}}^{1/2} \approx 0.0609 \gamma_\mu \Phi_0 / \lambda^2, \qquad (11)$$

where $B(\mathbf{r})$ is the field at position $\mathbf{r}$ and the averages are taken over all positions [70]. Thus the relaxation rate of the observed precession signal can be used to directly obtain the magnetic penetration depth. This principle has been applied to many different superconductors to extract both the penetration depth and its temperature dependence [71]. An advantage is that data are obtained from the bulk of the superconductor, in contrast to techniques involving microwaves which are only sensitive to effects at the surface.

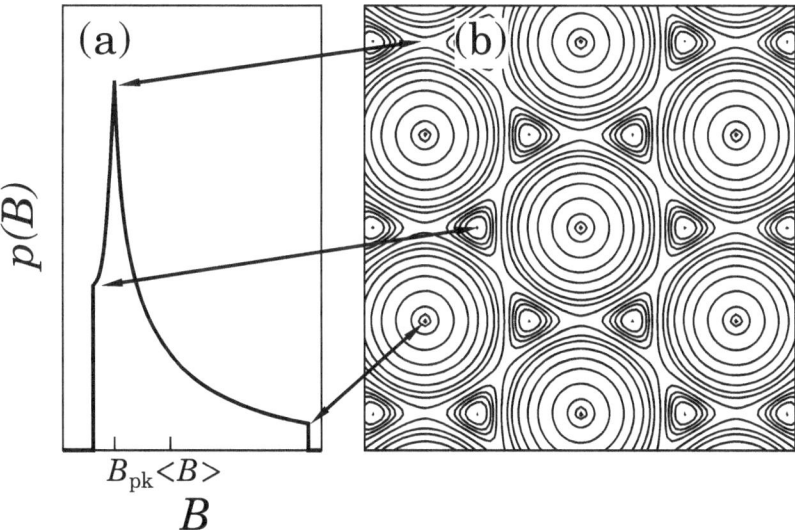

Fig. 14. (a) The field distribution $p(B)$ in the vortex lattice (contours of B shown in (b)).

5.1 Vortex states

A conventional type II superconductor exhibits three well-defined phases for $T < T_c$: (1) a Meissner phase for $B < B_{c1}$, (2) a mixed or Shubnikov phase for $B_{c1} < B < B_{c2}$ (in which the magnetic field enters the superconductor in the form of well defined flux lines or vortices arranged in a lattice) and (3) the normal metallic phase for $B > B_{c2}$. In highly anisotropic systems the vortex lattice is no longer a system of rigid rods but should be considered as a system of flexible interacting lines. A useful picture is that of a weakly coupled stack of quasi-two-dimensional (q2D) "pancake" vortices, each one confined to a superconducting plane [72, 73]. The phase diagram is thus substantially altered to take account of field and temperature dependent changes in the vortex lattice itself. At low T and low B the stacks resemble conventional vortex lines. Above a characteristic temperature T_b, but still below that at which superconductivity is destroyed, the vortex lattice is broken up by thermal fluctuations [73] (this is called vortex lattice melting). At low T, but this time increasing B, the energetic cost of interlayer deformations of the lattice (local tilting of the lines) is progressively outweighed by the cost of intralayer deformations within the superconducting plane (shearing). Above a crossover field B_{cr} the vortex lattice enters a more two-dimensional regime. Thus in anisotropic systems we may expect temperature and field dependent transitions in which the vortex lattice is destroyed. When muons are implanted into a superconductor in a field $B_{applied}$ one can directly measure the field distribution $p(B)$ which is given by $p(B) = \langle \delta(B - B(\mathbf{r})) \rangle_{\mathbf{r}}$ and is the probability that a randomly chosen point in the sample has field B [70]. This is shown in Fig. 14(a) for an ideal vortex line lattice. The distribution is highly asymmetric, the high field "tail" corresponding to regions of the lattice close to the vortex cores (see Fig. 14(b)). The maximum of the distribution occurs at B_{pk}, which lies below the mean field $\langle B \rangle$ (see Fig. 14(a)). Such lineshapes have been observed at low temperatures and fields in various anisotropic superconductors using μSR, including the high temperature superconductors [74] and, as will be discussed below, in organic superconductors [75]. In both cases it is found that the vortex lattice can be melted with temperature at T_b or can cross into a two-dimensional regime at fields above B_{cr}. Both transitions can be followed by measuring the field and temperature dependence of the $p(B)$ line shapes.

Early μSR measurements on κ-(BEDT-TTT)$_2$Cu(SCN)$_2$ [76, 77, 78] produced contradictory results concerning the low temperature behaviour of the penetration depth $\lambda(T)$. These were all performed at rather large transverse fields $\sim$ 300–400 mT (in order to obtain a larger number of precessions during the muon lifetime) but in fact this destroys the anisotropic vortex structure. Figure 15 is a μSR lineshape measured at 1.8 K for a sample of κ-(BEDT-TTF)$_2$Cu(SCN)$_2$ showing the measured lineshape cooled in a field of only 2.5 mT, which was derived from the muon time spectra using a maximum entropy technique [74]. The inset curve in Figure 15(a) is the probability dis-

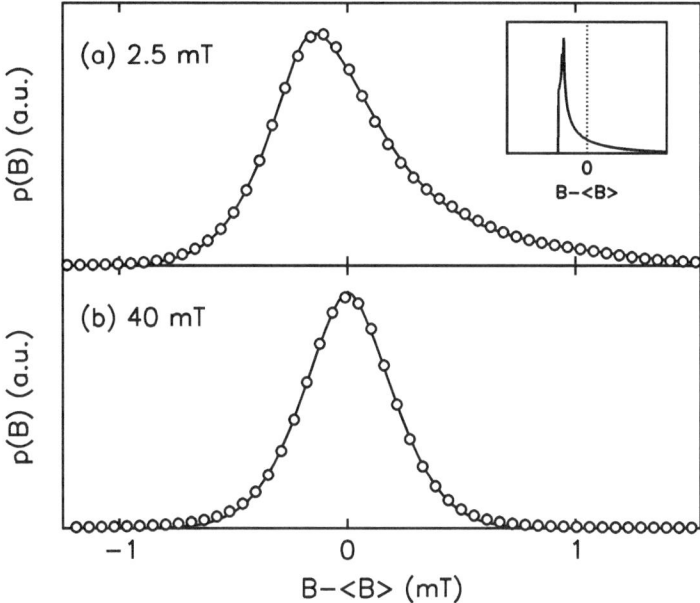

Fig. 15. κ-(BEDT-TTT)$_2$Cu(SCN)$_2$ showing the measured lineshape when field cooled to 1.8 K in a field applied at 45° to the superconducting planes. (a) At 2.5 mT the characteristic asymmetric vortex-line-lattice shape is seen. [Note the circles are data, solid line is simulation of ideal vortex-line-shape convoluted with instrument response. The inset shows the computed $p(B)$ before convolution.] (b) At 40 mT a symmetric line-shape is obtained due to the 2D arrangement of pancake vortices which results in a more symmetrical vortex field distribution [75].

tribution $p(B)$ from a numerical simulation of a vortex-line lattice in a uniaxial superconductor at an angle of 45° to the superconducting planes. The solid curve in the main part of Fig. 15(a) is the convolution of this field distribution with instrumental and dipolar broadening of 0.23 mT, which describes the data well. The long penetration depth in ET superconductors means that the lineshape is very narrow and the dipolar broadening very significant. Moreover the very low field used to ensure the existence of the vortex lattice means that it is essential to use a pulsed muon source such as ISIS since the very long time window available reduces the contribution from instrumental broadening [79].

Figure 15(b) shows the μSR-lineshape taken for the same sample cooled in a larger field of 40 mT. The lineshape is highly symmetric, with a mode at B_{pk} close to the average field $\langle B \rangle$. This change of lineshape with field is very similar to changes observed in the high-T_c superconductor Bi$_2$Sr$_2$CaCu$_2$O$_{8+\delta}$ (BSCCO), and indicates the loss of short-range correlations of the pancake vortices along the field direction [74]. It is attributed to the effective smearing-out of the core fields due to the local tilt deformations of the pancake stacks [74, 80, 81]. As a function of field we find that there is a broad crossover centred around $B_{cr} \sim 7$ mT from the asymmetric lineshape of Fig. 15(a) to the almost symmetric lineshape of Fig. 15(b) [75].

For $B < B_{cr}$ there is a dramatic increase in linewidth below $T^* \sim 5$ K and the *lineshape* also changes around T^*. An estimate of the characteristic temperature for the thermally-induced breakup of an electromagnetically-coupled pancake stack is given by $T_b = \phi_o^2 s / k_B \mu_o (4\pi)^2 2\lambda_\parallel^2 \approx 4.5$ K [73] where s is the interlayer spacing. The reduction in the positional correlations of pancake vortices at $T^* \sim 5$ K is thus consistent with this prediction [75, 79]. Flux decoration studies have independently confirmed this interpretation [82, 83].

The crossover field B_{cr} is closely related to the "second-peak" effect which has been observed in magnetization hysteresis loops [84]. For a Josephson coupled superconductor ($\lambda_\parallel \gg \gamma s$) the dimensional crossover is expected at a field $B_J \sim \phi_o/(\gamma s)^2$ when the width of the Josephson vortex core γs equals the vortex separation. When the anisotropy is very large $\gamma s \gg \lambda_\parallel$ the rigidity of the vortex line is controlled by the tilt modulus of the lattice and is dominated by a highly dispersive electromagnetic interaction so that despite long wavelength stiffness the vortices are subject to short wavelength fluctuations [85, 75]. Electromagnetic coupling is believed to dominate in BSCCO [86, 87] and κ-(BEDT-TTF)$_2$Cu(SCN)$_2$ [75, 79] and taking the layer separation for κ-(BEDT-TTF)$_2$Cu(SCN)$_2$ as $s \sim 1.6$ nm yields an estimate $B_{cr} \sim 7$mT which is in agreement with the μSR experiment [75]. Recent measurements have mapped out this phase diagram in more detail [88].

5.2 Scaling relations

There is a linear scaling between the superconducting transition temperature (T_c) and the superfluid stiffness ($\rho_s = c^2/\lambda^2$, where λ is the London penetration depth), which was first identified by Uemura *et al* for the underdoped cuprates [89]. Recently, scaling relations between ρ_s and the normal state conductivity σ_0 have also been suggested and a linear relation between ρ_s and the product $\sigma_0(T_c)T_c$ was demonstrated for cuprates and some elemental superconductors [90]. F. L. Pratt and I have recently found that these specific linear scaling relations do not hold for molecular superconductors, but a different form of power-law scaling is found to link ρ_s, $\sigma_0(T_c)$ and T_c [91]. These scaling properties hold as T_c varies over several orders of magnitude for materials with differing dimensionality and contrasting molecular structure, and the scaling is dramatically different from that of the cuprates. systems.

We considered values of ρ_s/c^2 ($= ne^2/m^*\epsilon_0 c^2$) and T_c derived from μSR measurements in the vortex state [71], and $\sigma_0(T_c)$ in the highest conductivity direction for a series of molecular superconductors. The materials range from a highly anisotropic quasi-one-dimensional (q1D) organic superconductor ((TMTSF)$_2$ClO$_4$), through systems of two-dimensional (2D) layered organic superconductors (BETS and ET salts) to examples of three-dimensional (3D) fulleride superconductors. Fig. 16 shows the Uemura plot of T_c against

$1/\lambda^2$ in log–log form where it can be seen that T_c follows ρ_s^m with the fitted value $m = 1.44(3)$. This approximate scaling of T_c with $\rho_s^{3/2}$, or equivalently λ^{-3}, in 2D organic superconductors was noted previously and discussed in terms of the 2D physics of layered superconductors [88, 92, 93]. However, it now appears that the scaling relations between T_c, ρ_s and σ_0 are more universal, encompassing examples of q1D and 3D molecular superconductors alongside the 2D systems. The non-linear scaling between T_c and ρ_s in the molecular case is much harder to understand than the linear scaling seen in the cuprates. In the cuprates the carrier density n is directly controlled by the doping level; in the underdoped regime ρ_s is directly proportional to n and T_c has been suggested to be linked to ρ_s either through Bose-Einstein condensation of preformed pairs [94] or through a mechanism in which phase fluctuations of the superconducting order parameter determine T_c [95]. In contrast, for the molecular systems n is fixed by the unit cell size and sto-ichiometry of the crystal structure and varies only little across the range of materials, whose superconducting parameters are nevertheless varying across several orders of magnitude. Differences in the superconducting properties must then arise entirely from differences in the details of the electronic many–body interactions.

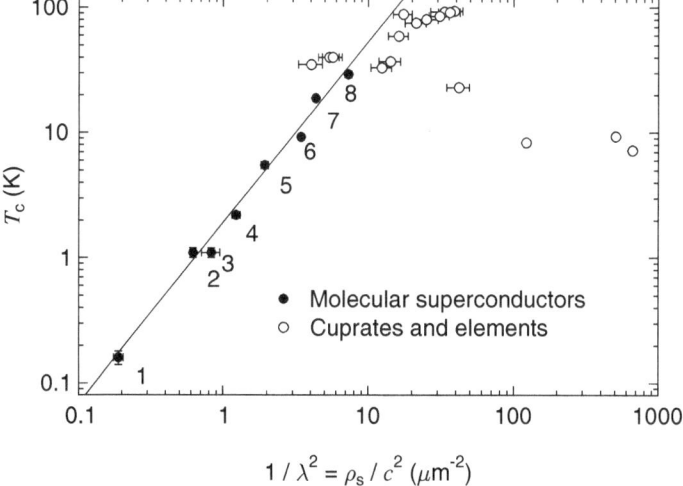

Fig. 16. Log-log Uemura plot of T_c against $1/\lambda^2$. Data for the cuprates and elemental superconductors tabulated by Homes *et al* [90] are also shown for comparison. For the molecular systems a scaling close to $\rho_s^{3/2}$ is observed, rather than the linear ρ_s scaling seen for the cuprates. The molecular superconductors are (1) κ-BETS$_2$GaCl$_4$, (2) (TMTSF)$_2$ClO$_4$, (3) α-ET$_2$NH$_4$Hg(SCN)$_4$, (4) β-ET$_2$IBr$_2$, (5) λ-BETS$_2$GaCl$_4$, (6) κ-ET$_2$Cu(NCS)$_2$, (7) K$_3$C$_{60}$, (8) Rb$_3$C$_{60}$. (After [91].)

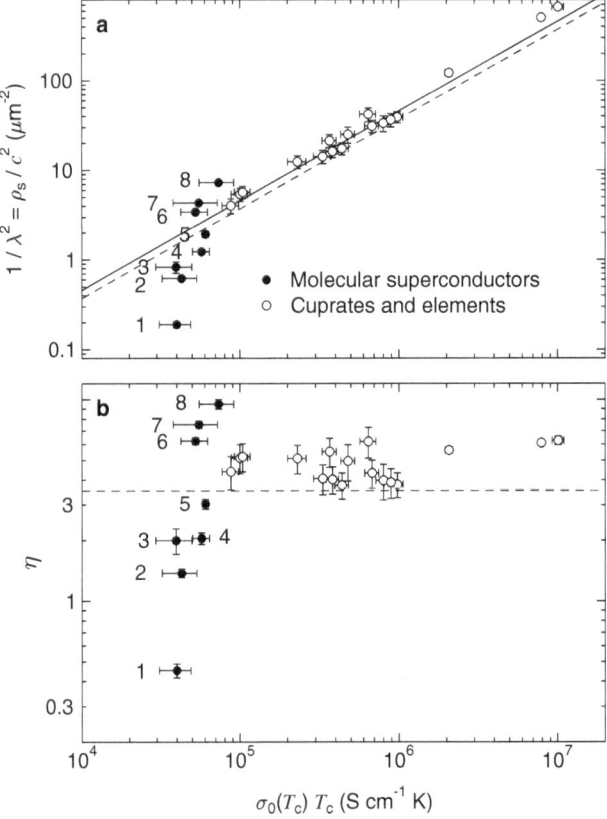

Fig. 17. (a) Plot of $1/\lambda^2$ against the product of T_c and $\sigma_0(T_c)$, following Homes *et al* [90]. For the molecular superconductors the data collapse onto a narrow ordinate range due to the inverse scaling between T_c and σ_0 demonstrated in Fig. 1(a). Open circles show the data of Homes *et al* [90] along with the linear fit (solid line). The dashed line shows the scaling expected for a weak coupling BCS superconductor in the high scattering rate limit (Eqn.13). (b) The data expressed as an effective gap parameter $\eta = (2\Delta/k_B T_c)(n_s/n_n)(m_n^*/m_s^*)$. Whereas the cuprates and elements are grouped around a value of η just above the BCS limit (dashed line) and comparable to the gap ratios seen using other techniques, the molecular systems cover a wide range of η values, both above and below the BCS limit. (After [91].)

Further evidence for fundamentally different behaviour between molecular and non-molecular superconductors is seen when an attempt is made to search for linear scaling between ρ_s and the product $\sigma_0(T_c)T_c$, of the form that was recently demonstrated by Homes *et al* [90]. Fig. 17(a) shows that such a simple linear scaling does not occur for the molecular superconductors. The linear behaviour seen for the non-molecular systems can be understood from applying the Ferrell-Glover-Tinkham sum rule for the real part of the frequency dependent conductivity [96],

$$\frac{2}{\pi} \int_0^\infty \sigma(\omega) \, d\omega = \frac{ne^2}{m^*} = \epsilon_0 \rho_s, \tag{12}$$

where $\sigma(\omega)$ takes the Drude form $\sigma_0/(1 + (\omega/\Gamma)^2)$, with $\sigma_0 = ne^2/(m^*\Gamma)$ and Γ being the scattering rate. In the case where Γ is significantly smaller than the frequency corresponding to the superconducting energy gap $2\Delta/\hbar$, the whole free carrier spectrum is redistributed to zero frequency to give the superfluid response peak, i.e. $\epsilon_0 \rho_s = \sigma_0(T_c)\Gamma$. If, on the other hand, $2\Delta/\hbar$ is significantly smaller than Γ, then the normal state conductivity is independent of frequency in the gap region, i.e. $\sigma(\omega, T_c) = \sigma_0(T_c)$; in this case, as the superconducting gap forms, an area of the conductivity spectrum with frequency width $2\Delta/\hbar$ and height σ_0 is redistributed to zero frequency to give the superfluid response peak. This leads to the following expression for $\epsilon_0 \rho_s$:

$$\epsilon_0 \rho_s = \frac{2}{\pi} \sigma_0(T_c) \frac{2\Delta}{\hbar} = \frac{2k_B}{\pi\hbar} \eta \, \sigma_0(T_c) T_c, \tag{13}$$

where $\eta = 2\Delta/k_B T_c$. In Fig. 17(a) the dashed line shows Eqn. 13 plotted taking the weak-coupling BCS limit $\eta = 3.53$ as a reference; this is seen to describe the general behaviour of the non-molecular data quite well. The effective value of η derived from the data using Eqn. 13 is shown in Fig. 17(b), which reveals the considerable variation among the molecular systems. Note that Eqn. 13 was derived assuming that the ratio of carrier density to effective mass is the same in the normal and superconducting states. If, however, this assumption is relaxed then the effective gap ratio observed in this plot becomes

$$\eta = \left(\frac{2\Delta}{k_B T_c}\right) \left(\frac{n_s}{n_n}\right) \left(\frac{m_n^*}{m_s^*}\right), \tag{14}$$

where the subscripts s and n refer to the superconducting and normal states respectively. Strong coupling can increase η over the BCS value via the first term of Eqn. 14, however if Eqn. 13 is applicable to the molecular systems then the reduced values of η seen for many cases would require a contribution from at least one of the other two terms, i.e. the superconducting carrier density would need to be less than normal state carrier density or the effective mass of superconducting carriers would have to be larger than that of the carriers in the normal state.

Another way to look at the data is to plot the ratio of the coherence length ξ (estimated using $\xi = \hbar v_F/\pi\Delta$, and writing a BCS-type gap $\Delta = 1.76 k_B T_c$) and the mean free path $\ell = v_F \gamma$. Thus $\xi/\ell = 0.181(\hbar\Gamma/k_B T_c)$ and this is shown in Fig. 18. The solid line shows a fit to $\xi/\ell = (T_c/T^*)^x$ and yields $T^* \sim 30\,\text{K}$ and $x \approx 0.5$. This implies that the scattering rate $\Gamma \propto \Gamma^y$ where $y \approx 1.5$. This is broadly consistent with the temperature dependence of the scattering rate deduced from the temperature dependent resistance of individual examples of the molecular metals; measurements for molecular metals just above T_c generally show power-law exponents in the region 1.5 to 2 [97] The scaling relations for the molecular superconductors [91] suggest

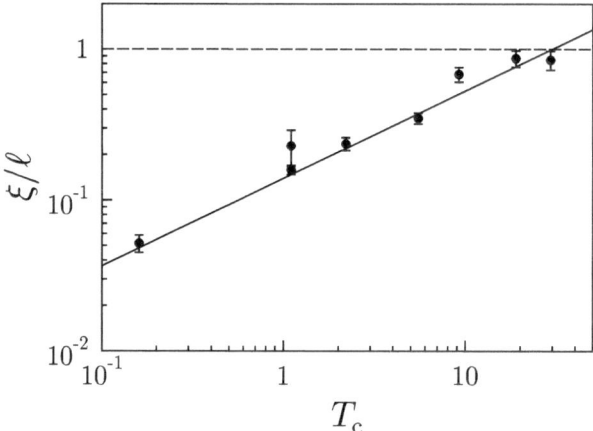

Fig. 18. The ratio ξ/ℓ as a functio of T_c. The horizontal dashed line shows $\xi = \ell$.

that there are features of their electronic properties that are common, despite the various materials having quite different dimensionality and Fermi surface topology. The simplicity of the scaling also suggests that it is being controlled by a single dominant parameter, such as the ratio of the electron correlation energy on a molecule U to the electronic bandwidth W. Identification of new theoretical models (see e.g. [98]) that match the observed scaling behaviour is clearly necessary; finding such models should lead to significant progress in understanding superconductivity in molecular systems.

6 Conclusion

This chapter has shown that muons can be used in a variety of ways to understand the magnetic, conducting and superconducting properties of a variety of organic and molecular systems. In each case, the muon is able to provide unique information which is unavailable using other techniques. However, the most information can be obtained when muons are used in tandem with other experimental probes.

I would like to thank the members of my research group and research collaborators at other institutions who have contributed to this work in many ways, and particularly Tom Lancaster and Francis Pratt.

References

1. O. Kahn: *Molecular Magnetism* (Wiley-VCH, Weinheim, 1993); J. S. Miller and M. Drillon (eds): *Magnetism: Molecules to Materials* volumes 1–4 (Wiley-VCH, Weinheim, 2001)
2. S. J. Blundell, F. L. Pratt: J. Phys.: Condens. Matter **16**, R771, (2004)
3. P. J. Brown, A. Capiomont, B. Gillon, J. Schweizer: J. Mag. Magn. Mat. **14**, 289 (1979)
4. M. Saint-Paul, C. Veyret: Phys. Lett. **45A**, 362 (1973)
5. A. Benoit, J. Flouquet, B. Gillon, J. Schweizer: J. Magn. Magn. Mater. **31–34**, 1155 (1983)
6. G. Chouteau, C. Veyret-Jeandey: J. de Phys. **42**, 1441 (1981)
7. A. Capiomont: Acta Cryst. B **28**, 2298 (1972)
8. S. J. Blundell, A. Husmann, T. Jestadt, F. L. Pratt, I. M. Marshall, B. W. Lovett, M. Kurmoo, T. Sugano, W. Hayes: Physica B **289**, 115 (2000)
9. S. T. Bramwell, P. C. W. Holdsworth: J. Phys.: Condens. Matter **5**, L53 (1993)
10. T. Sugano, S. J. Blundell, F. L. Pratt, T. Jestadt, B. W. Lovett, W. Hayes, P. Day: Mol. Cryst. Liq. Cryst. **334**, 477 (1999)
11. M. Tamura, Y. Nakazawa, D. Shiomi, K. Nozawa, Y. Hosokoshi, M. Ishikawa, M. Takahashi, M. Kinoshita: Chem. Phys. Lett. **186**, 401 (1991)
12. Y. Nakazawa, M. Tamura, N. Shirakawa, D. Shiomi, M. Takahashi, Mand. Ishikawa. M. Kinoshita: Phys. Rev. B **46**, 8906 (1992)
13. S. J. Blundell, P. A. Pattenden, F. L. Pratt, R. M. Valladares, T. Sugano, W. Hayes: Europhys. Lett. **31**, 573 (1995)
14. L. P. Le, A. Keren, G. M. Luke, W. D. Wu, Y. J. Uemura, M. Tamura, M. Ishikawa, M. Kinoshita: Chem. Phys. Lett. **206**, 405 (1993)
15. H. M. McConnell: J. Chem. Phys. **39**, 1910 (1963)
16. M. Kinoshita: Phil. Trans. R. Soc. Lond. A **357**, 2855 (1999)
17. E. Coronado, J.R. Galán-Mascaros, C.-J. Gómez-Garcia,J. Ensling, P. Gütlich: Chem. Eur. J. **6**, 552 (2000)
18. T. Lancaster, S.J. Blundell, F.L. Pratt, E. Coronado, J.R. Galán-Mascaros: J. Mater. Chem. **14**, 1518 (2004)
19. S. J. Blundell, T. Lancaster, M. L. Brooks, F. L. Pratt, E. Coronado, J. R. Galan-Mascaros, J. L. Manson, C. Cadiou, R. E. P. Winpenny: Synth. Met. **152**, 481 (2005)
20. L. J. de Jongh, A. R. Miedama: Adv. Phys. **23**, 1 (1974)
21. J. C. Bonner, M. E. Fisher: Phys. Rev. **135** A640 (1964)
22. S. Eggert, I. Affleck, M. Takahashi: Phys. Rev. Lett. **73**, 332 (1994)
23. T. Ami, M. K. Crawford, R. L. Harlow, Z. R. Wang, D. C. Johnston, Q. Huang, R. W. Erwin: Phys. Rev. B **51**, 5994 (1995)
24. K. M. Kojima, Y. Fudamoto, M. Larkin, G. M. Luke, J. Merrin, B. Nachumi, Y. J. Uemura, N. Motoyama, H. Eisaki, S. Uchida, K. Yamada, Y. Endoh, S. Hosoya, B. J. Sternlieb, G. Shirane: Phys. Rev. Lett. **78**, 1787 (1997)
25. C. Yasuda, S. Todo, K. Hukushima, F. Alet, M. Keller, M. Troyer, H. Takayama: Phys. Rev. Lett. **94**, 217201 (2005)
26. P. Sengupta, A. W. Sandvik, R. R. P. Singh: Phys. Rev. B **68**, 094423 (2003)
27. P. R. Hammar, M. B. Stone, D. H. Reich, P. J. Gibson, M. M. Turnbull, C. P. Landee, M. Oshikawa: Phys. Rev. B **59**, 1008 (1999)
28. N. Motoyama, H. Eisaki, S. Uchida: Phys. Rev. Lett. **76**, 3212 (1996)

29. M. B. Stone, D. H. Reich, C. Broholm, K. Lefmann, C. Rischel, C. P. Landee, M. M. Turnbull: Phys. Rev. Lett. **91**, 037205 (2003)

30. G. Mennenga, L. J. de Jongh, W. J. Huiskamp, J. Reedijk: J. Magn. Magn. Mater. **44**, 89 (1984)

31. T. Lancaster, S. J. Blundell, M. L. Brooks, P. J. Baker, F. L. Pratt, J. L. Manson, C. P. Landee, C. Baines: Phys. Rev. B **73**, R020410 (2006)

32. T. Lancaster, S. J. Blundell, M. L. Brooks, P. J. Baker, F. L. Pratt, J. L. Manson, C. Baines: Phys. Rev. B **73**, 172403 (2006)

33. J. L. Manson, M. M. Conner, J. A. Schlueter, T. Lancaster, S. J. Blundell, M. L. Brooks, F. L. Pratt,T. Papageorgiou, A. D. Bianchi, J. Wosnitza, M.-H. Whangbo: Chem. Comm. 4894 (2006)

34. F. L. Pratt, S. J. Blundell, T. Lancaster, C. Baines, S. Takagi: Phys. Rev. Lett. **96**, 247203 (2006)

35. S. J. Blundell, Chem. Rev. **104**, 5717, (5736)

36. F. L. Pratt: Phil. Mag. Lett. **75**, 371, (1997)

37. S. F. J. Cox, M. C. R. Symons: Chem. Phys. Lett. **126**, 516 (1986)

38. B. D. Patterson: Rev. Mod. Phys. **60**, 69 (1988)

39. S. F. J. Cox: J. Phys. C **20**, 3187 (1987)

40. K. H. Chow, B. Hitti, R. F. Kiefl: Semiconductors and Semimetals **51A**, 137 (1998)

41. S. F. J. Cox: J. Phys. Condens. Matter **15**, R1727 (2003)

42. E. Roduner: Chem. Soc. Rev. **22**, 337 (1993).

43. I. D. Reid, T. Azuma, E. Roduner: Nature **345**, 328, (1990)

44. B. W. Lovett, S. J. Blundell, J. Stießberger, F. L. Pratt, T. Jestädt, W. Hayes, S. P. Cottrell, I. D. Reid: Phys. Rev. B **63**, 054204, (2001)

45. Th. Jestädt: Thesis, University of Oxford (1999)

46. F. L. Pratt, S. J. Blundell, T. Jestädt, B. W. Lovett, A. Husmann, I. M. Marshall, W. Hayes, A. Monkman, I. Watanabe, K. Nagamine, R. Martin, A. B. Holmes: Physica B **289**, 625, (2000)

47. F. L. Pratt, S. J. Blundell, I. M. Marshall, T. Lancaster, A. Husmann, C. Steer, W. Hayes, C. Fischmeister, R. E. Martin and A. B. Holmes, Physica B **326**, 34, (2003)

48. W. P. Su, J. R. Schrieffer, A. J. Heeger: Phys. Rev. B **22**, 2099 (1980)

49. A. J. Heeger, S. Kivelson, J. R. Schrieffer, W. P. Su: Rev. Mod. Phys. **60**, 781 (1988)

50. A. J. Fisher, W. Hayes, D. S. Wallace: J. Phys.: Condens. Matter **1**, 5567 (1989)

51. K. Nagamine, K. Ishida, T. Matsuzaki, K. Nishiyama, Y. Kuno, T. Yamazaki: Phys. Rev. Lett. **53**, 1763 (1984)

52. F.L. Pratt, W. Hayes, G. R. Mitchell, B. Rossi, M. S. Kiani, B. D. Malhotra, S. S. Mandey, A. Milton, A. P. Monkman: Synth. Met. **55**, 677 (1993)

53. W. Hayes: Phil. Trans. R. Soc. Lond. A **350**, 249 (1995)

54. K. Nishiyama, K. Ishida, K. Nagamine, T. Matsuzaki, Y. Kuno, H. Shirakawa, R. Kiefl, J. H. Brewer: Hyp. Int. **32**, 551, (1986)

55. A. J. Fisher, W. Hayes, F. L. Pratt: *J. Phys.: Condens. Matter* **3**, 9823 (1991)

56. T. Jestadt, D. S. Sivia, S. F. J. Cox: Hyp. Int. **106**, 45, (1997)

57. F. L. Pratt, R. M. Valladares, P. A. Pattenden, S. J. Blundell, W. Hayes, A. J. Fisher, A. P. Monkman, B. D. Malhotra, K. Nagamine: Synth. Met. **69**, 231, (1995)

58. F. L. Pratt, K. Ishida, K. Nagamine, P. A. Pattenden, T. Jestädt, K. H. Chow, S. J. Blundell, W. Hayes, A. P. Monkman: Synth. Met. **84**, 943, (1997)

59. F. L. Pratt, S. J. Blundell, P. A. Pattenden, W. Hayes, K. H. Chow, A. P. Monkman, T. Ishiguro, K. Ishida, K. Nagamine: Hyp. Int. **106**, 33, (1997)
60. K. Mizoguchi: Jpn. J. Appl. Phys.**34**, 1 (1995)
61. K. Mizoguchi: Synth. Met. **199**, 35 (2001)
62. F. L. Pratt, S. J. Blundell, W. Hayes, K. Ishida, K. Nagamine, A. P. Monkman: Phys. Rev. Lett. **79**, 2855, (1997)
63. R. Risch, K. W. Kehr: Phys. Rev. B **46**, 5246 (1992)
64. D. S. Bourdeaux, R. R. Chance, J. L. Bredas, R. Silbey: Phys. Rev. B **28**, 6297 (1983)
65. S. J. Blundell, F. L. Pratt, I. M. Marshall, C. A. Steer, W. Hayes, A. Husmann, C. Fischmeister, R. E. Martin, A. B. Holmes: J. Phys.: Condens. Matter **14**, 9987, (2002)
66. S. Jeyadev, E. M.Conwell: Phys. Rev. B **35**, 6523 (1987)
67. K. Maki: Phys. Rev. B **26**, 2181 (1982)
68. F. L. Pratt: J. Phys.: Condens. Matter **16**, S4779 (2004)
69. F. L. Pratt, T. Lancaster, M. L. Brooks, S. J. Blundell, T. Prokscha, E. Morenzoni, A. Suter, H. Luetkens, R. Khasanov, R. Scheuermann, U. Zimmermann, K. Shinotsuka and H. E. Assender: Phys. Rev. B **72**, 121401, (2005)
70. E. H. Brandt: Phys. Rev. B **37**, 2349, (1988)
71. J. E. Sonier, J. H. Brewer, R. F. Kiefl: Rev. Mod. Phys. **72**, 769, (2000)
72. G. Blatter, M. V. Feigel'man, V. B. Geshkenbein, A. I. Larkin, V. M. Vinokur: Rev. Mod. Phys.**66**, 1125 (1995)
73. J. R. Clem: Phys. Rev. B **43**, 7837 (1991)
74. S. L. Lee, P. Zimmermann, H. Keller, M. Warden, I. M. Savic, R. Schauwecker, D. Zech, R. Cubitt, E. M. Forgan, P. H. Kes, T. W. Li, A. A. Menovsky and Z. Tarnawski: Phys. Rev. Lett. **71**, 3862, (1993)
75. S. L. Lee, F. L. Pratt, S. J. Blundell, C. M. Aegerter, P. A. Pattenden, K. H. Chow, E. M. Forgan, T. Sasaki, W. Hayes and H. Keller, Phys. Rev. Lett. **79**, 1563, (1997)
76. D. R. Harshman, R. N. Kleiman, R. C. Haddon, S. V. Chichesterhicks, M. L. Kaplan, L. W. Rupp, T. Pfiz, D. L. Williams, D. B. Mitzi: Phys. Rev. Lett. **64**, 1293, (1990)
77. L. P. Le, G. M. Luke, B. J. Sternlieb, W. D. Wu, Y. J. Uemura, J. H. Brewer, T. M. Riseman, C. E. Stronach, G. Saito, H. Yamochi, H. H. Wang, A. M. Kini, K. D. Carlson, J. M. Williams: Phys. Rev. Lett. **68**, 1923, (1992)
78. D. R. Harshman, A. T. Fiory, R. C. Haddon, M. L. Kaplan, T. Pfiz, E. Koster, I. Shinkoda and D. L. Williams: Phys. Rev. B **49**, 12990, (1994)
79. S. J. Blundell, S. L. Lee, F. L. Pratt, C. M. Aegerter, T. Jestadt, B. W. Lovett, C. Ager, T. Sasaki, V. N. Laukhin, E. Laukhina, E. M. Forgan, W. Hayes: Synth. Met. **103**, 1925, (1999)
80. E. H. Brandt: Phys. Rev. Lett. **66**, 3213, (1991)
81. E. H. Brandt: J. Low Temp. Phys. **73**, 355, (1988)
82. L. Y. Vinnikov, F. L. Barkov, M. V. Kartsovnik, N. D. Kushch: Phys. Rev. B **61**, 14358 (2000)
83. F. L. Barkov, L. Y. Vinnikov, M. V. Kartsovnik, N. D. Kushch: Physica C **385**, 568 (2003).
84. T. Nishizaki, T. Sasaki, T. Fukase, N. Kobayashi: Phys. Rev. B **54**, R3760 (1996)
85. G. Blatter, V. Geshkenbein, A. Larkin, H. Nordborg: Phys. Rev. B **54**, 72 (1996)

86. C. M. Aegerter, S. L. Lee, H. Keller, E. M. Forgan, S. H. Lloyd: Phys. Rev. B **54**, 15661, (1996)
87. S. L. Lee, C. M. Aegerter, H. Keller, M. Willemin, B. StaublePumpin, E. M. Forgan, S. H. Lloyd, G. Blatter, R. Cubitt, T. W. Li, P. Kes: Phys. Rev. B **55**, 5666, (1997)
88. F. L. Pratt, S. J. Blundell, T. Lancaster, M. L. Brooks, S. L. Lee, N. Toyota, T. Sasaki: Synth. Met. **152**, 417, (2005)
89. Y. J. Uemura, L. P. Le, G. M. Luke, B. J. Sternlieb, W. D. Wu, J. H. Brewer, T. M. Riseman, C. L. Seaman, M. B. Maple, M. Ishikawa, D. G. Hinks, J. D. Jorgensen, G. Saito, H. Yamochi: Phys. Rev. Lett. **66**, 2665, (1991)
90. C.C. Homes, S. V. Dordevic, M. Strongin, D. A. Bonn, R. Liang, W. N. Hardy, S. Komiya, Y. Ando, G. Yu, N. Kaneko, X. Zhao, M. Greven, D. N. Basov, T. Timusk: Nature **430**, 539 (2004)
91. F. L. Pratt, S. J. Blundell: Phys. Rev. Lett. **94**, 097006, (2005)
92. F. L. Pratt, S. J. Blundell, I. M. Marshall, T. Lancaster, C. A. Steer, S. L. Lee, A. Drew, U. Divakar, H. Matsui, N. Toyota, Polyhedron **22**, 2307, (2003)
93. B.J. Powell, R.H. McKenzie: J. Phys.: Condens. Matter **16**, L367 (2004)
94. Y.J. Uemura: Physica C**282-287**, 194 (1997)
95. V.J. Emery, S.A. Kivelson: Nature **374**, 434 (1995)
96. R.A. Ferrell, R.E. Glover: Phys. Rev. **109**, 1398 (1958)
97. H. Kobayashi, A. Kobayashi, P. Cassoux: Chem. Soc. Rev. **29**, 325 (2000)
98. B.J. Powell, R.H. McKenzie: J. Phys.: Condens. Matter **18**, R827 (2006)

Stochastic Landau-Zener model, and experimental estimates of dephasing time in molecular magnets.

Amit Keren

Technion-Israel Institute of Technology, Physics Department, Haifa 32000, Israel
keren@physics.technion.ac.il

Quantum tunneling of the magnetization in magnetic molecules (MM) with high spin value is a fascinating subject which contrasts clean and accurate experimental data with sophisticated theoretical models. At the heart of these models stands the Landau [1] and Zener [2] (LZ) derivation of quantum tunneling between levels, which at resonance have a tunnel splitting Δ, but are brought in to and out off resonance by a time-dependent field. This theory predicts transition probabilities, however, it has not been able to account for the size of the magnetization jumps in molecular magnets. In fact, the discrepancy between Δ deduced from LZ experiments [3] and the one calculated from spectroscopic data is more than three orders of magnitudes [4]. In these circumstances it might be essential to analyze experiments using a broader LZ theory which includes stochastic fluctuations produced by the environment. Such a theory was developed by Shimshoni and Stern (SS) [5]. The SS theory takes into account the dephasing effect due to stochastic field fluctuations. Combining this theory with measurements of dephasing times for MM could lead to a revision in tunnel splitting calculations. But, as far as we know, there are no estimates for dephasing time of MM. The purpose of the present work is to highlight the importance of dephasing in tunneling experiments and to measure the dephasing time.

First we revisit the SS theory, limiting the discussion to the parameter range relevant to MM. This will be done in a tutorial manner in the hope that even a non experienced reader will be able to follow the calculations. In order to produce a formula we assume, as did SS, a weak-coupling between the spins and the environment, in a sense that will be defined below. We show how ignoring dephasing time can lead to erroneous estimates of Δ. Second we review muon spin relaxation experiment on isotropic ($\Delta = 0$) MM with varying spin values. These experiments provide information on the source of dephasing and an estimate of dephasing times. In light of our finding we recognize that weak-coupling is not the correct assumption for MM. Therefore, we set the stage for further theoretical developments.

1 Landau-Zener model in the presence of stochastic field fluctuations

We use the simplest Hamiltonian appropriate for the Landau-Zener problem. It is the Hamiltonian of a spin $1/2$ which has a resonance tunnel splitting Δ at $t = 0$, and a time-dependent magnetic field αt in the z direction. The Hamiltonian is given by

$$\mathcal{H}_0 = \alpha t S_z + \Delta S_x \tag{1}$$

where $S_z = \sigma_z/2$, $S_x = \sigma_x/2$, and the σs are the Pauli matrixes. The Schrödinger equation could be written in a dimensionless form as

$$i\frac{t_T}{t_Z}\frac{\partial}{\partial y}|n\rangle = (yS_z + S_x)|n\rangle \tag{2}$$

where

$$t_z = \Delta/\alpha \tag{3}$$

is the Zener time [2],

$$t_T = \hbar/\Delta \tag{4}$$

is the tunneling time and

$$y = t/t_z \tag{5}$$

is dimensionless time.

Let us define the states $|+\rangle = [1,0]$ and $|-\rangle = [0,1]$. We are interested in the LZ probability that a spin prepared at time $t = -\infty$ in the low energy state $|+\rangle$ will be in the high energy state at $t = \infty$ which is again the $|+\rangle$ state. For this purpose we have to calculate the matrix element C_{LZ}

$$C_{LZ} = \langle +|U|+\rangle$$

where U is the time propagator operator. If the Hamiltonian had been time independent this operator would have been

$$\exp(-iH_0 t/\hbar). \tag{6}$$

But it does depend on time and a more complicated and approximated expression for U will be given soon. The probability of changing energy states is given by

$$P_{LZ} = |C_{LZ}|^2.$$

Sometimes a different definition for P_{LZ} is used where it is the probability of flipping energy states. In this case the spin stays in the low energy state throughout the field sweep. However, the two definitions sum up to 1 and extracting one from the other is trivial.

In the standard LZ model practically no transitions are taking place at very negative or very positive times. The transitions essentially take place within the Zener time scale t_z around $t = 0$. This is demonstrated in Fig. 1(a) which is a numerical solution of Eq. 2 taken from Ref. [6] as a function of time for three different values t_Z/t_T. Clearly most of the action is happening within t_z. The asymptotic case $t = \infty$ can be solved analytically [1, 2] and

$$P_{LZ} = \exp\left(-\frac{\pi \Delta^2}{2\hbar\alpha}\right) = \exp\left(-\frac{\pi t_Z}{2t_T}\right).$$ (7)

However, the solution involves reducing Eq. 2 to the Weber equation, which is not very well known in physics, and do not provide grate insight to the problem. We will take an approximation approach based on the SS theory.

The SS formulation of the problem gives a result similar to Eq. 7 and provides a natural platform for adding a fluctuating field. The solution starts by finding the instantaneous eigenstates and eigenvalues of the Hamiltonian in Eq. 1. These are

$$|\sigma_+(t)\rangle = \frac{\sqrt{2}}{2\sqrt{\Delta^2 + \alpha^2 t^2 + \alpha t \sqrt{\Delta^2 + \alpha^2 t^2}}}[\alpha t + \sqrt{\Delta^2 + \alpha^2 t^2}, \Delta]$$ (8)

with the eigenvalue

$$E_+ = +\frac{1}{2}\sqrt{\Delta^2 + \alpha^2 t^2}$$ (9)

and

$$|\sigma_-(t)\rangle = \frac{\Delta\sqrt{2}}{2\sqrt{\Delta^2 + \alpha^2 t^2 - \alpha t \sqrt{\Delta^2 + \alpha^2 t^2}}}[\alpha t - \sqrt{\Delta^2 + \alpha^2 t^2}, \Delta]$$ (10)

with the eigenvalue

$$E_- = -\frac{1}{2}\sqrt{\Delta^2 + \alpha^2 t^2}.$$ (11)

They are named according to their instantaneous energy. The two energy levels, normalized by Δ, are presented in Fig. 1(b) as a function of y (see Eq. 5) by the solid lines. We expect the instantaneous states and energies to be a useful concept in the adiabatic limit, namely, when the sweep rate is small. Looking at Eqs. 2 and 3 this means

$$\frac{t_T}{t_Z} < 1.$$ (12)

At $t \to -\infty$ we find that

$$|\sigma_-(-\infty)\rangle = |+\rangle$$

and at $t \to \infty$

$$|\sigma_+(\infty)\rangle = |+\rangle,$$

so that we need to calculate

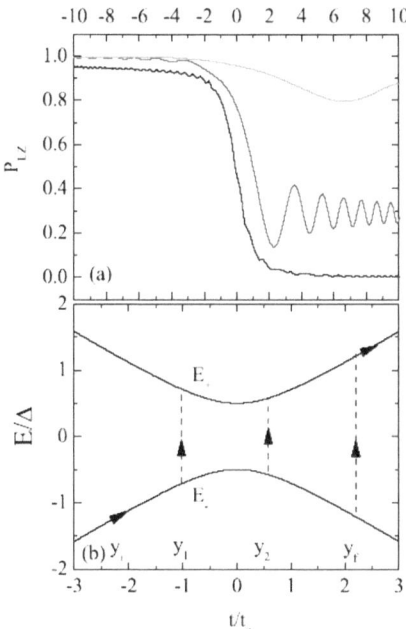

Fig. 1. (a) Landau-Zener transition probability as a function of normalized time reproduced from Ref. [6] for three values of t_Z/t_T: 0.05 (top), 0.4 (middle) and 5 (bottom). (b) Solid lines show the instantaneous energy levels as a function of normalized time in the Landau-Zener problem. Dashed lines are a schematic representation of the paths the spin can take when tunneling from the low energy state to the high energy state at times $t_j = y_j t_Z$. The transition amplitude is a sum of all paths with the appropriate matrix element as given by Eq. 16.

$$C_{LZ} = \langle \sigma_+(\infty)| \, U \, |\sigma_-(-\infty)\rangle \, .$$

To actually perform this calculation we divide the time into small segments and assume that we are allowed to use Eq. 6 (with $\hbar = 1$) in each segment, so that

$$U = \prod_{j=1}^{\infty} e^{-iH_0(t_j)\delta t} \, . \tag{13}$$

We also define the identity operator for each point in time

$$I_j = |\sigma_+(t_j)\rangle \, \langle \sigma_+(t_j)| + |\sigma_-(t_j)\rangle \, \langle \sigma_-(t_j)| \, . \tag{14}$$

We are now in position to perform the calculation. We do it by inserting this identity between every two exponents in Eq. 13. For example, let's divide the time between the initial time t_i and the final time $t_f = t_i + 2\delta t$ into two segment by introducing t_1 in between. In this case

$$C_{LZ} = \langle \sigma_+(t_f)| \, e^{-iH_0(t_1)\delta t} \, |\sigma_-(t_i)\rangle \tag{15}$$

and inserting Eq. 14 gives

$$C_{LZ} = \langle \sigma_+(t_f)| \, [|\sigma_+(t_1)\rangle \langle \sigma_+(t_1)| + |\sigma_-(t_1)\rangle \langle \sigma_-(t_1)|] \, e^{-iH_0(t_1)\delta t} \, |\sigma_-(t_i)\rangle .$$

Acting with the Hamiltonian backwards leavs

$$C_{LZ} = \langle \sigma_+(t_f)|\sigma_+(t_1)\rangle \langle \sigma_+(t_1)|\sigma_-(t_i)\rangle \, e^{-iE_+(t_1)\delta t}$$

$$+ \langle \sigma_+(t_f)|\sigma_-(t_1)\rangle \langle \sigma_-(t_1)|\sigma_-(t_i)\rangle \, e^{-iE_-(t_1)\delta t} .$$

We evaluate the bracket products to the lowest order in δt. For this purpose we approximate $|\sigma_-(t_1)\rangle$ using the time derivative

$$|\sigma_-'\rangle = \frac{d}{dt} |\sigma_-(t)\rangle$$

at the later time segment so that

$$\langle \sigma_+(t_f)|\sigma_-(t_1)\rangle \approx - \langle \sigma_+(t_f)|\sigma_-'(t_f)\rangle \, \delta t$$

and

$$\langle \sigma_+(t_1)|\sigma_-(t_i)\rangle \approx - \langle \sigma_+(t_1)|\sigma_-'(t_1)\rangle \, \delta t$$

since $\langle \sigma_+(t)|\sigma_-(t)\rangle = 0$. Similarly, to the lowest order in δt

$$\langle \sigma_-(t_1)|\sigma_-(t_i)\rangle = \langle \sigma_+(t_f)|\sigma_+(t_1)\rangle \approx 1.$$

Using these approximations in Eq. 15 gives

$$C_{LZ} = - \langle \sigma_+(t_f)|\sigma_-'(t_f)\rangle \, \delta t e^{-iE_-(t_1)\delta t} - \langle \sigma_+(t_1)|\sigma_-'(t_1)\rangle \, \delta t e^{-iE_+(t_1)\delta t} .$$

Now let's repeat the same exercise by dividing the time into three segments and introducing t_1 and t_2 in between t_i and t_f so that

$$C_{LZ} = \langle \sigma_+(t_f)| \, e^{-iH_0(t_2)\delta t} e^{-iH_0(t_1)\delta t} \, |\sigma_-(t_i)\rangle .$$

Again, inserting Eq. 14 behind the exponents gives

$$C_{LZ} = \langle \sigma_+(t_f)| \, [|\sigma_+(t_2)\rangle \langle \sigma_+(t_2)| + |\sigma_-(t_2)\rangle \langle \sigma_-(t_2)|] \, e^{-iH_0(t_2)\delta t}$$

$$\times [|\sigma_+(t_1)\rangle \langle \sigma_+(t_1)| + |\sigma_-(t_1)\rangle \langle \sigma_-(t_1)|] \, e^{-iH_0(t_1)\delta t} \, |\sigma_-(t_i)\rangle .$$

We look for cases where there is only one transition, since every transition introduces a factor δt. We also apply the Hamiltonians backwards and get

$$C_{LZ} = \langle \sigma_+(t_f)|\sigma_+(t_2)\rangle \langle \sigma_+(t_2)|\sigma_+(t_1)\rangle \langle \sigma_+(t_1)|\sigma_-(t_i)\rangle \, e^{-iE_+(t_2)\delta t} e^{-iE_+(t_1)\delta t}$$

$$+ \langle \sigma_+(t_f)|\sigma_+(t_2)\rangle \langle \sigma_+(t_2)|\sigma_-(t_1)\rangle \langle \sigma_-(t_1)|\sigma_-(t_i)\rangle \, e^{-iE_+(t_2)\delta t} e^{-iE_-(t_1)\delta t}$$

$$+ \langle \sigma_+(t_f)|\sigma_-(t_2) \rangle \langle \sigma_-(t_2)|\sigma_-(t_1) \rangle \langle \sigma_-(t_1)|\sigma_-(t_i) \rangle \, e^{-iE_-(t_2)\delta t} e^{-iE_-(t_1)\delta t}$$

Using the same rules as before we find

$$C_{LZ} = - \langle \sigma_+(t_1)|\sigma'_-(t_1) \rangle \, \delta t e^{-iE_+(t_2)\delta t} e^{-iE_+(t_1)\delta t}$$

$$- \langle \sigma_+(t_2)|\sigma'_-(t_2) \rangle \, \delta t e^{-iE_+(t_2)\delta t} e^{-iE_-(t_1)\delta t}$$

$$- \langle \sigma_+(t_f)|\sigma'_-(t_f) \rangle \, \delta t e^{-iE_-(t_2)\delta t} e^{-iE_-(t_1)\delta t}.$$

This expression has a graphical representation depicted in Fig. 1(b). Transitions are occurring at some time t. We take the exponent of $-i\delta t$ times the sum of the low energies at the t_js $[E_-(t_j)]$ until the transition. From the transition onward we do the same thing using the high energy $[E_+(t_j)]$ until the final time t_f. The exponential factor per transition is multiplied by $\langle \sigma_+(t)|\sigma'_-(t) \rangle$ at the time of the transition. Finally, we sum all contributions.

If we divide the time into an infinite number of segments and substitute $t_i \to -\infty$ and $t_f \to \infty$, we will find that

$$C_{LZ} = - \int_{-\infty}^{\infty} dt' \, \langle \sigma_+(t')|\sigma'_-(t') \rangle \, e^{-i \int_{t'}^{\infty} E_+(\tau)d\tau - i \int_{-\infty}^{t'} E_-(\tau)d\tau} \qquad (16)$$

where t' now represents the time at which a transition is taking place. Next using Eqs. 9 and 11 we replace $E_-(\tau)$ by $-E_+(\tau)$. To this we add zero in the form of

$$i \int_{-\infty}^{t'} E_+(\tau)d\tau + i \int_{t'}^{\infty} E_+(\tau)d\tau - i \int_{-\infty}^{\infty} E_+(\tau)d\tau$$

which leads to

$$-i \int_{t'}^{\infty} E_+(\tau)d\tau - i \int_{-\infty}^{t'} E_-(\tau)d\tau = 2i \int_{-\infty}^{t'} E_+(\tau)d\tau - i \int_{-\infty}^{\infty} E_+(\tau)d\tau.$$

We give the following name

$$A(t') = \langle \sigma_+(t')|\sigma'_-(t') \rangle$$

and find using Eqs. 10 and 8 that

$$A(t') = \frac{\Delta\alpha}{2(\Delta^2 + (\alpha t')^2)}. \qquad (17)$$

Similarly we name

$$\phi(t') = 2 \int_{-\infty}^{t'} E_+(\tau)d\tau = \int_{-\infty}^{t'} \sqrt{\Delta^2 + (\alpha\tau)^2} d\tau. \qquad (18)$$

Thus

$$C_{LZ} = -e^{-i\phi(\infty)/2} \int_{-\infty}^{\infty} dt' A(t') e^{i\phi(t')}. \tag{19}$$

In the adiabatic limit (Eq. 12) this integral can be solved using a saddle point approximation [7] as we demonstrate now. Since we are going to take the absolute value of C_{LZ} we can ignore all multiplying phases and express the integral in dimensionless form

$$D_{LZ} = \frac{1}{2} \int_{-\infty}^{\infty} dy \frac{1}{1+y^2} e^{i\frac{tz}{tT}w(y)} \tag{20}$$

where

$$w(y) = \int_0^y \sqrt{1+x^2} dx.$$

There are no poles between the real axis and an axis running parallel to it through, but avoiding from below, the point i. We therefore perform the integral along this new axis. We need to find y_c where $w(y)$ is stationary by taking

$$w' \equiv \frac{dw(y)}{dy} = \sqrt{1+y^2} = 0 \tag{21}$$

which gives $y_c = \pm i$. We expand $w(y)$ around y_c using small values of x so that

$$w(y) = w_c + \int_0^{y-y_c} \sqrt{1+(x+y_c)^2} dx \simeq w_c + \sqrt{2y_c} \frac{2}{3}(y-y_c)^{3/2} \tag{22}$$

where

$$w_c = \int_0^{y_c} \sqrt{1+x^2} dx = \pm i\pi/4.$$

The integral of Eq. 20 is done by changing variables $dw = w' dy$ and using Eq. 21

$$D_{LZ} = \frac{1}{2} \int_{-\infty}^{\infty} \frac{dw}{w'(1+y^2)} e^{i\frac{tz}{tT}w} = \frac{1}{2} \int_{-\infty}^{\infty} \frac{dw}{(w')^3} e^{i\frac{tz}{tT}w}.$$

From Eqs. 22 and 21

$$(w')^3 = (2y_c)^{3/2}(y-y_c)^{3/2} = 3y_c(w-w_c)$$

so that finally, by closing the path of integration in the upper part of the complex plane, we get

$$D_{LZ} = \frac{1}{6y_c} \int_{-\infty}^{\infty} \frac{dw}{(w-w_c)} e^{i\frac{tz}{tT}\phi} = \frac{\pi i}{3y_c} e^{-\frac{\pi tz}{4tT}}$$

yielding

$$P_{LZ} = \left(\frac{\pi}{3}\right)^2 \exp\left(-\frac{\pi \Delta^2}{2\hbar \alpha}\right).$$

This is only slightly different from the exact result in Eq. 7.

The calculation using the SS formalism clearly shows that P_{LZ} is a consequence of interference of different paths. If instead of summing transition amplitude we would sum transition probabilities $|A(t')|^2$ using discretization of time based on the uncertainty principle $\hbar/|2E_+(t')|$ we would find

$$\frac{1}{4} \int_{-\infty}^{\infty} |A(t')|^2 \frac{\hbar}{2E_+(t')} dt' = \frac{35\pi\hbar\alpha}{1024\Delta^2}$$

which is very different from P_{LZ}.

Now let's introduce dephasing. The starting point is the LZ transition probability obtained by taking the absolute value of Eq. 19

$$P_{LZ} = |C_{LZ}|^2 = \int_{-\infty}^{\infty} dt_1 \int_{-\infty}^{\infty} dt_2 A(t_1) A(t_2) e^{i[\phi(t_1)-\phi(t_2)]}.$$

We consider the simplest case where a fluctuating field fluctuates in the z direction, namely, the Hamiltonian is

$$\mathcal{H} = [\alpha t + 2B(t)]S_z + \Delta S_x$$

where $B(t)$ is a stochastic field. We include μ_B in the definition of B. This is equivalent to introducing noise to the sweep rate α which will affect both $A(t')$ and $\phi(t')$. The effect on $A(t')$ involves energy non-conserving transitions between states, which is equivalent to T_1 processes. The effect on $\phi(t')$ involves dephasing similar to T_2. Usually T_2 is shorter than T_1 so we will concentrate on the stronger effect. In other words we ignore the effect of B on the $A(t')$ and simply consider its impact on the phases. We now have to average P_{LZ} over different realizations of the noise $B(t)$. We rename this stochastic LZ transition probability as SS probability and write

$$\langle P_{SS} \rangle = \int_{-\infty}^{\infty} dt_1 \int_{-\infty}^{\infty} dt_2 A(t_1) A(t_2) \left\langle e^{i[\phi(t_1)-\phi(t_2)]} \right\rangle. \tag{23}$$

As a result of the fluctuations in B the phase ϕ will have an average part $\langle \phi(t) \rangle$ which is not random and given by Eq. 18, and a part $\delta\phi$ which is different for different realizations of B, therefore,

$$\left\langle e^{i[\phi(t_1)+\delta\phi(t_1)-\phi(t_2)-\delta\phi(t_2)]} \right\rangle = e^{i[\langle\phi(t_1)\rangle-\langle\phi(t_2)\rangle]} \left\langle e^{i[\delta\phi(t_1)-\delta\phi(t_2)]} \right\rangle.$$

We assume that $\langle \exp(i[\delta\phi(t_1) - \delta\phi(t_2)]) \rangle$ is a function of $t_1 - t_2$ and introduce a phenomenological dephasing time τ_ϕ which is defined as

$$\left\langle e^{i[\delta\phi(t_1)-\delta\phi(t_2)]} \right\rangle \simeq e^{-|t_1-t_2|/\tau_\phi}. \tag{24}$$

SS related τ_ϕ to the dynamic properties of $B(t)$. We will not give this derivation here and simply mention that if

$$\langle B(t)B(0)\rangle = B^2 \exp(-\nu t) \tag{25}$$

then

$$\frac{1}{\tau_\phi} = \frac{B^2}{\hbar^2\nu}. \tag{26}$$

For the dynamic fluctuations to be interesting the field must fluctuate several times before the tunneling process is over, therefore the interesting case is the fast fluctuation limit

$$\nu t_Z > 1. \tag{27}$$

The important point to carry to the next section is that the environment alone determines the dephasing time. If we can determine the dephasing time for one kind of molecule in a given environment, and if we believe the environment does not change between different molecules, then we know the dephasing time for other molecules.

Inserting Eqs. 17, 18 and 24 into 23 we find

$$\langle P_{SS}\rangle = \frac{1}{4} \int\limits_{-\infty}^{\infty} du \int\limits_{-\infty}^{\infty} ds \frac{1}{1+u^2} \frac{1}{1+s^2} \exp\left(i\frac{t_Z}{t_T}\int_s^u \sqrt{1+v^2} dv\right) \exp\left(-\frac{|u-s|}{u_\phi}\right) \tag{28}$$

where

$$u_\phi = \tau_\phi/t_Z.$$

Evaluation of this integral is not simple. To date SS have done this analytically for weak dephasing only where the dephasing time τ_ϕ is long compared to the resonance tunneling time t_T given in Eq. 4. In this case we have

$$\frac{t_T}{\tau_\phi} = \frac{B^2}{\Delta\hbar\nu} < 1. \tag{29}$$

Since this case means that B is small, it is named the weak-coupling limit. Another requirement for analytic evaluation of the integral is that the dephasing is taking place within the Zener time, namely, $\tau_\phi/t_Z < 1$. One can call this requirement long LZ time, and it is expressed as

$$\frac{\tau_\phi}{t_Z} = \frac{\hbar^2\nu\alpha}{B^2\Delta} < 1. \tag{30}$$

Put in other words, analytical evaluation of Eq. 28 is provided only for the time scales order

$$t_Z > \tau_\phi > t_T. \tag{31}$$

In the adiabatic (Eq. 12), fast fluctuation (Eq. 27), weak coupling (Eq. 29), and long LZ time (Eq. 30) limits, the integral in Eq. 28 can be evaluated [5]; the answer is

$$\langle P_{SS} \rangle \simeq \frac{\tau_\phi \alpha}{\Delta} \left(e^{\frac{2\Delta}{\alpha \tau_\phi}} P_{LZ} + \left\{ \frac{\hbar}{\Delta \tau_\phi} \right\}^2 \right). \tag{32}$$

It is clear from this expression that for a proper evaluation of Δ from a tunneling experiment τ_ϕ must be determined. If one insists on fitting experimental data with a LZ type expression, as in Eq. 7, using an observed tunnel splitting Δ_{obs} one finds, in terms of the three time scales, that

$$\frac{\Delta_{obs}^2}{\Delta^2} = -\frac{2t_T}{\pi t_Z} \ln \frac{\tau_\phi}{t_Z} \left(e^{\frac{2t_z}{\tau_\phi}} P_{LZ} + \left\{ \frac{t_T}{\tau_\phi} \right\}^2 \right).$$

Interestingly, due to the dephasing, the observed tunnel splitting becomes sweep rate dependent.

When considering experimental difficulties, the four limits leave a very narrow range of parameters in which Eq. 32 is valid. For example, for $\Delta = 10^{-7}$ K, $\alpha/\mu_B = 10^{-4}$ T/s, $B/\mu_B = 0.1$ G, and $\nu = 5 \times 10^8$ sec^{-1} we find that $t_z = 1.5 \times 10^{-3}$ sec, $\tau_\phi = 6.4 \times 10^{-4}$ sec, and $t_T = 7.6 \times 10^{-5}$ sec, so that the order of time scales given in Eq. 31 holds, and $\nu t_Z > 1$. In this case $\Delta_{obs} = 0.4 \times 10^{-7}$ K which is smaller than Δ. Thus, dephasing increases the probability of moving from the low energy state to the higher one, or, decreases the probability of staying in the low energy state.

2 Experimental dephasing time estimates

We extract the dephasing time in real material using muons coupled to the electronic spins of isotropic MM that experience only the stochastic field. The leading terms for such an Hamiltonian are

$$\mathcal{H} = -2\mu_B \left[\mathbf{H} + \mathbf{B}(t) \right] \mathbf{S} + \hbar^\mu \gamma \left[\mathbf{H} + \mathbf{SA} \right] \mathbf{I}$$

where $\mathbf{S}$ is the electronic spin, $\mathbf{I}$ is the muon spin, $\mathbf{H}$ is the external field, $\mathbf{B}$ is the stochastic field, $\gamma_\mu = 85.162$ MHz/kG is the muon gyromagnetic ratio, μ_B is the Bohr magneton, and $\mathbf{A}$ is a coupling matrix. In this section we use a definition of $\mathbf{B}$ which does not include μ_B since the electronic spin and the muon spins have different gyromagnetic ratios. We ignore a term of the form $\mathbf{B}(t)\mathbf{I}$ since the field experienced by the muon from the molecular spins is greater than this term. Due to the fluctuating field $\mathbf{B}$, $\mathbf{S}$ will vary in time. The simplest assumption that one can make is that the correlation function $\langle \{ \mathbf{S}(t), \mathbf{S}(0) \} \rangle$, where $\{\}$ stands for anticomutator, decay exponentially. The decay rate is determined by the dynamic properties of $\mathbf{B}(t)$ which is produced by the environment of the molecules. Therefore we expect

$$\{\mathbf{S}(t), \mathbf{S}(0)\} = 2S^2 \exp(-t/\tau_\phi). \tag{33}$$

A priori it is possible that τ_ϕ will be H-dependent but we will show experimentally that this is not the case for $H < 2$ kG.

The muon, which is prepared with 100% polarization, will decay towards its equilibrium polarization with a decay rate

$$\frac{1}{^\mu T_1} = \frac{2A^2\tau_\phi}{1 + (^\mu\gamma H\tau_\phi)^2} \tag{34}$$

where we assumed for simplicity that $\mathbf{A}$ is diagonal and isotropic as well. By measuring $^\mu T_1$ as a function of H^2 and fitting the measurement to

$$^\mu T_1 = m + nH^2 \tag{35}$$

one can obtain τ_ϕ from

$$\tau_\phi = \left(\frac{n}{m^\mu\gamma^2}\right)^{1/2}. \tag{36}$$

Salman et $al.$ [8] performed such an experiment for three different isotropic MM with different spin value. In these systems no tunneling is observed due to the absence of a tunnel splitting Δ. However, spin dynamics is observed even at very low temperatures ($T = 50$ mK) with no temperature dependence over a wide temperature range [9].

The molecules were [Cr{(CN)Cu(tren)}$_6$](ClO$_4$)$_{21}$, [Cr{(CN)Ni(tetren)}$_6$](ClO$_4$)$_9$ [10, 11] and [Cr{(CN)Mn(tetren)}$_6$](ClO$_4$)$_9$ [12], which are labeled as CrCu$_6$, CrNi$_6$ and CrMn$_6$, respectively. In these molecules a Cr(III) ion is surrounded by six cyanide ions, each bonded to a Cu(II), Ni(II) or Mn(II) ion. Their magnetic moments of CrCu$_6$, CrNi$_6$ and CrMn$_6$ is $\frac{9}{2}$ $g\mu_B$, $\frac{15}{2}$ $g\mu_B$ and $\frac{27}{2}$ $g\mu_B$, respectively. High field ESR measurements (on CrNi$_6$) [5] and susceptibility measurements (on CrCu$_6$, CrNi$_6$ and CrMn$_6$) found no evidence for anisotropy. This is consistent with the octahedral character of the molecules.

In Fig. 2 we reproduce the data from Ref. [8] where the relaxation time $T_1^\mu(H)$ is plotted at $T \to 0$ as a function of H^2 for all compounds, and for fields up to 2 kG (note the axis break). T_1^μ obeys Eq. 35 in this range. This is consistent with the assumption that for low H the field experience by the muon, which stems from the molecules, and hence the molecular spin autocorrelation function, can be described by a single correlation time τ_ϕ, and that τ_ϕ is field-independent.

From the linear fits in Fig. 2, and using Eq. 36, it was found that $\tau_\phi = 7 \pm 1$ nsec for CrCu$_6$, $\tau_\phi = 10 \pm 1$ nsec for CrNi$_6$ and $\tau_\phi = 11 \pm 1$ nsec for CrMn$_6$. According to Eq. 26 this kind of dephasing time could be generated by a field $B \sim 1$ G fluctuating with a fluctuation rate of $\nu \sim 10^6$ sec^{-1}.

3 Conclusions

It is highly significant that τ_ϕ is nearly spin-independent. Since τ_ϕ is determined by the environment in which the molecules are embedded, its S-

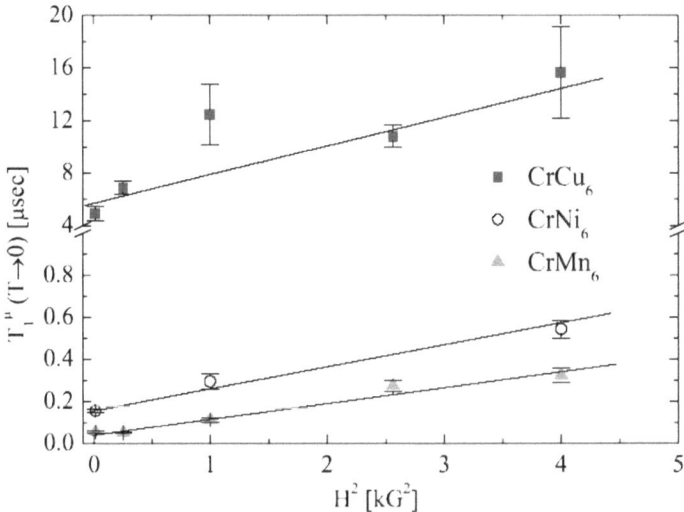

Fig. 2. The relaxation time at $T \sim 100$ mK as a function of H^2 for CrCu6, CrNi6 and CrMn6 with sping value $S = 9/2, 15/2, 27/2$. The solid lines are linear fits. Data are reproduced from Ref. [8]

independence means that coupling to other molecules or to phonons is not responsible for τ_ϕ. This can be understood by examining the Bloch equation which govern the spin motion. In zero external field this equation is given by

$$\frac{d\widehat{\mathbf{S}}}{dt} = {}^e\gamma[\widehat{\mathbf{S}} \times \mathbf{B}(t)]$$

where $\widehat{\mathbf{S}}$ is a unit vector in the direction of the magnetic moment. Only if $\mathbf{B}$ does not stem from other molecules or from spin-phonon coupling could the time scale for spin motion be S-independent, as found experimentally. We therefore conclude that at $T \to 0$ the stochastic field $\mathbf{B}(t)$ responsible for the MM spin motion most likely emanates from nuclear moments.

More importantly τ_ϕ is on the order of 10 nsec. As we argue before, we assume that this dephasing time is typical to high spin magnetic molecules made of transition metal ions embedded in a sea of other ions, including a large number of protons. Indeed, the three isotropic molecules reported here are different but have the same τ_ϕ. We have no reason to believe that τ_ϕ will be substantially different in Fe8 for example. In Fe8, the only molecule where as far as we know Δ_{obs} was measured, it was found to be on the order of 10^{-7} K and sweep rate dependent [3]. If this Δ_{obs} had been simply the intrinsic tunnel splitting Δ, then the tunneling time would have been $\hbar/\Delta \sim 7 \times 10^{-5}$ sec. This tunneling time is longer than the dephasing time $\tau_\phi = 10^{-8}$ sec. More over if, for example, $\alpha = 0.001$ T/sec, then the Zener time (for the 10 to -10 transition) $t_z = \Delta/(20\alpha) = 7 \times 10^{-5}$ sec. This makes the order of time scales $t_Z \sim t_T > \tau_\phi$. This order is very different from Eq. 31, and neither the LZ theory nor the SS theory are valid. However, it is conceivable that Δ_{obs} is not the intrinsic Δ, and that $\Delta > \Delta_{obs}$. If this is correct we might still be in the adiabatic limit and Δ_{obs} could be sweep rate dependent as was observed experimentally. However, to be in the $\tau_\phi > t_T$ range, it must be that $\Delta \gg \Delta_{obs}$ and in fact $\Delta > 10^{-3}$ K. It is more likely that in the MM the order of time scales is $t_Z > t_T > \tau_\phi$. For this case there is no theory available and a new approach is required.

4 Acknowledgments

The author is grateful for enlightening discussions with E. Shimshoni. The work is funded in part by the Russell Berrie Nanotechnology Institute.

References

1. L. D. Landau, Phys. Z. Sov. **2**, 46 (1932).
2. C. Zener, Proc. Roy. Soc. (London) A **137**, 696 (1932).
3. W. Wernsdorfer, R. Sessoli, A. Caneschi, and D. Gatteschi, A. Cornia, and D. Mailly, J. Appl. Phys. **87**, 5481 (2000).
4. S. Carretta, E. Liviotti, N. Magnani, P. Santini, and G. Amoretti, Phys. Rev. Lett. **92**, 207205 (2004).
5. E. Shimshoni and A. Stern, Phys. Rev. B **47**, 9523 (1993).
6. K. Mullen, E. Ben-Jacon, Y. Gefen, and Z. Schuss, Phys. Rev. Lett. **62**, 2543 (1989).
7. E. Shimshoni and Y. Gefen, Ann. of Phys. **210**, 16 (1991).
8. Z. Salman, A. Keren, P. Mendels, V. Marvaud, A. Suiller, and M. Verdaguer, Phys. Rev. B **65**, 132403 (2002).
9. Z. Salman, A. Keren, P. Mendels, A. Sciuller and M. Verdaguer, *Physica B* **289-290**, 106 (2000).
10. T. Mallah, C. Auberger, M. Verdaguer and P. Veillet, *J. Chem. Soc., Chem. Commun.*, **61** (1995).
11. T. Mallah, S. Ferlay, A. Sciuller and M. Verdaguer, *Molecular Magnetism: a Supramolecular Function*, NATO ASI series, C474, Reidel, Dordrecht, 597 (1996).
12. A. Sciuller, T. Mallah, M. Verdaguer, A. Nivorozkhin, J. L. Tholence and P. Veillet, *New J. Chem.* **20**, 1 (1996).
13. A. Barra and M. Verdaguer, Private communication.

Part III

Mössbauer Spectroscopy

Basic principles of Mössbauer spectroscopy and applications

George Filoti

National Institute for Materials Physics, P.O. Box MG-07, 77125
Bucharest-Magurele, Romania

1 The Method

The Mössbauer spectroscopy [1, 2, 3], know also as a recoilless nuclear gamma rays resonance, represents a very powerful tool by providing a large range of information via the modification of the nuclear level energy, derived from the interactions with external factors specific to condensed matter, mostly due to the electron contributions.

The Mössbauer effect comprises simultaneously two factors: a nuclear and a condensed matter state contribution in a related multiplication product and this feature will be present over the whole actual presentation. The "golden mine" of the Mössbauer spectroscopy is represented by the ^{57}Fe isotope (over 90% of publications) and most of the subsequent examples will refer to such related data.

2 Novelty of gamma recoilless phenomena: the classic and the quantum approach

There is widespread known that in most interactions a part of energy is lost by recoil according to the following formula

$$E_{recoil} = \frac{1}{2}mv^2 = \frac{p^2}{2m} = \frac{E_\gamma^2}{2mc^2} \tag{1}$$

where p is the momentum, m the mass and v the velocity of the nucleus. In the case of ^{57}Fe with an energy transition of 14.4 KeV, the recoil energy can be derived as follow:

$$E_{recoil} = \frac{E_\gamma^2}{2mc^2} = \frac{(14.4 keV)^2}{2(53.022 GeV)} = 0.002 eV \tag{2}$$

However, when the vibrations are not any more individual but collective, as best described by Debye model, one can replace m with the mass M for the whole crystal, exceeding the previous one at least by a factor equal to Avogadro's number, and the recoil vanishes. That was explained by Rudolf Mössbauer, when at low temperatures he discovered a huge emission line, compared with the room temperature one.

Because the crystal forces are very weak compared with the internal nuclear forces one can separate the matrix element for the lattice (expressing the change of its state from initial (n_i) to final (n_f)) from the matrix element describing the nuclear gamma transition, namely the Hamiltonian can be explicitly written with separate lattice and nuclear contributions [4]. Hence

$$M_L = (n_f \left| \exp(i\boldsymbol{K}\boldsymbol{X}_L) \right| n_i).(f \left| a(q) \right| i) \ , \tag{3}$$

where $\boldsymbol{K}$ is the momentum operator and $\boldsymbol{X}_L$ is the center of mass, both being related to the lattice, while $a(q))$ represents the nuclear properties operator. The quantum Mössbauer condition is fulfilled when the lattice conserves its initial state, i.e. there are no modifications of the vibrational spectrum and the initial n_i and final n_f states are the same.

In the early sixties of last century, it was of interest to search for all possible Mössbauer isotopes, looking especially for transitions with low free recoil energy and high "effective Mössbauer temperature" and on that search over all known gamma emitting nuclei the ^{57}Fe appeared then real era of Mössbauer spectroscopy as a powerful tool has started.

It is worth to mention that there are 46 elements involving 89 isotopes showing 104 Mössbauer transitions, but only up to ten or a dozen are mostly used in condensed matter, chemistry and biology or material science research.

3 Lamb-Mössbauer factor

The Mössbauer effect cannot be observed for freely moving atoms or molecules, i.e. in gaseous or liquid state. Only in the solid state, crystalline or non-crystalline, including frozen matter, there is possible a recoilless emission and absorption of a gamma quantum. The reason is related to much larger mass M of a solid particle as compared to that of an atom or a molecule, the linear recoil momentum created by emission and absorption of a gamma quantum practically vanishes, as mentioned above. The recoil energy caused by an emitting and absorbing atom tightly bound in the lattice is mostly transferred to the lattice in a vibration system. There is a certain probability f (called "Lamb-Mössbauer" factor , but also known, incorrectly, as Debye-Waller factor) that no lattice excitation (zero-phonon processes) takes place during γ-ray emission or γ-ray absorption. In this case f denotes the fraction of nuclear transitions which occur without recoil. The Mössbauer effect is observable only for this fraction.

The simplest model is due to Einstein (1907) and assumes a large number (via discrete approach) of independent linear harmonic oscillators at a frequency ω_E then the fraction of resonant events is given by Lamb-Mossbauer factor and the characteristic temperature is given by the relation: $k_B\Theta_E = \hbar\omega_E$ while a more adequate continuous model (ω_D frequency) was provided by Debye(1912) as expressed in the followings:

$$f = \exp\left\{-\frac{2R}{3N\hbar}\int_0^{\omega_{\max}}\frac{1}{\omega}\left[\frac{1}{2}+n\left[\omega\right]\right]\rho\left(\omega\right)d\omega\right\} ; n\left(\omega\right) = \frac{1}{\exp\left(\frac{\hbar\omega}{k_BT}\right)-1} \quad (4)$$

EinsteinModel :

$$\rho\left(\omega\right) = 3N\delta\left(\omega-\omega_E\right) \rightarrow f\left(T,\omega_E\right) = \exp\left[-\frac{R}{k_B\theta_E}cth\left(\frac{\hbar\omega_E}{2k_BT}\right)\right] \quad (5)$$

DebyeModel :

$$\rho\left(\omega\right) = \begin{cases} \frac{9N}{\omega_D^3}\omega^2 \text{ for } \omega < \omega_D \\ 0 \text{ for } \omega > \omega_D \end{cases} \quad (6)$$

$$\rightarrow f\left(T,\omega_D\right) = \exp\left\{-\frac{3R}{2\hbar\omega_D}\left[1+4\left(\frac{k_BT}{\hbar\omega_D}\right)^2\int_0^{\frac{\hbar\omega_D}{k_BT}}\frac{xdx}{e^x-1}\right]\right\} \quad (7)$$

R here is the photon energy. Within the Debye model for solids, f increases with decreasing transition energy E_γ, with decreasing temperature, and with increasing Debye temperature Θ_D. One can consider Θ_D as a measure for the strength of the bonds between the Mössbauer atom and the lattice. It is high (> room temperature=RT) for metallic materials and low (< RT) for soft compounds, including co-ordination and metal-organic ones.

Here is a typical disintegration scheme, exemplified by nuclear level of ^{57}Fe isotope:

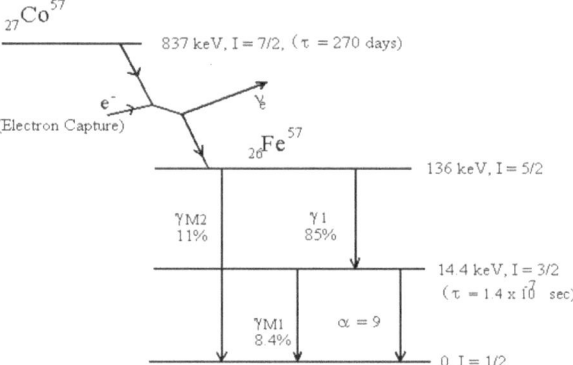

Fig. 1. Disintegration scheme for ^{57}Co.

4 Mössbauer Parameters

There are two types of parameters, one related with nuclear level energy modified by the interaction with external nucleus forces (called *energetical*) and the other (called *dynamic*) related to lattice vibrations and influenced by local surroundings, including impurities and defects (details about the following explanations can be found in classic books listed as Ref. [5]). The first category counts on three different parameters, namely Isomer Shift (IS), Quadrupole Splitting (QS) and Magnetic Splitting produced by an effective (or internal) field at nucleus(B_{int}).

4.1 Energetic Mössbauer Parameters

Isomer shift (IS) and second order Doppler effect

The interaction (mono-polar type) between the nuclear charge density and the electric potential $V(0)$ of the atomic electronic charge having a nonzero probability in the finite nuclear volume generates the isomer shift or center shift .

$$\delta = \left(\frac{Ze^2R^2c}{5\varepsilon_0 E_\gamma} \right) \cdot [\rho_a\,(0) - \rho_S\,(0)] \cdot \left[\frac{\Delta R}{R} \right] mm \cdot s^{-1} \qquad (8)$$

Z is the atomic number, e - electronic charge, R- nucleus radius, c -light velocity, ε_0- dielectric constant, E_γ-gamma quantum energy, ρ- charge density at the nucleus, in absorber (a) and, source (s), respectively.

In addition, there is also a shift, called the second order Doppler shift of the phonon spectrum parameters, depending either on the temperature or on pressure. In particular for the Einstein and Debye models this temperature dependence is:

Einstein model:

$$\delta_{SODS}(T,\theta_E) = \frac{E_R k_B \theta_E}{2Mc^2} cth \left(\frac{\theta_E}{2T} \right) ; k_B\theta_E = \hbar\omega_E \qquad (9)$$

Debye model :

$$\delta_{SODS}(T,\theta_D) = -\frac{9k_B E_0}{16Mc^2} \left[\theta_D + 8T \left(\frac{T}{\theta_D} \right)^3 \int_0^{\frac{\theta_D}{T}} \frac{x^3 dx}{e^x - 1} \right] ; k_B\theta_D = \hbar\omega_D \quad (10)$$

IS will decrease (when $\Delta R/R$ is negative, Fe-case) if there is:
a) a lone pair with low s character present,
b) a decrease in oxidation state of a metal centre,
c) a decrease in co-ordination number, such as for example tetrahedral sp^3 versus octahedral d^{2}sp^3

Quadrupole splitting (QS)

The expression of the isomer shift has been derived assuming a spherical charge, but the real nucleus has a non-spherical charge density. Consequently higher order interactions with a non-cubic distribution of ionic and/or electronic charges are possible, if the quadrupole moment Q of the nucleus is different from zero and if the extra-nuclear (atom co-ordination) symmetry is non cubic, the relationship is for iron as follow:

$$\Delta E_Q = e(QV_{zz}/2)(1 + \eta^2/3)^{1/2}$$

$$\eta = (V_{xx} - V_{yy})/V_{zz}$$

where eQ is the nuclear electric quadrupole moment, V_{zz} the principal component of the Electric Field Gradient (EFG) tensor and η is the asymmetry parameter, with specific definition of axis.

The contributions concern both the lattice and own electronic shell, therefore the quadrupole splitting show various values what represent "a good finger print" for both valence and co-ordination providing accurate and reliable assignments, enforced within correlation with the isomer shift and the magnetic field data

Magnetic Splitting (B)

The third hyperfine interaction corresponds to the nuclear Zeeman effect, reflecting interactions between the nuclear magnetic momentum and the extra-nuclear magnetic field. This interaction occurs if there is an effective magnetic field at the Mössbauer nucleus.

When nucleus possesses a magnetic moment (I) in the presence of a magnetic field B, the interaction is $\mu_N g_N I B$, where μ_N is the nuclear magneton and g_N is the nuclear gyromagnetic ratio.

In a magnetically ordered compound it contains three main contributions:

$$\boldsymbol{B}_{int} = \boldsymbol{B}_F + \boldsymbol{B}_L + \boldsymbol{B}_D$$

Here B_F is the Fermi contact field arising from the polarisation of the core electrons via the exchange interactions with the valence electrons. Generally this term is proportional to the net spin density at the nucleus. B_L is the magnetic field produced at the same nucleus by the orbital motion of the electrons from a partially filled shell. B_D is the dipolar field due to the spins of the valence shell.

There is largely agreed that in most cases the main contribution to B_{int} comes from B_F, specifically in the case of ions with a half-filled shell (for example Fe(III)). The detailed expressions of the three components could be written, for instance in the way presented in the followings. The Fermi contact field contribution is:

$$B_F = \frac{8}{3}\pi\mu_B \left[\rho_\uparrow(0) - \rho_\downarrow(0)\right] \tag{11}$$

where $\rho_i(0)$ is the electron density with spin up or spin down.

The orbital contribution provides a field:

$$B_L = -2\mu_B \left\langle r^{-3} \right\rangle_{3d} \left\langle L_z \right\rangle \tag{12}$$

While the dipolar field contribution is:

$$B_D = 2\mu_B q_{3d} \left\langle S_z \right\rangle \tag{13}$$

$q_{3d} = < r^{-3} > <3 \cos^2 \theta - 1 >$, where $<3 \cos^2 \theta - 1 >$ is the angular distribution of the electrons in partially filled shell, similar to one of quadrupole splitting.

A combined quadrupole (Q) and magnetic (M) interaction provide the

$$E = \mu_N g_N m_I B + (-1)^{|m_I|+1/2}(e^2 qQ/4)(3cos^2\phi - 1)/2$$

with the symbols and parameters having the same meaning as above, while m_I is the projection of nuclear spin I along the quantization axes.

The fitted parameters provide the angle ϕ between the B_{int} and the main Electrical Field Gradient (EFG) principal axis (V_{zz}), derived from the quadrupole splitting (QS) values observed in the magnetically ordered temperature region, ε, and non-ordered (paramagnetic) higher temperature region ΔE_Q, according to the formula:

$$\varepsilon = (\Delta E_Q/2)(3cos^2\phi - 1)$$

The relative intensity of the $\Delta m = 0$ lines of the sextet (specifically number 2 and 5) with respect to the $\Delta m = 1$ lines yields information on the angle θ between B_{int} and γ-rays. In the case when the quadrupole interaction is treated as a small perturbation to the magnetic one and the absorber thickness tends to zero, for ^{57}Fe, the relative intensities of the sextet lines, the outer, middle and inner pairs of the sextet pattern, are given by R_1 /R_2/ R_3 :

3 $(1+\cos^2 \theta)$: 4 $\sin^2 \theta$: $(1+\cos^2 \theta)$

The intensity ratio for random distribution is 3:2:1. The middle line ratio could vary between 0 (moments/ B direction and γ-ray are parallel) and 4 (when they are perpendicular). The values are playing an important role in defining the orientations of the magnetic moment directions per site.

In many cases of measurements on polycrystalline samples or powders, due to the non-vanishing absorber thickness, existing texture and to the competitive electronic mechanisms, the ratios show smaller values than the theoretical ones, calculated from Clebsch-Gordan coefficients. When spin re-orientation processes are present, the angle provided by the combined interaction points towards new directions.

Information derived from the energetic parameters:

- the number of un-equivalent positions,
- the number of order or non-ordered ones,
- the sites showing non-spherical distribution,
- their real individuality

    IS provides data on:
- the electron distribution,
- the valence,
- the spin state,
- the local co-ordination,
- the change of orbital populations,
- the type of transition (first or second order)
    QS supplies information about:
- the valence,
- the Crystal Field symmetry,
- the local distortions,
- the local defects,
- the spin re-orientations (joint Q and M interactions)
    B presents knowledge concerning
- electron configuration,
- magnetic moment,
- spin value,
- magnetic order,
- various contributions to magnetic field, spin and transition
- magnetic transition temperature,
- spin flop processes,
- magnetic relaxation*
    R_2 / R_3 ratio gives information
- related to spin reorientation and magnetic moment direction

4.2 Dynamic Mössbauer Parameters

Natural line width Γ

The uncertainty principle can be presented in the form:

$$\Delta E \Delta t \geq \hbar \tag{14}$$

In the following picture, the distribution of energies called a Lorentzian or a Breit-Wigner distribution, is presented:

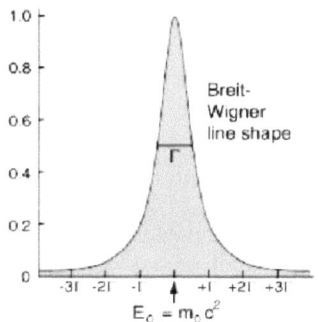

Fig. 2. Lorentzian distribution.

If the width of this distribution at half-maximum is labelled Γ, then the uncertainty in energy E, could be reasonably expressed as:

$$\Delta E = \Gamma = \frac{\hbar}{\tau} \tag{15}$$

Where τ is the life time of the nuclear level and Γ is normally referred as the "natural line width".

The Lamb-Mössbauer factor

As mentioned above, f_{LM} represents the fraction of gamma quanta absorbed or emitted without recoil. Here is re-iterated expression based on Debye model, with the same notation used above when comparing the Einstein and Debye model for lattice vibration:

$$f_{LM} = exp(-E_R/\hbar\omega_D) = exp(-E_R/k_B\Theta_D) \tag{16}$$

The factor appeared both in area and absorption parameters.

Absorption effect

The effect $\varepsilon(v)$ of the absorber, is defined (in the emission spectroscopy) by the following formula, related to maximum and minimum absorption counts as presented in the subsequent figure.

The most relevant parameters are the Lamb-Mössbauer factor f, the absorption section σ_0, the amount of isotope in the compound c_a and natural line width Γ.

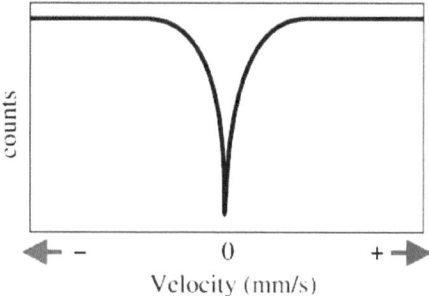

Fig. 3. Defined effect in absorption line.

$$\epsilon(\nu) = \frac{N_\infty - N_0}{N_0} = 1 - \frac{T(\nu)}{c_s.f_s}\alpha\frac{2c_A.f_A\sigma_0 Z}{\pi.\Gamma} \tag{17}$$

Area of one pattern

There is very important parameter providing relative population of different phases, sites or in-equivalent accommodations, being more complex but directly related to the absorption effect:

$$A = \int_{-v_{\min}}^{+v_{\max}} \epsilon(\nu)d\nu\alpha\epsilon(\nu_a)\Gamma \tag{18}$$

Information derived from dynamic parameters

- the population of various sites
- the relative ratios among phases
- the Debye temperature and Lamb-Mössbauer factor
- the local defects
- the solid state diffusion
- the relaxation phenomena
- the anisotropy (magnetic, electric, vibration)
- the dynamic Jahn-Teller effects (not the co-operative provided by energetic parameters)
- the nuclear after effects

A very short survey on relaxation processes in various cases

A- Fluctuating isomer shift
B- Fluctuating field in combined magnetic and quadrupole interactions

i) fluctuating Electric Field Gradient
ii) fluctuating magnetic field
C-   Paramagnetic relaxation
i) spin–phonon processes
ii) spin-spin mechanism
D-   Magnetic relaxation
i) spin-spin
ii) spin-lattice effects
E-   Spin relaxation in case of quadrupole interactions
F-   Super-paramagnetic limits
G-   Goldanskii- Karyagyn effect

5 Mössbauer set-up

Normally a set-up is composed from three elements (as observed from the following picture):
    a.   Driving system generator (pulse, ramp), amplifier (power + operational), transducer (with mounted source)
b.   Nuclear Gamma detection and signal selection chain (detector, preamplifier, single channel analyser)
c.   Acquisition/storage data device (multi-channel analyser or computer acquisition card)
    Additionally, there are devices providing external effects such as: furnaces, cryostats, pressure cell, irradiation facilities, solenoids for magnetic field, etc....The few examples presented hereafter are from basic books on Mössbauer spectroscopy but most of them are from references listed under Ref.[6].
    Fig. 4 present two alternatives: transmission with source-absorber and detector in line or back-scattering arrangement for detecting re-emitted or scattered gamma, X rays or conversion electron

    In the figure above specific transmission and back-scattered spectra are presented for the sake of comparison (one is truly absorption and the other is characteristic to emission/re-emission spectra).

Basic requirements for Mössbauer Source

-    emitting nuclei must be in a very symmetrical co-ordination and the matrix to show no splitting or enlargement of line width at very low temperature
-    accurate and homogeneous distribution of the emitting nuclei into the matrix
-    high Debye temperature of the matrix and large Mössbauer- Lamb fraction
-    narrow emitting line width close to natural one

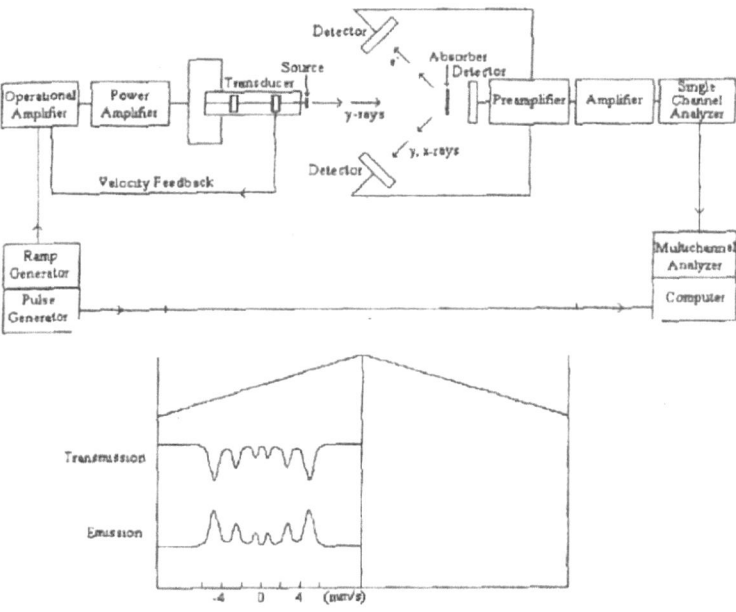

Fig. 4. Set-up description.

-    adequate selection of the source activity related to specific investigations and/or applications has to be considered

Nowadays most of (rather few) companies which are delivering Mössbauer sources take care to provide good sources.

Essential qualities for Detectors

-    Adequate efficiency for specific (for each isotope) gamma Mössbauer energy
-    High counting rate derived from the filling gas type and pulse shape
-    Very good energetic resolution
-    Endurance/life of use (related to number of scintillation/discharges or working ambient conditions)
-    elaborate shielding system against competitive scattered radiations

The detectors could be proportional counters (most widely used), crystals such as NaI (Tl) or Lithium drifted Ge or InP and their selection is related with type of Mössbauer source and specific experimental condition related to efficiency and resolution quality.

Gamma Spectrometric Chain requirements are

-    specific preamplifiers
-    pulse amplifiers with large energy range and pulse characteristics
-    stable single channels analysers
-    integrated devices (preamplifier +amplifier + single channel/ pulse discriminator)

The following figures present the effect of filling gas for proportional counters on emission line intensity for ^{57}Fe:

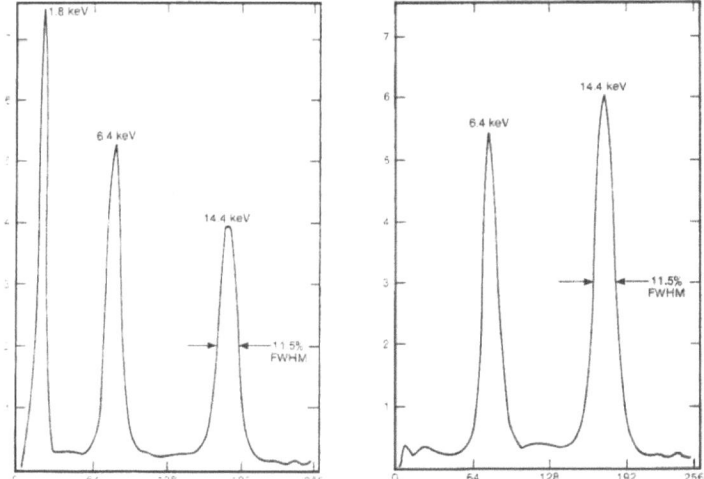

Fig. 5. Exemplification from a leaf-let of proportional counter producer: (left) One atmosphere (97% Krypton - 3% CO_2) and anode voltage of 1.8 kV. (right) One atmosphere (97% Xenon - 3% CO_2) and anode voltage of 2 kV.

Driving systems assuring the relative velocity, involve several characteristics

-    High quality electro-mechanical performances
-    Very precise and flexible electronic control
-    Attached calibration units, when possible
-    Warranted linearity for main types of driving wave form: sinusoidal, triangular- symmetrical, saw tooth
-    Easy settings optimal conditions

-    Adequate velocity range, depending on Mössbauer isotopes to be used
-    Precise protection against vibrations of building or incidentally induced by co-lateral daily activities

Acquisition systems options

-    independent multi channels acquisition systems
-    Computer plug-in card with various functions
-    Incorporated system of spectroscopic chain and acquisition
   The following picture describes the main elements of a performing driving system:

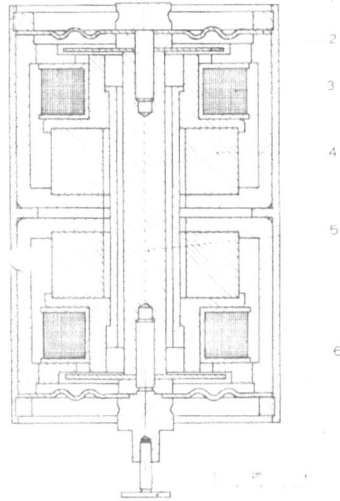

Fig. 6. Home made vibrator/transducer: 1.Diaphragm, 2. Plates, 3.Coils, 4.Permanent magnet, 5.Rod, 6.Source.

   One coil is the excitation signal and the second is the control signal one, the perfect movement being controlled by power amplifier (see above) which permanently compare and ensures the wave form accuracy.

Absorbers and thickness effects

This is a very important matter in the optimisation (acquisition time, large effects, avoiding the increase of experimental line width, etc) of a Mössbauer experiment. Normally the absorption of gamma rays is given by

$$\epsilon(\nu) = \frac{N_\infty - N_0}{N_0} \tag{19}$$

where $N_0 = N_\infty \, exp(-\mu x) + N_B$, N_B is the radiation background μ the linear absorption coefficient and x the thickness of the absorber.

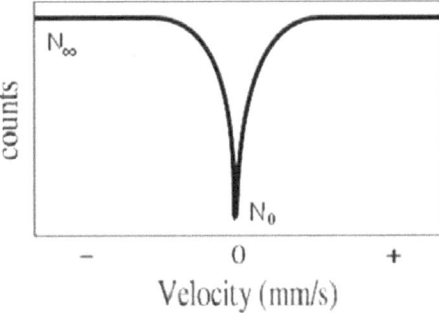

Fig. 7. Absorption related with transmission and radiation background.

This type of absorption is related with electronic configuration of ions or atoms and is always used in the form (μ/ρ) $(\rho.x)$ where the ratio could be found in tables of gamma absorptions per elements and the product is providing the thickness of the absorber in mg per cm^2.

The thickness could be derived from the product $\mu x = 1-2$, meaning that the absorption in the sample decrease the intensity of emergent radiations by a roughly factor between 3 and 7. Depending on specific arrangement the μx value can be taken around 1.3.

Another approach is related to restrict the experimental line width enlargement to maximum 20% from the natural line width value:

$$\Gamma_{app}/\Gamma = 2 + 0.27 n a f \sigma_0$$

(then restricted to additional maximum 20 %). Here n is the number of atoms, a the abundance, f- Lamb-Mössbauer factor and σ_0 the absorption section. From this relation another value for absorber thickness could be derived as well.

There is known an already custom recommendation of 10 mg/cm^2 of Fe related to the thickness absorber and the total amount (derived thickness) is calculated taking into account the holder area and the percentage of Fe in the sample molecular formula. This simple rule does not work when compounds contain heavy absorbing elements (Ba, Pb) or competitive X-ray emitting energy (close to 14.4 KeV for Fe) as in the case of Ga !!

6 Presentation of selected experiments

With isotopic sources (for example ^{57}Fe, ^{119}Sn or ^{151}Eu) two types of experiments could be realized (as already mentioned above) :

a.   in a transmission arrangement where the quanta are passing through absorber to reach detector placed on their way

b.   in a scattering geometry (mostly back scattered) when detector accommodate the absorber and the pulses are generated either by conversion electrons, X-rays or by gamma rays.

The back scattering is recommended for thin layers or when a depth selective investigation of surface formed phases is envisaged. For an optimized detection of one specific from various emerged radiations the gas filling of detector is of a great importance (see Fig.5 above).

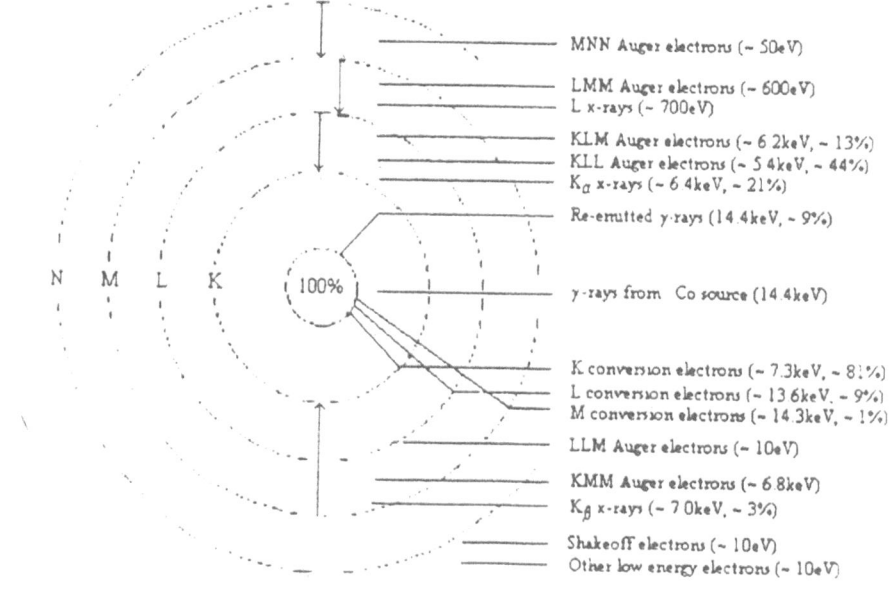

Fig. 8. Energies of various emissions of ^{57}Fe atom and nucleus.

In the next figure an example of a very simple CEMS (Conversion electron spectroscopy) counter is shown.

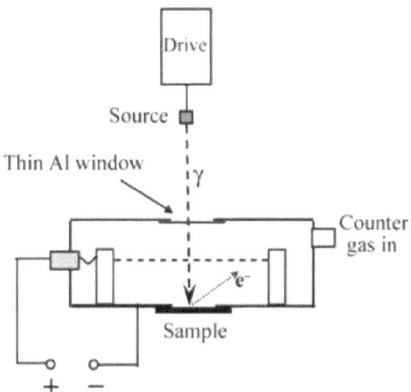

Fig. 9. Simple CEMS detector

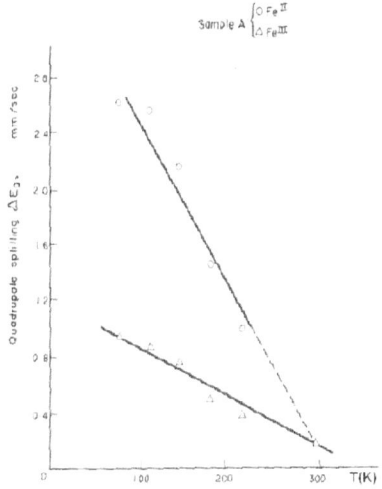

Fig. 10. Effect of electron delocalization on the quadrupole splitting

7 Selected examples for the influence of external factors on Mössbauer spectra

7.1 Temperature Effect

The temperature dependence Mössbauer spectra provided an almost exhaustive range of information concerning all types transition (crystallographic, order-disorder, magnetic, spin, electron flop), mechanisms (donor-acceptor, relaxation, anisotropic coupling or vibration, Jahn-Teller distortion) or change in orbital population and valence. This is obviously the one more consistently reported.

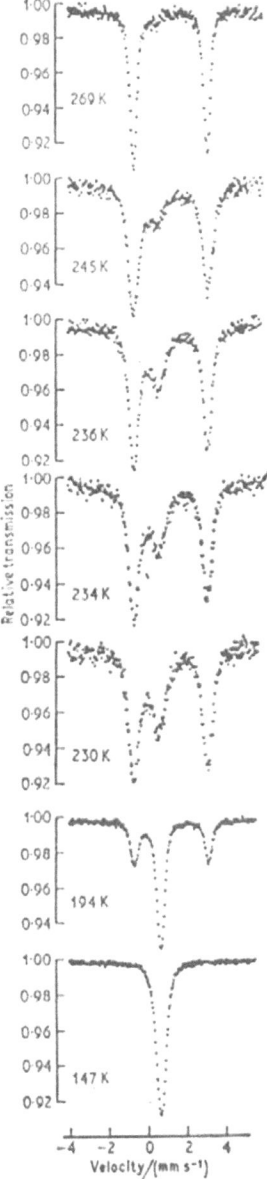

Fig. 11. Effect of spin crossover on Mössbauer spectra.

One particular example of an electronic process seen via Mössbauer spectroscopy is the delocalization of the electron among a tri-nuclear core of Fe_2^{3+}-Fe^{2+}. The electron hopping increases in frequency with temperature and modifies the electron population on the orbitals, which determines the individual quadrupole splitting of each iron species. Fig. 10 shows the quadrupole splittings of Fe^{3+} and Fe^{2+} decreasing with increasing temperature, until 300 K where only one species is seen in the Mössbauer time window.

A more currently mention of the temperature effects is related to spin crossover transition from high spin (HS)to low-spin (LS) states. Typical spectra showing transition from S=2 to S=0 for Fe^{2+} are presented in the next figure.

Another example proves the behaviour of three iron sites in the pyrochlore structure showing very distinct spin-spin relaxation versus temperature.

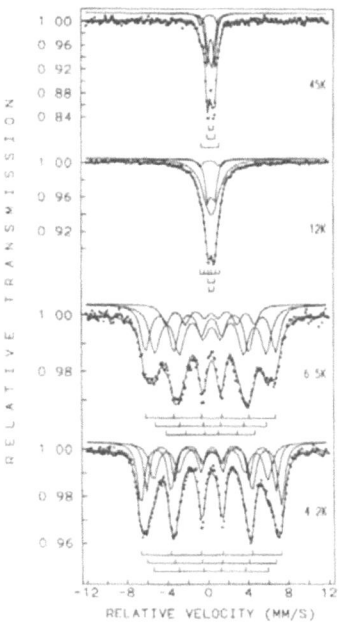

Fig. 12. Effect of spin-spin relaxation on Mössbauer spectra

7.2 Pressure effect

Pressure cells are often used in Mössbauer spectroscopy as they allow to obtain specific relevant data. The example in Fig. 13 shows the reduction of Fe ion valence under the applied pressure.

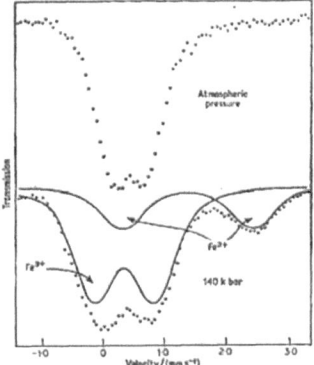

Fig. 13. Mössbauer spectra of iron(III) phosphate showing the electron transfer which takes place at high pressure.

7.3 Effect of an external magnetic field

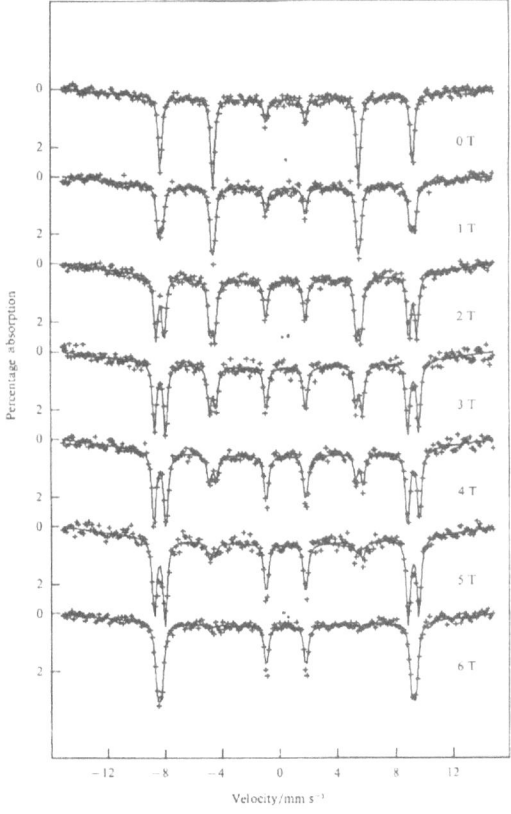

Fig. 14. Effect of spin re-orientation on Mössbauer spectra.

Spin reorientation can occur in a canted antiferromagnet when magnetic fields of increasing magnitude are applied, as shown in the next figure. At B =0 T the direction of magnetic moment is perpendicular to incident γ-rays, as derived from the value 4 for the ratio between middle and inner line pairs. The two canted sublattices (two sextets) are better evidenced with increasing field until there is a re-orientation to a direction parallel to the incident γ-rays, as proven by the absence of second and fifth lines (middle pair).

Another example of canting, this time in a ferromagnet, is provided by Fe-phthalocyanine (FePc) giving in the absence of applied field the largest value reported for the hyperfine field for Fe in an S=1 state. Another unusual aspect is that both sub-lattices appearing under applied field show an increase in the local field with the external field. This was explained by the fact that in that peculiar case the resulting magnetic field at nucleus is positive and coming from a huge orbital and dipolar contributions which surmount the negative spin contribution.

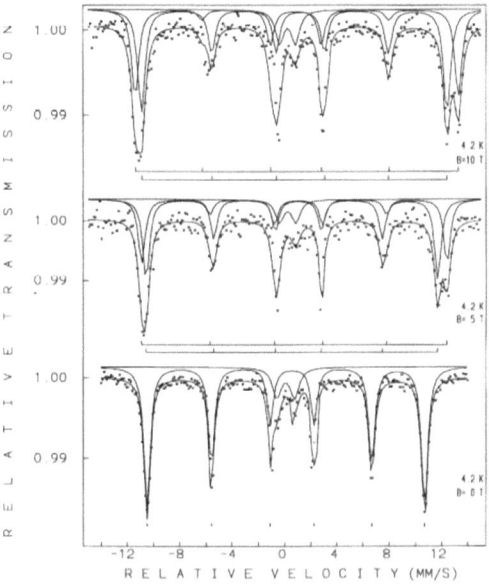

Fig. 15. Mössbauer spectra in Fe-phthalocyanine in external magnetic field.

A more striking effect is that the resulting field observed under external applied field is higher than algebraic sum of the two components, namely the hyperfine field on Fe and the applied magnetic field. The only reliable explanation is an increased orbital contribution under applied field.

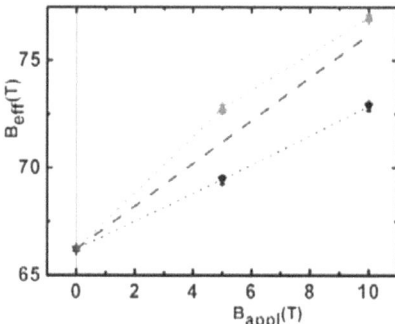

Fig. 16. Effective field at Fe nucleus vs. the external magnetic field in Fe-phthalocyanine.

7.4 Irradiation

Some of the more interesting phenomena are those produced by irradiation, either with particle (electron, neutron, heavy nuclei) or rays (gamma, monochromatic light or micro-waves). The present example is showing the rotation inside a molecular crystal of Fe with 8-hydroxyquinoline. Below are presented two possible packing of the molecule in a either "cis" or "trans" arrangement. The synthesis provides the cis version with inequivalent bonds, inducing an anisotropy in Lamb-Moossbauer factor. Accordingly different areas and linewidths are observed for the paramagnetic doublet.

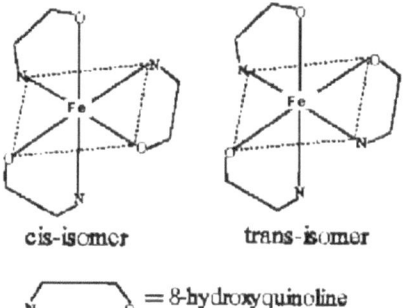

cis-isomer   trans-isomer

= 8-hydroxyquinoline

Under irradiation, the two lines become more and more equal in their parameters. The process is more evident for the irradiation with 2 MeV electrons than for gamma irradiation at 1.2 MeV from a ^{60}Co source. The following figures show the Mössbauer spectra and the evolution of area and line width of each of the two components line of the doublet. Finally, when the lines have the same parameters the molecule is completely rotated in the trans-configuration

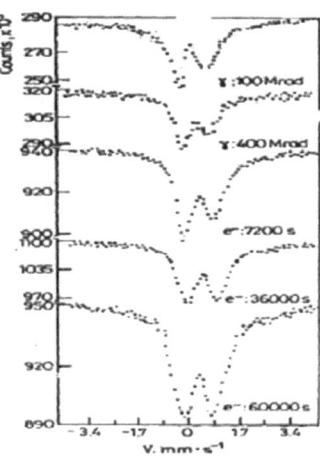

Fig. 17. Effect of irradiation on Mössbauer spectra.

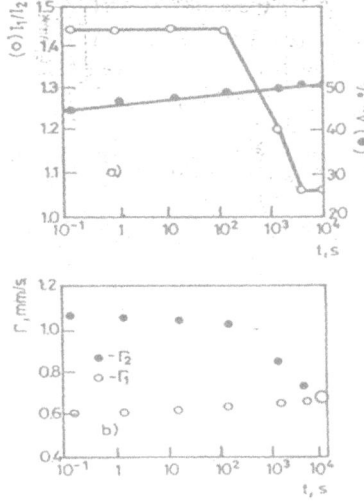

Fig. 18. Mössbauer parameters vs. the electron irradiation time: (top) I_1/I_2 ratio (open circles) and relative area (closed circles); (bottom) linewidths Γ_1 and Γ_2.

Acknwoledgements

I would like to thank MAGMANet Network of excellence, which provided the scientific and financial support for organizing and presenting the lectures in September 2006 in Pavia, Italy. I am deeply indebted to Alessandro Lasciafari for his precise hospitality and creative atmosphere and to Pietro Carretta for his care and support in editing.

References

1. R. L. Mössbauer, Naturwissenschaften 45, 538 (1958)
2. R.L. Mössbauer, Z.Physik 151, 124(1958)
3. R.L. Mössbauer, Z. Naturforsch. 14a, 211 (1959)
4. Harry Lipkin, Journal of Physics 9, 332 (1960)
5. hereafter a list of basic books on Mössbauer spectroscopy is presented: a) H. Frauenfelder in *The Mössbauer Effect* (Benjamin, New York (1962)).

   b) G. K. Wertheimin *Mössbauer Effect* (Academic, New York (1964)).

   c) V. I. Goldanskii and R. H. Herber in *Chemical Application of Mössbauer Spectroscopy*, (Academic Press (1964)).

   d) N. N. Greenwood and T. C. Gibb in *Mössbauer Spectroscopy* (Chapman and Hall, London (1971)).

   e) G. K. Shenoy and F. E. Wagner (Eds) in *Mössbauer Isomer Shifts* p. 106 (North Holland, Amsterdam (1978)).

   f) B. V. Thosar, J. K. Srivastava, P. K. Iyengar, S. C. Bhargava (Eds.) in *Advances in Mössbauer Spectroscopy* (Elsevier, Amsterdam (1983))

   g) D.P.E Dickson and F.J. Berry in *Mössbauer Spectroscopy* (Cambridge University Press, 1986)
6. For further details on the examples given above see: M. Bornaz, G. Filoti, A. Gelberg, V. Grabari, C. Nistor, Nucl. Instr. & Meth. 40, 61 (1966); D. Lupu, D. Barb, G. Filoti, M. Morariu, D. Tarina, J. Inorg. Nucl. Chem. 34, 2803 (1972); G. Filoti, C.N. Turcanu, C. Oproiu in Proc. ICAME Krakow-1975, Eds. A.Z. Hrynkiewicz and J.A. Sawicki, Vol.II, p.283; G. Filoti, M. D. Kuzmin and J. Bartolomé, Phys Rev. B 74, 134420 (2006)

Line shape of Mössbauer relaxation spectra

Gabriele Spina[12] and Luciano Cianchi[3]

[1] Dipartimento di Fisica - Università di Firenze, Via Sansone 1, 50019 Sesto Fiorentino (Firenze) - Italy `gabriele.spina@unifi.it`
[2] Affiliation: Centro Nazionale di Ricerca S3, INFM-CNR
[3] Istituto dei Sistemi Complessi (Sezione di Firenze), CNR, Via Madonna del Piano, 10, 50019 Sesto Fiorentino (Firenze) - Italy
`luciano.cianchi@isc.cnr.it`

1 Mössbauer cross section line shape

The resonant absorption of gamma rays, emitted by nucleus of particular isotopes embedded in solid state matrices, was discovered and explained in 1957 by Rudolph Mössbauer. From the time when it was discovered, scientists have used the Mössbauer effect to investigate valence states, structure and bonding properties in solid state materials. Moreover, vibrational properties and dynamics of the hyperfine fields have been investigated.[1][2][3]

Usually, the spectrum consists of a set of sharp lines. This is true only if, within the lifetime of the nuclear states, the hyperfine interaction is practically stationary. Two opposite situations give rise to such result.

The first case occurs when the lifetime of the electronic state is much greater than the nuclear one. The second one is characterized by the opposite situation. In fact, in the first case the nucleus sees the hyperfine fields produced by each single electronic level, in the second case it sees the mean value of hyperfine fields generated by all the electronic levels. In both cases the hyperfine fields are defined and time independent. In the intermediate situation the shape of the spectrum can be very complex and reflects the dynamics of the electronic levels.

Previous considerations give rise to the time window for the electronic dynamics that can be studied by Mössbauer spectroscopy. For the common case of the Fe^{57} we have $10^{-5} \div 10^{-11}$ sec [Section 5].

The functional dependence of the Mössbauer spectrum on the source velocity v is given by [2]

$$Y(v) = F(v)\left\{1 - \frac{f_s}{1+B}\int_{-\infty}^{\infty} L(\omega - v, \Gamma_s)[1 - \exp(-t_a\sigma(\omega))]d\omega\right\} \quad (1)$$

where $F(v)$ is the background function, not very dependent on v. Moreover $L(\omega, \Gamma_s)$ is the line shape of the source characterized by the linewidth value

Γ_s, $\sigma(\omega)$ is the Mössbauer absorption cross section, t_a indicates the sample thickness and f_s is the source Mössbauer f factor . Finally, B depends on the detected spurious counts.

This chapter is devoted to the evaluation of the Mössbauer spectrum line shape in presence of transitions among electronic states.

In the present section, we derive a general expression for the absorption cross section $\sigma(\omega)$. Subsequent sections, taking into account electronic states and involved interactions, will be devoted to the application of the derived expression in the case of time dependent hyperfine interactions.

The usual perturbation theory states that the probability amplitude for an absorption transition from the ground state $|m_g\rangle$ to a final excited state $|m_e\rangle$ is given by [4]

$$a_{m_e,m_g}(t) = -i \langle m_e | A^\dagger | m_g \rangle \int_0^t \exp[-i(\omega - \omega_{m_e} + \omega_{m_g})\tau - \frac{\Gamma_a}{2}\tau]d\tau \quad (2)$$

where Γ_a is the linewidth of the excited state and the energies are expressed in units $\hbar = 1$. Moreover $A^\dagger$ represents the operator describing the absorption of a gamma ray by the nuclear system. From this expression we obtain

$$a_{m_e,m_g}(\infty) = \frac{\langle m_e | A^\dagger | m_g \rangle}{-\omega + \omega_{m_e} - \omega_{m_g} + i\frac{\Gamma_a}{2}}$$

Consequently, the transition probability is expressed by

$$P = \frac{\langle m_g | A | m_e \rangle \langle m_e | A^\dagger | m_g \rangle}{\left(-\omega + \omega_{m_e} - \omega_{m_g} - i\frac{\Gamma_a}{2}\right)\left(-\omega + \omega_{m_e} - \omega_{m_g} + i\frac{\Gamma_a}{2}\right)} \quad (3)$$

Now by using the relation

$$\frac{1}{\left(\Delta\omega - i\frac{\Gamma_a}{2}\right)\left(\Delta\omega + i\frac{\Gamma_a}{2}\right)} = \frac{1}{i\Gamma_a}\left(\frac{1}{\Delta\omega - i\frac{\Gamma_a}{2}} - \frac{1}{\Delta\omega + i\frac{\Gamma_a}{2}}\right)$$

we obtain

$$P = \frac{2}{\Gamma_a}\Re\left\{\frac{\langle m_g | A | m_e \rangle \langle m_e | A^\dagger | m_g \rangle}{i(-\omega + \omega_{m_e} - \omega_{m_g}) + \frac{\Gamma_a}{2}}\right\}$$

where $\Re\{a\}$ indicates the real part of a.

By summing contributions from all possible ground and exited states $|m_g\rangle$ and $|m_e\rangle$ we obtain the absorption cross section.

$$\sigma(\omega) = \frac{2}{\Gamma_a}\Re\left\{\sum_{m_g} \rho_{m_g} \sum_{m_e} \frac{\langle m_g | A | m_e \rangle \langle m_e | A^\dagger | m_g \rangle}{i(-\omega + \omega_{m_e} - \omega_{m_g}) + \frac{\Gamma_a}{2}}\right\} \quad (4)$$

where ρ_{m_g} represents the thermal occupation probability for the generic ground state.

Posing $p = -i\omega + \frac{\Gamma_a}{2}$, using the relation

$$\frac{\langle m_g|A|m_e \rangle}{i(-\omega + \omega_{m_e} - \omega_{m_g}) + \frac{\Gamma_a}{2}} = \int_0^\infty e^{-pt} \langle m_g|A(t)|m_e \rangle dt$$

and the completeness of the base states, we finally find the general expression for the Mössbauer absorption cross section

$$\sigma(\omega) = \frac{2}{\Gamma_a} \Re \left\{ \sum_{m_g} \rho_{m_g} \int_0^\infty e^{-pt} \left\langle m_g \left| A(t)A^\dagger(0) \right| m_g \right\rangle dt \right\} =$$

$$= \frac{2}{\Gamma_a} \Re \left\{ \sum_{m_g} \int_0^\infty e^{-pt} \left\langle m_g \left| \rho A(t)A^\dagger(0) \right| m_g \right\rangle dt \right\} \tag{5}$$

We note that the derived expression contains a Trace; consequently any complete set of base states can be used to evaluate Eq. 5.

2 Time dependent hyperfine interactions and the relaxation superoperator

When transitions among electronic levels occur, we must consider the structure of the states and of the involved interactions.

The quantum system is composed of the gamma radiation field, the nucleus, the surrounding atomic electrons and, finally, of the states describing the thermal bath. The nucleus interacts with the gamma radiation field; moreover, it also interacts with the electrons through hyperfine interaction. Lastly, electronic states interact with the thermal bath.

The energy values connected with the hyperfine interaction are much smaller than the typical energy differences among the electronic levels; so that, it is usual to consider the hyperfine interaction operator as diagonal on the electronic states. In other terms, magnetic fields and E.F.G. tensors are considered as expectation values, evaluated for each electronic state, instead of quantum operators acting on the electronic states.

In the following we indicate the atomic Hamiltonian by

$$H_{at}(t) = H_{ne}(t) + H_{hyp}(t) = (H_n(t) + H_e(t)) + H_{hyp}(t) \tag{6}$$

where the subscripts n, e and hyp denote respectively the nuclear, electronic and hyperfine components. Moreover, indicating by H_b the thermal bath Hamiltonian and by H_{e-b} the electron-bath interaction, the total Hamiltonian takes the form

$$H(t) = H_{at}(t) + H_b(t) + H_{e-b}(t) \tag{7}$$

Taking into account that Eq. 5 contains a Trace, the base states of the system are usually chosen as the product of the eigenstates describing the various subsystems. A common choice is

$$|\psi\rangle = |I, I_z\rangle |\varphi_i\rangle |n\rangle \tag{8}$$

where $|I, I_z\rangle$ represents the nuclear quantum state, $|\varphi_i\rangle$ indicates the ith electronic level and finally $|n\rangle$ denotes the state of the thermal bath. We stress that the atomic base states $|I, I_z\rangle |\varphi_i\rangle$ are eigenstates of $H_{ne}(t)$, but not necessarily they are also eigenfunctions of $H_{at}(t)$.

The Eq. 5 contains the density matrix that depends on the whole Hamiltonian. It is however possible to forget contributions to ρ due to H_{hyp} and H_{e-b}. Such approximation is physically correct if electron-bath and hyperfine interactions do not significantly change the occupation probabilities. Under such hypothesis, we write $\rho \simeq \rho_{at}\rho_b$, where ρ_{at} and ρ_b respectively depend on the atomic and bath Hamiltonians.

The operator $A^\dagger(0)$ does not act on the thermal bath states, so the absorption cross section can be rewritten as

$$\sigma(\omega) = \frac{2}{\Gamma_a} \Re \left\{ \int_0^\infty e^{-pt} \mathcal{T}r_{at}[\rho_{at}\langle A(t)\rangle_b A^\dagger(0)]dt \right\} \tag{9}$$

where $\langle A(t)\rangle_b$ and $\mathcal{T}r_{at}$ respectively represent the average of the $A(t)$ operator over the thermal bath and the Trace over the atomic states.

In order to treat in a simple and condensed way the time dependence of $A(t)$, is it useful to introduce the Liouville operator [Appendix 1] related to the Hamiltonian of the system (Eq. A.10). This way we can write the matrix elements of the operators $A(t)$ and $A(p)$ as linear combinations of those of $A(0)$. We obtain $A(t) = e^{iH^x t}A(0)$ and $(p - iH^x)A(p) = A(0)$, where $A(p)$ is for the Laplace transform of $A(t)$.

Eq. 9 contains the Laplace transform of $\langle A(t)\rangle_b$. To evaluate $A_{at}(p) = \int_0^\infty e^{-pt}\langle A(t)\rangle_b dt$, we divide the Hamiltonian in two parts, by separating the term H_{e-b}. Moreover we introduce the projection operator P which, applied to a generic operator M returns its thermal average over the bath states $|n\rangle$.

$$H = H_{at} + H_b + H_{e-b} = H_1 + H_{e-b}$$

$$PM = \langle M\rangle_b = \sum_{n,m} \rho_{n,m} \langle m|M|n\rangle \tag{10}$$

Obviously $P^2 = P$ and $P(1 - P) = 0$.

Applying P and $(1 - P)$ to $(p - iH^x)A(p) = A(0)$ we obtain the equation system

$$\begin{cases} P(p - iH^x)A(p) = A(0) \\ (1 - P)(p - iH^x)A(p) = 0 \end{cases} \tag{11}$$

The thermal average of the Laplace transform of the nuclear-radiation interaction Hamiltonian can be written as $A_{at}(p) = PA(p)$; where the index denotes the system on which the operator acts. Moreover, defining the *difference* operator $A_d(p) = A(p) - A_{at}(p) = (1 - P)A(p)$ and resolving the system for $A_{at}(p)$, we obtain

$$\{p - iPH^x P + PH^x[p - i(1 - P)H^x]^{-1}(1 - P)H^x P\}A_{at}(p) = A(0) \quad (12)$$

Since $[H_b, A_{at}(p)]_- = 0$, we have $H_b{}^x A_{at}(p) = 0$; moreover $[P, H_{at}{}^x]_- = 0$ because $H_{at}{}^x$ does not operate on the thermal bath. Using these properties, the Eq. 12 is rewritten as

$$\{p - i(H_{at}{}^x + \langle H_{e-b}\rangle_b{}^x) + PH^x[p - i(1 - P)H^x]^{-1}(1 - P)H_{e-b}{}^x\}A_{at}(p) = A(0) \quad (13)$$

Thus we have for $A_{at}(p)$:

$$A_{at}(p) = [p - i(H_{at}{}^x + \langle H_{e-b}\rangle_b{}^x) + R(p)]^{-1}A(0) \quad (14)$$

where we introduced the *relaxation operator* $R(p)$, defined as

$$R(p) = PH^x[p - i(1 - P)H^x]^{-1}(1 - P)H_{e-b}{}^x \quad (15)$$

To obtain useful expressions for the relaxation superoperator, we consider the series expansion $[p - i(1 - P)H^x]^{-1} = \frac{1}{p}(1 + \sum_n (\frac{i}{p})^n[(1 - P)H^x]^n)$, and we observe that $PH^x(1 - P) = P(H_{at}{}^x + H_b{}^x + H_{e-b}{}^x)(1 - P) = PH_{e-b}{}^x(1 - P)$
So that, we obtain

$$PH^x[p - i(1 - P)H^x]^{-1}(1 - P)H_{e-b}{}^x A_{at}(p) =$$
$$= PH^x(1 - P)[p - i(1 - P)H^x]^{-1}(1 - P)H_{e-b}{}^x A_{at}(p) =$$
$$= PH_{e-b}{}^x(1 - P)[p - i(1 - P)H^x]^{-1}(1 - P)H_{e-b}{}^x A_{at}(p)$$

and, consequently

$$R(p) = PH_{e-b}{}^x(1 - P)[p - i(1 - P)H^x]^{-1}(1 - P)H_{e-b}{}^x P \quad (16)$$

The absorption cross section is then given by

$$\sigma(\omega) = \frac{2}{\Gamma_a}\Re\left\{\mathcal{T}r_{at}[\rho_{at}(G(p)^{-1}A(0))A(0)^\dagger]\right\} \quad (17)$$

where $G(p)^{-1} = [p - i(H_{at}{}^x + \langle H_{e-b}\rangle_b{}^x) + R(p)]^{-1}$. If the diagonal elements of the electron-bath interaction Hamiltonian are zero, we finally obtain

$$G(p)^{-1} = [p - iH_{at}{}^x + R(p)]^{-1} \quad (18)$$

Eq. 14 expresses the matrix elements of $A_{at}(p)$ as linear combination of those of $A(0)$ and Eq. 18 allows us to evaluate the involved coefficients. The Mössbauer cross section, Eq. 17, depends on the dynamical properties of the electronic states through $R(p)$ that makes $A_{at}(p)$ able to operate on electronic levels; in fact $A(0)$ does not operate on them.

3 Evaluation of the relaxation superoperator matrix elements

In order to evaluate the derived expressions, we use the base states expressed by Eq. 8. In particular we write $|\psi_g\rangle = |\mu\rangle|i\rangle$ and $|\psi_e\rangle = |\nu\rangle|i\rangle$ for atomic states respectively characterized by nuclear ground and exited levels. Where $|i\rangle$ denotes the electronic state and $|\mu\rangle$, or $|\nu\rangle$, represents the nuclear eigenfunction. This way, Eq. 17 is rewritten as

$$\sigma(\omega) = \frac{2}{\Gamma_a}\Re\left\{\sum_{\mu,i}\rho_{at}(\mu,i)\sum_{\nu}\langle\mu,i|G(p)^{-1}A(0)|\nu,i\rangle\langle\nu,i|A(0)^{\dagger}|\mu,i\rangle\right\} \quad (19)$$

where we also considered that the nuclear-radiation operator $A(0)^{\dagger}$ does not act on the electronic states.

Moreover, denoting by $|\mu',i';\nu',i'\rangle$ the generic decay transition $|\mu',i'\rangle\langle\nu',i'|$, one has

$$\langle\mu,i|G(p)^{-1}A(0)|\nu,i\rangle = \sum_{\mu',\nu',j}\langle\mu,i;\nu,i|G(p)^{-1}|\mu',j;\nu',j\rangle\langle\mu',j|A(0)|\nu',j\rangle$$

$$(20)$$

By ordering the transitions, the elements of $A(0)$ can be written as a column vector and the superoperator G^{-1} takes the form of a square matrix whose order is equal to the total number of transitions. For example, in the case of Fe^{57}, one has eight transitions for each electronic state.

Instead of directly evaluating the matrix elements of G^{-1}, the more suitable procedure consists of the evaluation of the eigenvalues and eigenvectors of G and $G^{\dagger}$ and in the subsequent reconstruction of G^{-1}[5][6].

For this reason let us now consider the superoperator $G(p) = p - iH_{at}{}^x + R(p)$.

It is formed by three terms. The first one, containing the natural linewidth and the independent variable ω, is diagonal. The second one describes the hyperfine interaction; as mentioned, we assume that it does not connect transitions between base states having different electronic parts, so that it is characteristic of each electronic state and depends also on eventually applied external fields.

Finally the last term, depends both on the electron-bath interaction and on the total Hamiltonian. Moreover it depends on the natural linewidth Γ_a and on the independent variable ω.

In order to evaluate the matrix elements of the relaxation operator, a few approximations must be introduced.

To discuss them it is useful to write

$$R(p) = PH_{e-b}{}^x(1-P)[p - i(1-P)H^x]^{-1}(1-P)H_{e-b}{}^xP = \int_0^{\infty}e^{-pt}R(t)dt$$

where

$$R(t) = PH_{e-b}{}^x(1-P)e^{(i(1-P)H^x t)}(1-P)H_{e-b}{}^x P \tag{21}$$

To deduce the matrix elements of $R(t)$, we use the general rule of Eq. A.5 by applying the superoperator to the general operator $|\mu, i; n\rangle\langle\nu, j; m|$, where n and m denote generic bath states. The application of P to $|\mu, i; n\rangle\langle\nu, j; m|$ gives an operator that works only on the electronic and nuclear parts.

In fact $P|\mu, i; n\rangle\langle\nu, j; m| = \sum_{m,n} \rho_{nm}|\mu, i; n\rangle\langle\nu, j; m| = |\mu, i\rangle\langle\nu, j|$

Now, by applying $H_{e-b}{}^x$ one has

$$H_{e-b}{}^x|\mu, i\rangle\langle\nu, j| = H_{e-b}|\mu, i\rangle\langle\nu, j| - |\mu, i\rangle\langle\nu, j|H_{e-b}$$

and the subsequent application of $(1-P)$ gives

$$(1-P)H_{e-b}{}^x|\mu, i\rangle\langle\nu, j| = H_{e-b}|\mu, i\rangle\langle\nu, j| - |\mu, i\rangle\langle\nu, j|H_{e-b} - \\ -P(H_{e-b}|\mu, i\rangle\langle\nu, j| - |\mu, i\rangle\langle\nu, j|H_{e-b})$$

If H_{e-b} does not contain diagonal elements on the bath states

$$P(H_{e-b}|\mu, i\rangle\langle\nu, j| - |\mu, i\rangle\langle\nu, j|H_{e-b}) = 0$$

so that

$$(1-P)H_{e-b}{}^x|\mu, i\rangle\langle\nu, j| = H_{e-b}|\mu, i\rangle\langle\nu, j| - |\mu, i\rangle\langle\nu, j|H_{e-b}$$

The subsequent application of $e^{(i(1-P)H^x t)}$ means that we must consider the temporal evolution of the operator $(1-P)H_{e-b}{}^x|\mu, i\rangle\langle\nu, j|$ due to the total Hamiltonian that, we note, also contains electron-bath terms. It is however physically correct to introduce the lifetime τ_b of the bath states and to not consider the contribution of H_{e-b} to the temporal evolution. In fact, bath states interact not only with the considered ion, but with all the ions existing in the compound. Moreover the description of the bath states is in every case approximate. Consequently it is correct to introduce a decay time for the bath states; furthermore H_{e-b} can be omitted because the time constants involved in the electron-bath term are obviously much greater than τ_b.

Thus, eliminating H_{e-b} from the argument of the exponential and using the usual power expansion we have

$$e^{(i(1-P)H^x t)} \simeq e^{(i(1-P)(H_{at}+H_b)^x t)} = 1 + \sum_n \frac{(i(1-P)(H_{at}+H_b)t)^n}{n!}$$

Considering the general term of the series, it is easy to see that

$$(1-P)\, e^{(i(1-P)(H_{at}+H_b)^x t)}(1-P)H_{e-b}{}^x \, |\mu, i\rangle\,\langle\nu, j| = e^{(iH_{at}{}^x t)}H_{e-b}{}^x \, |\mu, i\rangle\,\langle\nu, j|$$

where H_{at} contains both electronic and nuclear terms. Consequently one has

$$R(t)\,|\mu,i;n\rangle\,\langle\nu,j;m| = PH_{e-b}{}^{x}e^{(\mathrm{i}H_{at}{}^{x}t)}H_{e-b}{}^{x}\,|\mu,i\rangle\,\langle\nu,j| =$$
$$= e^{-i\omega_{\mu,\nu}t}P[H_{e-b},H_{e-b}(t)\,|\mu,i\rangle\,\langle\nu,j| - |\mu,i\rangle\langle\nu,j|H_{e-b}(t)]_{-} =$$
$$= e^{-i\omega_{\mu,\nu}t}P\{H_{e-b}H_{e-b}(t)|\mu,i\rangle\langle\nu,j| - H_{e-b}|\mu,i\rangle\langle\nu,j|H_{e-b}(t) -$$
$$- H_{e-b}(t)|\mu,i\rangle\langle\nu,j|H_{e-b} + |\mu,i\rangle\langle\nu,j|H_{e-b}(t)H_{e-b}\}$$

where $\omega_{\mu,\nu} = \omega_\mu - \omega_\nu$ and the temporal evolution of $H_{e-b}(t)$ is due to the electronic Hamiltonian.

Let us now consider the interaction Hamiltonian H_{e-b} as a product of two terms[7]; the first one, denoted by S, acts on the electronic states and the second one, indicated by F, works on the bath:

$$H_{e-b} = S \cdot F \tag{22}$$

Under such hypothesis and, moreover, supposing that $|\omega_j - \omega_i| \simeq 0$ we write

$$R(t)\,|\mu,i;n\rangle\,\langle\nu,j;m| =$$
$$= e^{-i\omega_{\mu,\nu}t}\,\{\langle F(0)F(t)\rangle_b[S(0)S(t)\,|\mu,i\rangle\,\langle\nu,j\,|-S(0)\,|\mu,i\rangle\,\langle\nu,j\,|S(t)] -$$
$$- \langle F(0)F(t)\rangle_b{}^{*}[S(t)\,|\mu,i\rangle\,\langle\nu,j\,|S(0) - |\mu,i\rangle\,\langle\nu,j\,|S(t)S(0)]\}$$

where $\langle F(0)F(t)\rangle_b = P(F(0)F(t)) = \langle F(0)^2\rangle_b e^{-t/\tau_b}$ depends exponentially on the time and the electronic Hamiltonian H_e determines the time dependence of $S(t)$.

Since the S operator does not act on the nuclear states, we have

$$R(t)|\mu,i;n\rangle\langle\nu,j;m| =$$
$$= e^{-i\omega_{\mu,\nu}t}\{\langle F(0)F(t)\rangle_b[S(0)S(t)|i\rangle\langle j| - S(0)|i\rangle\langle j|S(t)] -$$
$$- \langle F(0)F(t)\rangle_b{}^{*}[S(t)|i\rangle\langle j|S(0) - |i\rangle\langle j|S(t)S(0)]\}\,|\mu\rangle\langle\nu|$$

and, introducing the completeness, we obtain:

$$R(t)|\mu,i;n\rangle\langle\nu,j;m| =$$
$$= e^{-i\omega_{\mu,\nu}t}\{\langle F(0)F(t)\rangle_b[\sum_k S(0)|k\rangle\langle k|S(t)|i\rangle\langle j| - S(0)|i\rangle\langle j|S(t)] -$$
$$- \langle F(0)F(t)\rangle_b{}^{*}[S(t)|i\rangle\langle j|S(0) - \sum_k |i\rangle\langle j|S(t)|k\rangle\langle k|S(0)]\}\,|\mu\rangle\langle\nu|$$

Consequently, we get the following expression for the matrix element of the relaxation superoperator $R(t)$

$$\langle \mu',i';\nu',j'\,|R(t)|\mu,i;\nu,j\rangle = \delta_{\mu,\mu'}\delta_{\nu,\nu'}e^{-i\omega_{\mu,\nu}t}\,.$$
$$\cdot\{\langle F(0)F(t)\rangle_b[\sum_k \langle i'|S(0)|k\rangle\langle k|S(t)|i\rangle\delta_{j,j'} - \langle i'|S(0)|i\rangle\langle j|S(t)|j'\rangle]-$$

$$-\langle F(0)F(t)\rangle_b{}^* [\langle i^{'}|S(t)|i\rangle\langle j|S(0)|j^{'}\rangle - \sum_k \delta_{i,i'}\langle j|S(t)|k\rangle\langle k|S(0)|j^{'}\rangle]\} \quad (23)$$

Let us discuss the obtained result. We first note that the Relaxation super-operator only connects transitions characterized by identical changes in the nuclear part of the base states.

Let us now consider the time dependence of the superoperator matrix elements. They are composed of a sum of terms whose general time dependence is given by $e^{-i\omega_{\mu,\nu}t}e^{-t/\tau_b}e^{i\omega_e t}$, where ω_e is the energy difference between two electronic states. The Laplace transform of the general term is given by

$$\int_0^\infty e^{-pt}e^{-t/\tau_b}e^{-i\omega_{\mu,\nu}t}e^{i\omega_e t}dt = \frac{1}{-i(\omega - \omega_{\mu,\nu} + \omega_e) + \frac{1}{\tau_b} + \Gamma_a} \quad (24)$$

and contains the difference $\omega - \omega_{\mu,\nu}$, where ω is associated with the value of the source velocity and $\omega_{\mu,\nu}$ depends on the particular nuclear transition. To introduce useful approximations, the orders of magnitude for the quantities appearing in Eq. 24 must be considered.

For the case of the Fe^{57}, we have[3][1][7][8]:

$$\omega - \omega_{\mu,\nu} \approx (11.62\text{MHz} \cdot \sec/\text{mm}) \cdot 10\text{mm}/\sec \approx 10^8\text{Hz}$$

$$\Gamma_a \simeq (11.62\text{MHz} \cdot \sec/\text{mm}) \cdot 0.1\text{mm}/\sec \approx 10^6\text{Hz}$$

$$\omega_e < K_B 300 \approx 10^{13}\text{Hz}$$

$$\frac{1}{\tau_b} \sim 10^{11} \div 10^{13}\text{Hz}$$

It is clear that Γ_a can be surely omitted. Moreover, considering the value for the lifetime of the bath states, the dependence of the Relaxation operator on $\omega - \omega_{\mu,\nu}$ can also be neglected: *white noise approximation* (WNA). In such a case, the relaxation operator depends on the Laplace transform of the electron-bath interaction, evaluated at frequencies equal to the energy separations among the electronic levels.

$$\int_0^\infty e^{-pt}e^{-t/\tau_b}e^{-i\omega_{\mu,\nu}t}e^{i\omega_e t}dt \simeq \frac{1}{-i\omega_e + \frac{1}{\tau_b}} = \frac{i\omega_e + \tau_b{}^{-1}}{\omega_e{}^2 + \tau_b{}^{-2}}$$

The WNA approximation is of great practical importance and it has been applied in the interpretation of virtually every experimentally measured Möss-bauer spectrum, although its limitations have been realized[8]. In the framework of this approximation, the value of the Relaxation superoperator becomes independent of p and, consequently, of the velocity values associated to the distinct channels of the Mössbauer spectrum.

$$G(p) \simeq p - iH_{at}{}^x + R(0) \quad (25)$$

Finally, if the energy values of the electronic states are close each other, it is also possible to neglect the ω_e terms with respect to $\tau_b{}^{-1}$; so that the relaxation operator depends on the Laplace transform of the electron-bath interaction evaluated at zero frequency

$$\int_0^\infty e^{-\text{p}t} e^{-t/\tau_b} e^{-i\omega_{\mu,\nu}t} e^{i\omega_e t} dt \simeq \tau_b$$

This situation is very usual, so that we restrict our discussion to such a case, obtaining:

$$\langle \mu', i'; \nu', j' | R(p) | \mu, i; \nu, j \rangle = \int_0^\infty e^{-\text{p}t} \langle \mu', i'; \nu', j' | R(t) | \mu, i; \nu, j \rangle dt \simeq$$

$$\simeq \delta_{\mu,\mu'} \delta_{\nu,\nu'} \cdot \tau_b \langle F^2 \rangle_b \{ [\sum_k \langle i' | S | k \rangle \langle k | S | i \rangle \delta_{j,j'} \quad - \langle i' | S | i \rangle \langle j | S | j' \rangle] -$$

$$- [\langle i' | S | i \rangle \langle j | S | j' \rangle - \sum_k \delta_{i,i'} \langle j | S | k \rangle \langle k | S | j' \rangle] \}$$

Eq. 20 does not contain transitions among different electronic states; consequently the relevant elements of the relaxation operator are

$$\langle \mu, i; \nu, i | R(p) | \mu', j; \nu', j \rangle \simeq \delta_{\mu,\mu'} \delta_{\nu,\nu'} \cdot \tau_b \langle F^2 \rangle_b \cdot$$

$$\cdot 2 \{ \sum_k \langle i | S | k \rangle \langle k | S | i \rangle \delta_{i,j} \quad - | \langle j | S | i \rangle |^2 \} \tag{26}$$

In particular, the diagonal elements on the electronic states, are given by

$$\langle \mu, i; \nu, i | R(p) | \mu', i; \nu', i \rangle \simeq \delta_{\mu,\mu'} \delta_{\nu,\nu'} \cdot \tau_b \langle F^2 \rangle_b \cdot 2 \sum_{j \neq i} | \langle j | S | i \rangle |^2 \tag{27}$$

Moreover the non diagonal ones are expressed by

$$\langle \mu, i; \nu, i | R(p) | \mu', j; \nu', j \rangle \simeq -\delta_{\mu,\mu'} \delta_{\nu,\nu'} \cdot \tau_b \langle F^2 \rangle_b \cdot 2 | \langle j | S | i \rangle |^2 \tag{28}$$

In the general case, id est for $|\omega_j - \omega_i| \geq K_B T$, the second member of previous equations contains the multiplicative factor $\exp \left(-\frac{(\omega_j - \omega_i)}{2 K_B T} \right)$. Moreover $\langle F^2 \rangle_b$ depends on $|\omega_j - \omega_i|$ and the approximation $\omega_e \ll \tau_b^{-1}$ must be reconsidered. Finally, we note that the following *sum rule* is verified

$$\langle \mu, i; \nu, i | R(p) | \mu', i; \nu', i \rangle = - \sum_{j \neq i} \langle \mu, i; \nu, i | R(p) | \mu', j; \nu', j \rangle \tag{29}$$

The temperature dependence of the elements of $R(p)$ is mainly due to $\langle F^2 \rangle_b$. The mean value of F^2 depends on the temperature for two reasons : first of all, it contains the occupation probability for the various states of the thermal bath, secondly the matrix elements of the F operator depend on the temperature thought the occupation numbers of the involved states. It must be also noted that an additional dependence on the temperature is due to the bath state lifetime τ_b. The non diagonal elements of $R(p)$ are proportional to the transition probabilities among the electronic states $|i\rangle$ and $|j\rangle$. They exhibit identical temperature dependence and differ each other by the matrix elements $\langle i|S|j\rangle$ of the spin part of the interaction Hamiltonian; so that they can be considered as proportional to each other.

The diagonal elements, expressed by Eq. 27, are proportional to the sum of the transition probabilities from the considered level to a different generic state, so that they express the total probability of leaving the considered electronic level.

Usually, the fitting procedure of the Mössbauer spectrum uses $\tau_b\langle F^2\rangle_b$ as the free parameter connected with dynamical properties. Alternatively, one of the non diagonal elements of $R(p)$ may be used.

4 Cross section evaluation

The expression for the absorption cross section is given by Eq. 19 and Eq. 20, where the matrix elements of the G superoperator, defined by Eq. 25, are given by

$$\langle\mu,i;\nu,i|G(p)|\mu^{'},j;\nu^{'},j\rangle \simeq$$

$$\simeq p\delta_{\mu,\mu'}\delta_{\nu,\nu'}\delta_{i,j'} - i\langle\mu,i|(H_{at}{}^{x}|\mu^{'},j\rangle\langle\nu^{'},j|)|\nu,i\rangle + \langle\mu,i;\nu,i|R(0)|\mu^{'},j;\nu^{'},j\rangle =$$

$$= p\delta_{\mu,\mu'}\delta_{\nu,\nu'}\delta_{i,j} - i\ (\langle\mu,i|H_{at}|\mu^{'},i\rangle\delta_{\nu,\nu'}\delta_{i,j} - \langle\nu^{'},i|H_{at}|\nu,i\rangle\delta_{\mu,\mu'}\delta_{i,j}) +$$

$$+\delta_{\mu,\mu'}\delta_{\nu,\nu'}\cdot\tau_b\langle F^2\rangle_b 2\Re\{\sum_{k}\langle i|S|k\rangle\langle k|S|j\rangle\delta_{i,j}\ - |\langle i|S(t)|j\rangle|^2\}$$

By ordering the transitions, first considering the electronic levels and then the nuclear states, the whole matrix is divided into square blocks having dimensions equal to the number of nuclear ground states multiplied by the number of excited ones. Each block corresponds to an ordered couple of electronic levels. In the case of Fe^{57} the dimensions of the square blocks are $8 \cdot 8$.

The superoperator G consists of three parts. The first and second parts only contribute to diagonal blocks. The third part contributes to the diagonal elements of all blocks, both diagonal and not.

Table 1 shows the structure of the G superoperator in the case of three electronic levels. $p = (-i\omega + \Gamma_a/2)\delta_{\mu,\mu'}\delta_{\nu,\nu'}\delta_{i,j}$ denotes diagonal $8 \cdot 8$ blocks. Moreover $H_{at}^x(i) = \langle\mu,i|H_{at}|\mu^{'},i\rangle\delta_{\nu,\nu'} - \langle\nu^{'},i|H_{at}|\nu,i\rangle\delta_{\mu,\mu'}$ denotes the $8 \cdot 8$ square blocks containing the contributions from the hyperfine interaction corresponding to the ith electronic level and to external equipments. Finally $R_{i,j}$ indicates diagonal $8 \cdot 8$ blocks, defined by $R_{i,j} = \delta_{\mu,\mu'}\delta_{\nu,\nu'}\cdot\tau_b\langle F^2\rangle_b\cdot 2\Re|\langle i|S|j\rangle|^2$.

The ordering of the electronic level is obviously arbitrary, nevertheless it can be useful to utilize an electronic level sequence based on the energy values.

Table 1. Structure of the G matrix in the case of three electronic states. See text for symbols.

1	2	3
$p - iH_{at}^x(1) + R_{1,2} + R_{1,3}$	$-R_{1,2}$	$-R_{1,3}$
$-R_{2,1}$	$p - iH_{at}^x(2) + R_{2,1} + R_{2,3}$	$-R_{2,3}$
$-R_{3,1}$	$-R_{3,2}$	$p - iH_{at}^x(3) + R_{3,1} + R_{3,2}$

The explicit expression for the atomic hamiltonian depends on the chosen quantization direction. In our case the electric interaction does not depend on the time; consequently, the principal axes of the E.F.G. tensor appear to be good candidates for the direction of quantization. This way, the atomic hamiltonian is written as

$$H_{at}(i) = H_n + H_e + H_{hyp} = H_n + H_e(i) + \left(\delta_\nu^{I.S.}(i)\delta_{n,\nu} + \delta_\mu^{I.S.}(i)\delta_{n,\mu}\right) +$$

$$+ \frac{\Delta(i)}{6}\left[3I_z^2 - I(I+1) + \frac{\eta(i)}{2}\left(I_+^2 + I_-^2\right)\right] -$$

$$- (g_\nu\delta_{n,\nu} + g_\mu\delta_{n,\mu})\,\mu_n\left[B_z(i)I_z + \frac{1}{2}\left(B_-(i)I_+ + B_+(i)I_-\right)\right] \qquad (30)$$

where $\left(\delta_\nu^{I.S.}(i)\delta_{n,\nu} + \delta_\mu^{I.S.}(i)\delta_{n,\mu}\right)$ gives rise to the Isomer Shift. Moreover the Electric Quadrupolar Interaction is expressed in terms of $\Delta = \frac{e^2qQ}{2}$ and of the E.F.G. asymmetry parameter η. Furthermore B_z, B_+ and B_- denote the components of the magnetic field acting on the nucleus, containing contributions arising from the ith electronic state and from external sources. Finally μ_n indicates the nuclear Bohr magneton; g_g , g_e are the nuclear gyromagnetic factors for $|\mu\rangle$ and $|\nu\rangle$ nuclear states and $\delta_{n,\nu}$, $\delta_{n,\mu}$ respectively identify the proper terms for exited and ground nuclear states.

Obviously, H_n and H_e do not contribute to the relevant matrix elements of H_{at}^x so that the structure of the $H_{at}^x(i)$ matrix is as shown in Table 2 and Table 3, where we also omitted the I.S. terms.

Tables 2 and 3 report the quantities $\gamma_{i,j}$, defined as $\gamma_{i,j} = \gamma_{2\mu,2\nu} = (g_g\mu - g_e\nu)$ where μ, ν indicate the I_z components for $|\mu\rangle$ and $|\nu\rangle$ nuclear states.

As mentioned, to obtain the matrix elements of G^{-1}, the G matrix must be diagonalized. In the following we denote by VL_α and VR_α the left and right eigenvectors of G for $\omega = 0$; moreover ω_α indicates the corresponding eigenvalues. As the G matrix is non hermitian, left eigenvectors differ from the right ones and eigenvalues ω_α are complex numbers[6][9]. From these quantities it is possible to reconstruct the G^{-1} operator.

We have $G^{-1} = \sum_\alpha \frac{VR_\alpha \times VL_\alpha^*}{-i\omega + \omega_\alpha}$

Table 2. Structure of the iH^x_{at} matrix (first four columns). See text for symbols.

$|\frac{1}{2},\frac{3}{2}\rangle$	$|\frac{1}{2},\frac{1}{2}\rangle$	$|\frac{1}{2},-\frac{1}{2}\rangle$	$|\frac{1}{2},-\frac{3}{2}\rangle$
$i\frac{\Delta}{2}+i\mu_n B_z\gamma_{1,3}$	$-i\mu_n g_e\frac{\sqrt{3}}{2}B_+$	$i\frac{\Delta}{6}\eta\sqrt{3}$	0
$-i\mu_n g_e\frac{\sqrt{3}}{2}B_-$	$-i\frac{\Delta}{2}+i\mu_n B_z\gamma_{1,1}$	$-i\mu_n g_e B_+$	$i\frac{\Delta}{6}\eta\sqrt{3}$
$i\frac{\Delta}{6}\eta\sqrt{3}$	$-i\mu_n g_e B_-$	$-i\frac{\Delta}{2}+i\mu_n B_z\gamma_{1,-1}$	$-i\mu_n g_e\frac{\sqrt{3}}{2}B_+$
0	$i\frac{\Delta}{6}\eta\sqrt{3}$	$-i\mu_n g_e\frac{\sqrt{3}}{2}B_-$	$i\frac{\Delta}{2}+i\mu_N B_z\gamma_{1,-3}$
$+i\mu_n g_g\frac{1}{2}B_+$	0	0	0
0	$+i\mu_n g_g\frac{1}{2}B_+$	0	0
0	0	$+i\mu_n g_g\frac{1}{2}B_+$	0
0	0	0	$+i\mu_n g_g\frac{1}{2}B_+$

Table 3. Structure of the iH^x_{at} matrix (last four columns). See text for symbols.

$|-\frac{1}{2},\frac{3}{2}\rangle$	$|-\frac{1}{2},\frac{1}{2}\rangle$	$|-\frac{1}{2},-\frac{1}{2}\rangle$	$|-\frac{1}{2},-\frac{3}{2}\rangle$
$+i\mu_n g_g\frac{1}{2}B_-$	0	0	0
0	$+i\mu_n g_g\frac{1}{2}B_-$	0	0
0	0	$+i\mu_n g_g\frac{1}{2}B_-$	0
0	0	0	$+i\mu_n g_g\frac{1}{2}B_-$
$i\frac{\Delta}{2}+i\mu_n B_z\gamma_{-1,3}$	$-i\mu_n g_e\frac{\sqrt{3}}{2}B_+$	$i\frac{\Delta}{6}\eta\sqrt{3}$	0
$-i\mu_n g_e\frac{\sqrt{3}}{2}B_-$	$-i\frac{\Delta}{2}+i\mu_n B_z\gamma_{-1,1}$	$-i\mu_n g_e B_+$	$i\frac{\Delta}{6}\eta\sqrt{3}$
$i\frac{\Delta}{6}\eta\sqrt{3}$	$-i\mu_n g_e B_-$	$-i\frac{\Delta}{2}+i\mu_n B_z\gamma_{-1,-1}$	$-i\mu_n g_e\frac{\sqrt{3}}{2}B_+$
0	$i\frac{\Delta}{6}\eta\sqrt{3}$	$-i\mu_n g_e\frac{\sqrt{3}}{2}B_-$	$i\frac{\Delta}{2}+i\mu_n B_z\gamma_{-1,-3}$

Consequently, the absorption cross section, expressed by Eq. 19, is rewritten as:

$$\sigma(\omega) = \frac{2}{\Gamma_a}\Re\ \{\sum_{\mu,i}\rho_{at}(\mu,i)\sum_{\nu}\sum_{\mu',\nu';j}\sum_{\alpha}\frac{VR_\alpha(\mu,\nu;i)\times VL^*_\alpha(\mu',\nu';j)}{-i\omega+\omega_\alpha}\cdot$$

$$\cdot\langle\mu',j|A(0)|\nu',j\rangle\langle\nu,i|A(0)^\dagger|\mu,i\rangle\ \} \tag{31}$$

Now let us consider the matrix elements of the nucleus-radiation interaction A. The elements are independent of the electronic state, contain Clebsch-Gordan coefficients $\langle j_1,j_2,m_1,m_2|j_1,j_2,J,M\rangle$ [10] [11] and magnetic vectorial spherical harmonics $\boldsymbol{Y}^m_1(\vartheta,\varphi)$ defined as [12]

$$\boldsymbol{Y}^m_j(\vartheta,\varphi) = \frac{\boldsymbol{n}\times\boldsymbol{\nabla}_n Y^m_j(\vartheta,\varphi)}{\sqrt{j(j+1)}} = \frac{1}{\sqrt{j(j+1)}}\left(\boldsymbol{a}_\varphi\frac{\partial}{\partial\vartheta}-\boldsymbol{a}_\vartheta\frac{1}{\sin(\vartheta)}\frac{\partial}{\partial\varphi}\right)Y^m_j(\vartheta,\varphi)$$

where $\boldsymbol{n}$, $\boldsymbol{a}_\varphi$ and $\boldsymbol{a}_\vartheta$ represent the three orthogonal directions for a spherical system of coordinates and the angles (ϑ,φ) identify the gamma photon direction in the chosen system of coordinates. Table 4 shows the involved Clebsch-Gordan coefficients for the common Fe^{57} case.

Table 4. Clebsch-Gordan coefficients $\langle 1, \frac{1}{2}, m_1, m_2 | 1, \frac{1}{2}, \frac{3}{2}, M \rangle$

m_1	m_2	M	$\langle 1, \frac{1}{2}, m_1, m_2 | 1, \frac{1}{2}, \frac{3}{2}, M \rangle$
1	$\frac{1}{2}$	$\frac{3}{2}$	1
1	$-\frac{1}{2}$	$\frac{1}{2}$	$\sqrt{\frac{1}{3}}$
0	$\frac{1}{2}$	$\frac{1}{2}$	$\sqrt{\frac{2}{3}}$
0	$-\frac{1}{2}$	$-\frac{1}{2}$	$\sqrt{\frac{2}{3}}$
-1	$\frac{1}{2}$	$-\frac{1}{2}$	$\sqrt{\frac{1}{3}}$
-1	$-\frac{1}{2}$	$-\frac{3}{2}$	1

Moreover the explicit expressions for the vectorial armonical functions are

$$\boldsymbol{Y}_1^1(\vartheta, \varphi) = \sqrt{\frac{3}{16\pi}} \left(i\,\boldsymbol{a}_\vartheta - \cos(\vartheta)\,\boldsymbol{a}_\varphi \right) e^{i\varphi}$$

$$\boldsymbol{Y}_1^0(\vartheta, \varphi) = -\sqrt{\frac{3}{8\pi}} \sin(\vartheta)\,\boldsymbol{a}_\varphi$$

$$\boldsymbol{Y}_1^{-1}(\vartheta, \varphi) = \sqrt{\frac{3}{16\pi}} \left(i\,\boldsymbol{a}_\vartheta + \cos(\vartheta)\,\boldsymbol{a}_\varphi \right) e^{-i\varphi}$$

Consequently we have

$$\langle \mu', j | A(0) | \nu', j \rangle \langle \nu, i | A(0)^\dagger | \mu, i \rangle =$$

$$= \langle 1, \frac{1}{2}, \nu - \mu, \mu | 1, \frac{1}{2}, \frac{3}{2}, \nu \rangle \langle 1, \frac{1}{2}, \nu' - \mu', \mu' | 1, \frac{1}{2}, \frac{3}{2}, \nu' \rangle \boldsymbol{Y}_1^{\nu' - \mu'}(\vartheta, \varphi) \cdot (\boldsymbol{Y}_1^{\nu - \mu}(\vartheta, \varphi))^*$$

(32)

where the selection rules $|\nu - \mu| \leq 1$ and $|\nu' - \mu'| \leq 1$ must also be considered. It is evident that the products $\langle \mu', j | A(0) | \nu', j \rangle \langle \nu, i | A(0)^\dagger | \mu, i \rangle$ form a square matrix of order $8n_e \cdot 8n_e$ where n_e is the number of electronic states. Such a matrix is composed by identical n_e^2 square blocks.

We also note that $\rho_{at}(\mu, i)$ is the occupation probability of the ith electronic state divided by the degeneration of the nuclear ground level; for example $\rho_{at}(\mu, i) = \frac{\rho_{at}(i)}{2}$ in the iron case.

Shortly, the Mössbauer absorption cross section is expressed by

$$\sigma(\omega) = \frac{2}{\Gamma_a} \Re \left\{ \sum_\alpha \frac{\sum_{k,l} \text{VRL}_{k,l}^\alpha B_{l,k}}{-i\omega + \omega_\alpha} \right\}$$

(33)

where the indices k and l respectively localize the transitions $|\mu, \nu; i\rangle$ and $|\mu', \nu'; j\rangle$ within rows and columns of superoperator matrices. Moreover $\text{VRL}_{k,l}^\alpha = \text{VR}_\alpha(k) \times \text{VL}_\alpha^*(l)$ and $B_{l,k} = \rho_{at}(\mu, k) \langle \mu', l | A(0) | \nu', l \rangle \langle \nu, k | A(0)^\dagger | \mu, k \rangle$

5 Result discussion

First of all, we stress that the cross section, as expressed by Eq. 33, does not have a Lorentian shape. In fact the $\mathrm{VRL}_{k,l}^{\alpha}$ terms are complex numbers and, consequently, dispersive terms also appear in the cross section expression. The imaginary parts of the eigenvalues ω_{α} define the central positions of both Lorentian and dispersive terms. Whereas the real parts, that contain Γ_a and contributions from $R(p)$, determine linewidth values.

In order to discuss the evolution of the spectra, we consider three physical situations; namely $R_{i,j} \approx \Gamma_a$, $R_{i,j} \gg H_{at}(i)$ and $R_{i,j} \approx H_{at}(i)$. The first and second situations are respectively called *slow* and *fast relaxation limits*. The third is the general case.

In the *slow relaxation limit*, as a first approximation, we may omit, out of the structure of the G matrix, all the non diagonal blocks $R_{i,j}$; id est those with $i \neq j$. This way the diagonal elements $R_{i,i}$ act as additional terms to the linewidth value Γ_a. Therefore the cross section can be approximated by a set of structures, described by the various Hamiltonians $H_{at}(i)$, characterized by lines having lorentian shape and linewidth given by $\Gamma_i \simeq \Gamma_a + R_{i,i}$. In the slow relaxation limit, the dispersive terms act as second order contributions and give rise to non symmetric lines[7]. The Γ_a value is connected to the minimum frequency value of the electronic dynamics, detectable by the experimental technique. Practically $R^{Min} \approx \frac{\Gamma_a}{10}$. For the iron case $R^{Min} \approx 10^5 \div 10^6 Hz$

In the *fast relaxation limit*, the effect of the $R_{i,j}$ terms result in the superposition of the various physical situations associated with the set of electronic levels. Consequently, the cross section is connected to the mean value, with respect to the electronic levels, of the atomic Hamiltonians: $\langle H_{at}\rangle_{\mathrm{el}} = \sum_i \varrho_i(T) H_{at}(i)$[7].

The linewidth values are greater than the natural one and, as a consequence of the *sum rule* (Eq. 29), the growing of R makes them go back to the natural Γ_a value. In fact, for high R values, $\Re(\omega_{\alpha}(R)) \simeq \Gamma_a + \frac{[\Im(\omega_{\alpha}(0))]^2}{2R}$ where $\Im(\omega_{\alpha}(0))$ denotes the corresponding line position for $R = 0$. From the above relation, the upper limit for the experimental window is determined. We have $R^{Max} \approx 10 \frac{[\Im(\omega_{\alpha}(0))]^2}{\Gamma_a}$. For the iron case $R^{Max} \approx 10^{11} Hz$.

In the *fast* and in the *slow relaxation limits* the Mössbauer spectrum depends on the value of the time constants but does not depend on the relaxation mechanism. In fact the spectrum is characterized only by the linewidth value.

In the general case, $R_{i,j} \approx H_{at}(i)$, the evolution of the cross section is complex and strongly dependent not only on the relaxation times but also on the relaxation mechanism itself. To demonstrate it, let us consider a simple case in two different situations.

The simplest case that can be considered consists of transitions between two electronic levels. For the sake of simplicity we also suppose that there is no quadrupolar interaction and the two atomic Hamiltonians only differ for the direction of the hyperfine magnetic field. In this situation the static spectra, belonging to each electronic level, are identical and consist of six lines, that we number, as usual, in order of increasing energy, id est $\omega_1 < \omega_2 < \cdots < \omega_6$.

If the two possible directions for the magnetic fields are opposite each other, the nuclear eigenstates are the same and the corresponding eigenvalues are opposite each other. Considering the structure of the G superoperator (Table 1, Table 2 and Table 3), it is easy to see that the two H_{at} matrices are diagonal. Consequently line 1 will only *interact* with line 6. For the same reason line 2 will interact with line 5 and line 3 with line 4. It is also clear that no other interaction will occur. The interaction will result in a collapse of each couple of lines to a central structure. Such collapse will occur for $R_{i,j} \geq |\omega_j - \omega_i|$, consequently the evolution of the shape of the cross section, as a function of the temperature, is quite simple. We will first observe the collapse, to a central structure, of the two internal lines. Next, the intermediate lines will collapse and, finally, we will observe the collapse of the external part of the spectrum.

Figures 1, 2 and 3 show the evolution of the spectrum when a small quadrupolar interaction is also present. The electric interaction is introduced to make evident that the inner lines collapse before the external ones; moreover we note that, in the high temperature limit, there is no trace of magnetic interactions.

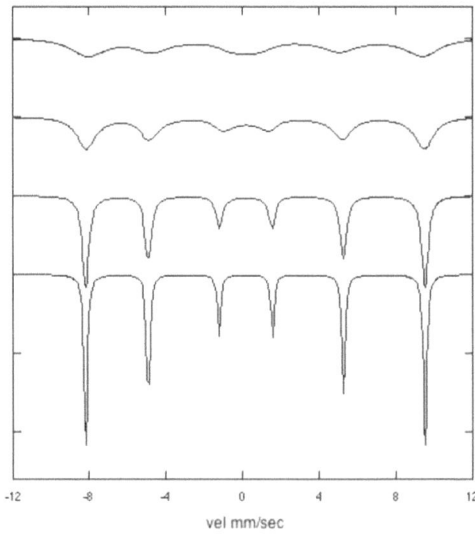

Fig. 1. Mössbauer spectra in the "Slow Relaxation Limit". Relaxation parameter $\tau_b \langle F^2 \rangle_b$ increases from bottom to top.

Very different, and much more complex will be the evolution of the spectrum if the two magnetic fields are not opposite each-other. In this case the nuclear transition states diagonalizing the first H_{at} matrix does not diagonal-

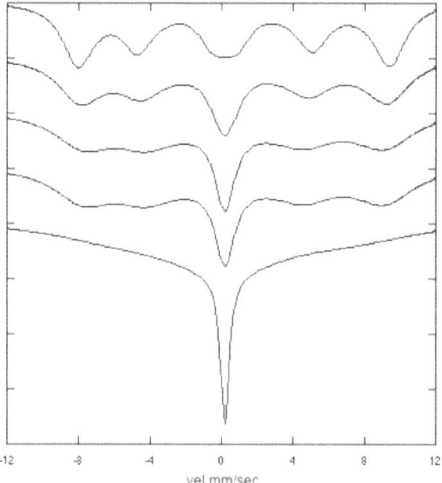

Fig. 2. Mössbauer spectra for intermediate relaxation times. Relaxation parameter $\tau_b \langle F^2 \rangle_b$ increases from top to bottom.

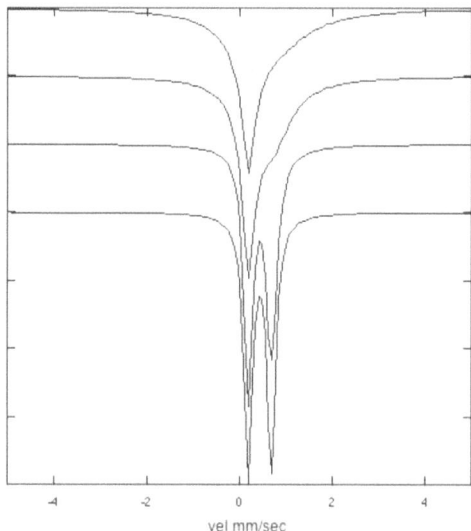

Fig. 3. Mössbauer spectra in the "Fast Relaxation Limit". Relaxation parameter $\tau_b \langle F^2 \rangle_b$ increases from top to bottom. The right line originates from the collapse of external lines of Fig.1

ize the second one. For this reason each line will interact with all the others and the strength of the interaction will depend on the values of the non diagonal coefficients. In particular, if the angle between the two magnetic fields is 90°degrees, it can be seen that the first line will interact much more with the second one than with the sixth [13]. Moreover, in the high temperature limit, the hyperfine interaction is not completely quenched and the spectrum will exhibit magnetic splitting and narrow lines.

Consequently the first and the second situation will give rise to very different evolution of the spectrum.

Another particular situation can arise when the system is characterized by a set of non degenerate electronic states, whose associated spin values monotonically increase with the energy. When the temperature increases, the spectrum may show an apparent slowing down of the electronic dynamics, Fig. 4 . Such behavior is simply due to the increase in occupation probability of electronic exited levels, characterized by high S values and giving rise to strong hyperfine interactions. It must be noted that the mean number of quantum jumps necessary to reverse the spin of the Mössbauer atom is proportional to the S^2 value. Therefore there are two quantities that increase with temperature: namely $\langle S^2 \rangle_{at}$ and $\tau_b \langle F^2 \rangle_b$. If the increase of $\langle S^2 \rangle_{at}$ predominates, the apparent slowing down of the dynamics will occur[14].

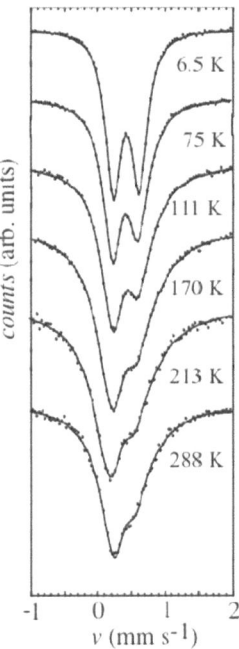

Fig. 4. Example of an apparent slow down of the electronic dynamics as a function of the temperature increase[14]

Generally speaking, the evolution of a relaxing Mössbauer spectrum permits not only to determine the time constants but also to identify the actual relaxation mechanism.

6 Saturation, texture and Goldanskii effects

Dynamical properties of the electronic states affect the shape of the cross section, so that the fitting procedure of the experimental data must reproduce not only peak positions but also the whole shape of the spectrum. Consequently, all experimental and theoretical causes affecting the line shape must be also considered.

As reported by Eq. 1 , the spectrum shape is connected to the absorption cross section by the relation $Y(v) = F(v)\{1 - \frac{f_s}{1+B} \int_{-\infty}^{\infty} L(\omega - v, \Gamma_s)[1 - \exp(-t_a \sigma(\omega))] d\omega\}$ where $t_a = n\beta\sigma_0 f_a$ is the thickness of the absorber and the source line shape $L(\omega - v, \Gamma_s)$ is usually assumed to have a Lorentian profile. Occasionally, Voigt profiles are also used.

The cross section expression contains the angular terms $Y_1^{\nu' - \mu'}(\vartheta, \varphi) \cdot (Y_1^{\nu - \mu}(\vartheta, \varphi))^*$ that make $\sigma(\omega)$ to depend on the local orientation of surroundings of the Mössbauer atom with respect to the gamma ray direction: $\sigma = \sigma(\omega, \vartheta, \varphi)$. Usually the sample is constituted by a set of differently oriented microcrystals, so that the argument of the exponential function is obtained by considering all possible orientations.

$$t_a \sigma(\omega) = \int_{4\pi} \varrho_{t_a}(\vartheta, \varphi) \sigma(\omega, \vartheta, \varphi) d\Omega \tag{34}$$

where $\varrho_{t_a}(\vartheta, \varphi) = \varrho_n(\vartheta, \varphi)\beta\sigma_0 f_a$ depends on the direction through $\varrho_n(\vartheta, \varphi) = nD(\vartheta, \varphi)$, where $\int_{4\pi} D(\vartheta, \varphi) d\Omega = 1$ and n is the number of nucleus per square centimeter.

The simplest case corresponds to a random distribution over all the possible directions. Id est: $D(\vartheta, \varphi) = \frac{1}{4\pi}$

In this situation, the thickness does not depend on the direction, so that we have

$$t_a \sigma(\omega) = t_a \langle \sigma(\omega, \vartheta, \varphi) \rangle$$

where $t_a = n\beta\sigma_0 f_a$, and consequently

$$Y(v) = F(v) \left\{ 1 - \frac{f_s}{1 + B} \int_{-\infty}^{\infty} L(\omega - v, \Gamma_s)[1 - \exp(-t_a \langle \sigma(\omega, \vartheta, \varphi) \rangle)] d\omega \right\} \tag{35}$$

There are three items to discuss, connected with the absorber thickness, that affect the spectrum shape. The first one is connected with the order of magnitude of t_a, the other two are connected with possible dependencies of the thickness on spatial directions.

If the value for the absorber thickness is smaller than one, it is possible to substitute the exponential function present in Eq. 35 by its series development, obtaining

$$Y(v) \simeq F(v) \left\{ 1 - \frac{f_s t_a}{1+B} \int_{-\infty}^{\infty} L(\omega - v, \Gamma_s) \langle \sigma(\omega, \Gamma_a, \vartheta, \varphi) \rangle \, d\omega \right\}$$

In this case, if the cross section can be expressed as a sum of lorentzian shapes, the spectrum is composed by a superposition of lorentzian lines.

$$Y(v) \simeq F(v) \left\{ 1 - \frac{f_s t_a}{1+B} \langle \sigma(v, \Gamma_a + \Gamma_s, \vartheta, \varphi) \rangle \right\} \qquad (36)$$

In other words, Mössbauer spectrum and cross section exhibit the same functional shape (*thin sample approximation*). Line intensities are proportional to the sample thickness and moreover to the square values of the corresponding Clebsch-Gordan coefficients. Finally, linewidths are given by $\Gamma = \Gamma_a + \Gamma_s$. The saturation of the spectra makes the differences among the line intensities to be smaller than those previewed by Eq. 36. Moreover it increases the apparent line width value. Therefore the use of the *thin sample approximation* for the fitting of saturated spectra gives rise to incorrect evaluations of the electronic transition probabilities.

Usually, in our case, we cannot use Eq. 36. First of all, the absorption cross section does not have Lorentzian shape. Moreover the Mössbauer f_a factor of molecular cluster systems, and consequently the sample thickness t_a, is usually heavily dependent on the temperature. To study dynamical mechanisms, a set of Mössbauer spectra, as a function of the temperature, must be collected; consequently, if the high temperature value for the target thickness is high enough to produce spectra characterized by low noise to signal ratio, the low temperature value for t_a can be high enough to require the use of the correct integral expression. Finally the dependence of the cross section on the system dynamics (Fig. 1) strongly enhances such necessity.

The use of the integral expression (Eq. 35) is a time consuming task, because the convolution integral must be evaluated for each value of the source velocity. Moreover the source factor f_s and the B parameter of Eq. 35 must be known. In particular, the evaluation of B requires the collection of *Total* and *Anticoincidence PHA* (Pulse Height Analysis) spectra. Anticoincidence spectra are collected using photons giving rise to voltage pulses whose values lie outside the range chosen to select the Mössbauer nuclear transition. Fig. 5 shows *Total* and *Anticoincidence* spectra for a Fe^{57} source. From the anticoincidence spectrum the lower and upper limits of the voltage range are determined and the amount of spurious counts under the Mössbauer line can be evaluated by standard fitting procedures of the total PHA spectrum.

Alternatively, the ulterior fitting parameter $\frac{f_s}{1+B}$, strongly correlated with t_a, must be introduced.

Two effects act on the spectrum shape if the thickness depends on the direction. They are the *Goldanskii*[15] and *texture*[16] effects.

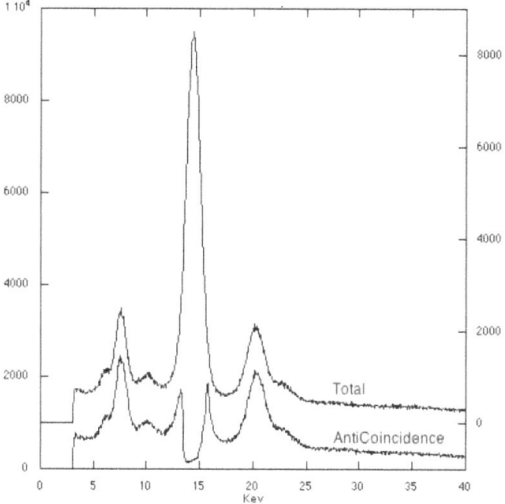

Fig. 5. Normal and AntiCoincidence PHA spectra for a Fe^{57} source

The first one describes the dependence of the thickness on the direction caused by anisotropies of the mean square displacement of the vibrating nucleus: $\langle x^2 \rangle = \langle x^2 \rangle(\vartheta, \varphi)$. In fact, the Mössbauer f factor depends on the atomic mean square displacement through the relation $f_a = \exp[-k^2 \langle x_a^2 \rangle]$, where k is the photon impulse; consequently, the sample thickness t_a will also depend on the direction of the gamma ray with respect to the crystal cell axes.

Indicating by $\langle f_a \rangle$ the mean values of the absorber f value over all the possible orientations, we have $f_a(\vartheta, \varphi) = \langle f_a \rangle g(\vartheta, \varphi)$, where $g(\vartheta, \varphi)$ is the Goldanskii function normalized by the relation $\frac{1}{4\pi} \int_{4\pi} g(\vartheta, \varphi)d\Omega = 1$. The thickness parameter is therefore expressed as $t_a(\vartheta, \varphi) = \langle t_a \rangle g(\vartheta, \varphi)$, where $\langle t_a \rangle = n\beta\sigma_0\langle f_a \rangle$, and the argument of the exponential function (Eq.34) is given by

$$t_a\sigma(\omega) = \langle t_a \rangle \int_{4\pi} g(\vartheta, \varphi)\sigma(\omega, \Gamma_a; \vartheta, \varphi)d\Omega$$

In solid state physics, the *Goldanskii* effect is rarely observed. In the present case of molecular cluster the situation is different. In fact, the low symmetry, that may characterize molecular cluster systems, can give rise to dependencies of the $\langle x^2 \rangle$ value on the direction.

Moreover the f factor value is much smaller than those common in solid state physics; consequently the effect of mean square displacement anisotropies on the Mössbauer f factor is expected to be enhanced.

We speak of *texture effects* when the simple assumption $D(\vartheta, \varphi) = \frac{1}{4\pi}$ is not adequate.

First of all, it is possible that the target construction procedure from the chemical compound, may cause preferred orientations. In fact certain types of microcrystallytes are particularly prone to partial orientation on compacting and the Mössbauer target is shaped as a thin flat disk, so that the distribution function D may depend on the angle between the normal to the disk surface and a particular crystallographic axis. The safest construction procedure of preparing unoriented absorbers appears to be by grinding with large bulk of fine-grained Al_2O_3 or other abrasive materials [17]. However we note that preferred orientations can be also desired, because spectra of strong orientated samples may be similar to those of monocrystals. Consequently, from the spectra of oriented samples, magnetic anisotropy directions, with respect to the crystal structure, can be extracted [18].

It is evident that preferred orientations and Goldanskii effect influence the Mössbauer spectrum in similar ways. It is however possible to distinguish between them.

First of all the dependence of the Goldanskii function $g(\vartheta, \varphi)$ on the direction cannot be very strong. In fact all $f(\vartheta, \varphi)$ values are constrained into the 0-1 range; moreover, in order to obtain a well defined spectrum, the $\langle f_a \rangle$ medium value must be not very small. Consequently, the maximum and minimum values for $f(\vartheta, \varphi)$ must be close to each other.

For this reason if, in order to fit the spectrum, strong angular dependencies are to be introduced, they are surely related to texture effects.

Moreover, in the case of the Goldanskii effect, the shape of Mössbauer spectrum is independent from the relative orientation between target and experimental optical axis. On the contrary, the effect produced by preferred orientations depends on the target axis direction.

A technique useful to see if texture is present and moreover able to determine the $D(\vartheta, \varphi)$ distribution function, is based on the collection of five independent spectra characterized by different spatial orientations of the sample.

The first one with the target axis parallel to the experimental optical one, and the other four with the target axis forming the *magic angle* of 54.7° degrees with the optical one. The last four experimental dispositions differentiate among them by subsequent rotations of 90° degrees of the sample around the target axis [19] [Fig. 6].

In order to avoid the influence of the magnetization dynamics, it is necessary to make such measurements either in the low or high temperature range where the magnetic interaction is, respectively, static or completely quenched.

In the second case, the interpretation of the data requires the knowledge of the quadrupolar parameters V_{zz} and $\eta = \frac{V_{xx} - V_{yy}}{V_{zz}}$. In particular a non zero value for η produces a mixing of the nuclear $|I, I_z\rangle$ states, so that it can greatly affect the intensities of the various transitions. For this reason the experimental work must go along with molecular orbital calculations.

In order to show the influence of the η value, let us consider the simple case of a quadrupolar Mössbauer spectrum. The intensities of the cross section lines are respectively given by [20]

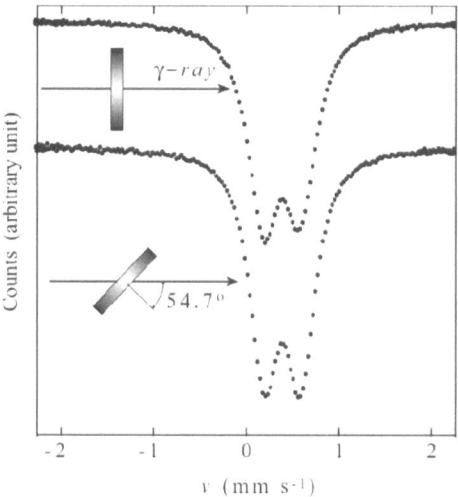

Fig. 6. Quadrupolar Mössbauer spectra for two different orientations of the sample: the former with the normal to the sample parallel to the gamma rays direction and the latter with the normal forming the *magic angle* of 54.7°.

$$I_1 = \sigma_0 [4\sqrt{\left(1 + \frac{\eta^2}{3}\right)} + \left(3\cos^2(\vartheta) - 1 + \eta\sin^2(\vartheta)\cos(2\varphi)\right)]$$

$$I_2 = \sigma_0 [4\sqrt{\left(1 + \frac{\eta^2}{3}\right)} - \left(3\cos^2(\vartheta) - 1 + \eta\sin^2(\vartheta)\cos(2\varphi)\right)]$$

where the polar angles define the gamma ray direction with respect the principal axes of the E.F.G. tensor.

Introducing the texture function $D(\vartheta, \varphi)$ one has

$$\frac{I_1}{I_2} = \frac{4\sqrt{\left(1 + \frac{\eta^2}{3}\right)} + \int_{4\pi} D(\vartheta, \varphi)\left(3\cos^2(\vartheta) - 1 + \eta\sin^2(\vartheta)\cos(2\varphi)\right) d\Omega}{4\sqrt{\left(1 + \frac{\eta^2}{3}\right)} - \int_{4\pi} D(\vartheta, \varphi)\left(3\cos^2(\vartheta) - 1 + \eta\sin^2(\vartheta)\cos(2\varphi)\right) d\Omega}$$

Let us now consider two extreme situations: the first one characterized by $V_{zz} = a$, $V_{xx} = V_{yy} = -\frac{a}{2}$ and $D(\vartheta, \varphi) \simeq \delta_{\vartheta=0}$; the second one characterized by $V_{zz} \simeq 0$, $V_{xx} \simeq a$, $V_{yy} \simeq -a$ and $D(\vartheta, \varphi) \simeq D(\vartheta)$.

It is easy to see that, in the first case, we have $\eta = 0$ and $\frac{I_1}{I_2} = 3$. In the second situation we obtain $\eta \simeq \infty$ and $\frac{I_1}{I_2} \simeq 1$; moreover the intensity ratio results to be independent of the texture dependence on the ϑ angle, so that $\frac{I_1}{I_2} \simeq 1$ also holds for $D(\vartheta, \varphi) \simeq \delta_{\vartheta=0}$.

Fig. 6 shows Mössbauer spectra of a sample affected by texture. It is evident that the upper spectrum, collected with the normal of the sample parallel to the gamma ray direction, is quite similar to the second from bottom of Fig. 3 where the intensity difference between the two lines is due to dynamical processes.

We conclude that texture, Goldanskii and saturation effects must be taken into account in order to extract, from the Mössbauer spectra, information on the electronic dynamics.

Appendix: Liouville Operators

Let us consider a complete set of base states $|\mu\rangle$ describing a physical system and a generic operator A. The application of the operator to a base state gives rise to a quantum state expressed as a linear combination of all the base states

$$|\psi\rangle = A\,|\mu\rangle = \sum_{\nu} \langle \nu\,|A\,|\mu\rangle\,|\nu\rangle \tag{A.1}$$

If the system is characterized by a limited number n of base states, it is possible to order the n^2 generic operators $|\nu\rangle\langle\mu|$ corresponding to all possible transitions. That allows us to introduce the *transition base states* $|\mathbf{m}\rangle = |\nu,\mu\rangle = |\nu\rangle\langle\mu|$ [21].

Let us now consider a linear combination of them, expressed by $\sum_{\mu,\nu} C_{\mu,\nu}|\nu\rangle\langle\mu| = \sum_{\mathbf{m}} C_{\mathbf{m}}|\mathbf{m}\rangle$. From the formal identity with the second term of Eq. A.1 it is possible to regard $\sum_{\mathbf{m}} C_{\mathbf{m}}|\mathbf{m}\rangle$ as resulting from the application of a *"superoperator"* to a specified *transition base state* $|\mathbf{n}\rangle$.

So that, we write

$$\mathcal{C}\,|\mathbf{n}\rangle = \sum_{\mathbf{m}} \mathcal{C}_{\mathbf{m}}\,|\mathbf{m}\rangle \tag{A.2}$$

where $\mathcal{C}_{\mathbf{m}} = \langle \mathbf{m}|\mathcal{C}|\mathbf{n}\rangle$. The term *"superoperator"* simply indicates an operator that operates on operators, instead of on quantum states, and generates new operators. From the analogy between Eq. A.1 and Eq. A.2 it follows that *superoperators* and standard operators obey the same algebra[21].

The matrix elements $\langle \mathbf{m}|\mathcal{C}|\mathbf{n}\rangle$ define the particular *superoperator*; to derive them we apply $\mathcal{C}$ to a generic operator

$$\mathbf{B} = \sum_{\mathbf{n}} \mathbf{B_n}|\mathbf{n}\rangle = \sum_{\mu',\nu'} \langle \nu'\,|\mathbf{B}|\mu'\rangle|\nu'\rangle\langle\mu'| \tag{A.3}$$

obtaining

$$\mathcal{C}\mathbf{B} = \mathcal{C}\sum_{\mathbf{n}} \mathbf{B_n}|\mathbf{n}\rangle = \sum_{\mathbf{m},\mathbf{n}} \mathbf{B_n}\langle \mathbf{m}|\mathcal{C}|\mathbf{n}\rangle|\mathbf{m}\rangle =$$

$$= \sum_{\mu,\nu}\sum_{\mu',\nu'} \langle \nu'\,|B|\mu'\rangle\langle\nu,\mu|\mathcal{C}|\nu',\mu'\rangle|\nu,\mu\rangle$$

and, consequently

$$\langle \nu | \mathcal{C} B | \mu \rangle = \sum_{\mu',\nu'} \langle \nu' | B | \mu' \rangle \langle \nu, \mu | \mathcal{C} | \nu', \mu' \rangle \tag{A.4}$$

Eq. A.4 shows that the matrix elements of $\mathcal{C}\mathbf{B}$ are linear combinations of those of $\mathbf{B}$. The coefficients of the linear combinations give a representation for $\mathcal{C}$.

Now, taking into account Eq. A.3, we obtain the general rule that permits us to determine the representation

$$\langle \nu, \mu | \mathcal{C} | \nu', \mu' \rangle = \langle \nu | (\mathcal{C} | \nu' \rangle \langle \mu' |) | \mu \rangle \tag{A.5}$$

In the following, we restrict our discussion to the so called *Liouville superoperator* $\mathbf{A}^x$ associated with a standard operator A. It is defined as the *superoperator* that, applied to a generic operator B, returns the *commutator* between A and B[22].

$$\mathbf{A}^x B = [A, B]_- \tag{A.6}$$

The great importance of *Liouville superoperators* is mainly due to the fact that the time dependence of operators is expressed through commutators. In fact, posing $\hbar = 1$,

$$\frac{dB(t)}{dt} = i [H, B]_- \tag{A.7}$$

So that

$$\frac{dB(t)}{dt} = i \mathbf{H}^x B(t) \tag{A.8}$$

and

$$B(t) = e^{i\mathbf{H}^x t} B(0) \tag{A.9}$$

where

$$\mathbf{H}^x = [H, \ ]_- \tag{A.10}$$

identifies the *Liouville superoperator* associated, through Eq. A.6, with the Hamiltonian of the system.

Using Eq. A.5, we obtain the expression for the matrix elements of the *Liouville superoperator* $\mathbf{A}^x$.

$$\langle \nu, \mu | \mathbf{A}^x | \nu', \mu' \rangle = \langle \nu | [A, | \nu' \rangle \langle \mu' |]_- | \mu \rangle = \langle \nu | A | \nu' \rangle \delta_{\mu,\mu'} - \langle \mu' | A | \mu \rangle \delta_{\nu,\nu'} \tag{A.11}$$

In the particular case of $\mathbf{H}^x$, choosing the eigenvector base state, we have

$$\langle \nu, \mu | \mathbf{H}^x | \nu', \mu' \rangle = (\omega_\nu - \omega_\mu) \, \delta_{\mu,\mu'} \delta_{\nu,\nu'}$$

and, from Eq. A.6,

$$\mathbf{H}^x | \nu \rangle \langle \mu | = (\omega_\nu - \omega_\mu) | \nu \rangle \langle \mu |$$

where ω_ν and ω_μ are the eigenvalues corresponding to the states $|\nu\rangle$ and $|\mu\rangle$.

In other words, transitions and corresponding frequencies are the eigenvectors and eigenvalues of H^x, and consequently the Liouville superoperator is closely related to the information immediately extracted from an experimental spectrum.

This statement gives rise to the physical importance of the Liouville supeoperator formalism.

As a concluding remark, let us consider the general function $f(z)$, which can be expressed as a Laurent series:

$$f(z) = \sum_{n=-\infty}^{\infty} c_n z^n$$

It is interesting to give a representation for the two operators $f(H^x)$ and $[f(H)]^x$.

We have:

$$f(H^x)|\nu\rangle\langle\mu| = \sum_{n=-\infty}^{\infty} c_n (H^x)^n |\nu\rangle\langle\mu| =$$

$$= \sum_{n=-\infty}^{\infty} c_n (\omega_\nu - \omega_\mu)^n |\nu\rangle\langle\mu| = f(\omega_\nu - \omega_\mu)|\nu\rangle\langle\mu|$$

and

$$[f(H)]^x |\nu\rangle\langle\mu| = \left[\sum_{n=-\infty}^{\infty} c_n H^n\right]^x |\nu\rangle\langle\mu| =$$

$$= \sum_{n=-\infty}^{\infty} c_n (\omega_\nu{}^n - \omega_\mu{}^n) |\nu\rangle\langle\mu| = [f(\omega_\nu) - f(\omega_\mu)]|\nu\rangle\langle\mu|$$

References

1. C. Janot in *Leffet Mössbauer et ses applications* (Masson, Paris, first edition, 1972).
2. B. Kolk in *Dynamical properties of solids* vol. 5 (Nort-Holland, Amsterdam, 1984).
3. A. Vértes, L. Korecz, and K. Burger in *Mössbauer spectroscopy* (Elsevier, Amsterdam, 1979).
4. L. Landau and E. Lifchitz in *Theorie quantique relativiste* Number 1 (Mir, Moscow, 1972), pag.283-293.
5. G. K. Shenoy and B.D. Dunlap, Phys. Rev. B 13, 1353 (1976).
6. M. J. Clauser, Phys. Rev. B 3, 3748 (1971).
7. L. Cianchi, P. Moretti, M. Mancini, and G. Spina, Rep. Prog. Phys. 49, 1243 (1986).

8. S. Dattagupta, G. K. Shenoy, B. D. Dunlap, and L. Asch, Phys. Rev. B 16, 3893 (1977).
9. G. K. Shenoy, B. D. Dunlap, S. Dattagupta, and L. Asch, Phys. Rev. Lett. 37, 539542 (1976).
10. http://pdg.ge.infn.it/2004/reviews/clebrpp.pdf.
11. W.-M. Yao, C. Amsler, D. Asner, and Alii, Journal of Physics G 33, 1 (2006).
12. L. Landau and E. Lifchitz in *Theorie quantique relativiste* Number 1 (Mir, Moscow, 1972), pag. 33-45.
13. L. Cianchi, F. Gulisano, and G. Spina, J. Physics : Condensed Matter 6, 2269 (1994).
14. L. Cianchi, F. Del Giallo, M. Lantieri, P. Moretti, G. Spina, and A. Caneschi, Phys. Rev. B 69, 014418 (2004).
15. V. I. Goldanskii, E. F. Makarov, and V. V. Khrapov, Phys. Letters 3, 344 (1963).
16. H. D. Pfannes and H. Fisher, Applied Physics 317325 (1977).
17. N. N. Greenwood and T. C. Gibb in *Mössbauer Spectroscopy* (Chapman and Hall Ltd., London, 1971),
18. L. Cianchi, F. Del Giallo, M. Lantieri, P. Moretti, G. Spina, and A. Caneschi. To be published, 2006.
19. J. M. Greneche and F. Varret, J. Phys. C Solid State Phys.15, 5333 (1982).
20. Peter Zory, Phys. Rev. 140, A1401 (1965).
21. R. Zwanzig, Physica 30, 1109 (1964).
22. Frédéric Schuller in *The Liouville Space Formalism in Atomic Spectroscopy* (Heron Press Science Series, 2002).

Studies of spin fluctuations in single molecular magnets by using Mössbauer spectroscopy

Luciano Cianchi[1] and Gabriele Spina[2]

[1] Istituto dei Sistemi Complessi (Sezione di Firenze), CNR,
   via Madonna del Piano, 10, 50019 Sesto Fiorentino, Firenze, Italy.
   `luciano.cianchi@isc.cnr.it`
[2] Dipartimento di Fisica Università di Firenze, Via G. Sansone,1,50019 Sesto
   Fiorentino, Firenze, Italy.
   Affiliation:Centro Nazionale di Ricerca S3, INFM-CNR.
   `gabriele.spina@unifi.it`

Potential applications of single molecular magnets (SMMs) in high density storage devices have been a major driving force for researchs in this field [1, 2, 3]. An hindrance to applications of this kind lies in the mobility of spins. That is, uncontrolled spin inversions can arise which destroy any magnetic order. Therefore, we need to understand the microscopic mechanisms on which such interactions are based, so that their effects can be removed or reduced in the new generations of magnetic molecules.

As far as the experimental techniques are concerned, Mössbauer spectroscopy is an excellent tool for the study of the magnetic properties of molecules containing iron ion, since Mössbauer probes are the nuclei of these ions, i.e. the ions which determine the magnetic properties of the molecules.

Spin fluctuations influence the spectrum shapes through the hyperfine interaction, according to the theory developed in the previous lesson. However, the Mössbauer spectra do not directly depend on the molecule spin. In fact, they consist of the superposition of sub-spectra corresponding to the different sites of the ^{57}Fe(III) ions, which depend on the spins of these ions.

Furtherly, we observe that the iron-spin changes are due to a hierarchic series of interactions. First of all, the iron spins are coupled through the isotropic exchange interaction, for which only the total spin S of the molecule is conserved. Their fluctuation frequencies are of the order of J/h, where the exchange constant J is in the range between some tens and few hundreds kelvins [3], corresponding to $10^{12}s^{-1}$. This value is behind the top border of the Mössbauer temporal window and one can assume that each iron nucleus sees a mean spin $\langle s^{(i)} \rangle$ where $i = 1, 2, \ldots$ denotes the different sites. Moreover, for the presence of the axial magnetic anisotropy (z-axis) – which is usually much larger than the hyperfine interaction –, the iron nucleus only "sees" the mean z-component $\langle s_z^{(i)} \rangle$ of the iron-ion spins, so that $S_z = \sum_i \langle s_z^{(i)} \rangle$. Lastly, the transverse terms of magnetic anisotropy, if any, and the spin-thermal bath interactions, which produce spin fluctuations with frequencies usually within the Mössbauer window, must be considered. These latter interactions induce transitions between S_z states and consequently between $\langle s_z^{(i)} \rangle$ states; they are thus the cause of the hyperfine-field fluctuations affecting the spectrum shapes.

Usually, it is not simple to establish the relations between the single spins $\langle s^{(i)} \rangle$ and the molecule spin S, as they depend on the exchange constants in a cumbersome way. In any case, we have $\langle s^{(i)} \rangle = C^{(i)} S$ [4] where $C^{(i)}$ are parameters that can be determined by the spectrum fits.

Here as follows, we report the properties of the magnetization dynamics of some species of Fe(III)-molecules studied by using Mössbauer spectroscopy.

1 Molecules with $S \neq 0$ ground spin state

Two molecules with spin ground state different from zero will be considered: the one (Fe4) containing four $s = 5/2$ Fe(III) ions and the other (Fe8) eight. In both the molecules the magnetic ions are antiferromagnetically coupled through bridging atoms.

First, a detailed Mössbauer analysis of the spin fluctuations in Fe4 will be reported, as this case is rather simple and particularly suitable in order to show the potentialities of the Mössbauer spectroscopy in spin fluctuations studies.

1.1 Ground state of Fe4 molecula

This molecule has formula $Fe_4(OCH_3)_6(dpm)_6$ and its structure is shown in Figure 1.

The trend of χT versus T was found to be characteristic of a antiferromagnet with spin $S \neq 0$ [5]. The antiferromagnetic coupling between the central ion Fe1 and the peripheral ions Fe2, Fe3, and Fe3' leads to a ground state $S = 5$. The corresponding Heisenberg Hamiltonian H_H is given by:

$$H_H = J s_1 \cdot (s_2 + s_3 + s_{3'})   \tag{1}$$

However, in order to obtain a good fit of χT one has to suppose the presence of a small ferromagnetic coupling between peripheral ions. The best fit parameters are $J = 21.1 \, \text{cm}^{-1} = 30.4 \, k_B$ with $g = 1.97$ [5]. Moreover, the first excited level is a doubly degenerate $S = 4$ state and its energy with respect to the ground state is about $60 \, \text{cm}^{-1} = 86.4 \, k_B$. Here, we consider spin fluctuations at low temperatures, so that only the ground states will be considered. The spin Hamiltonian for the ground state has the form:

$$H_S = D S_z^2 + \frac{E}{2}(S_+^2 + S_-^2) + H_{HO}   \tag{2}$$

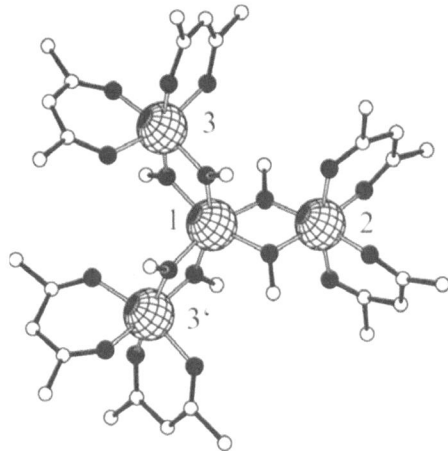

Fig. 1. Schematic structure of the Fe4 molecule. Globes denote Fe(III) ions; black and white circle denote oxygen and carbon atoms, respectively. Hydrogen atoms are not indicated.

where HO denotes fourth-order terms in the spin components; the constants D and E are relative to the axial anisotropy and the transverse one, respectively. These constants were determine by means of HF-EPR and INS thecniques [6, 7]. Three different isomers are present in the crystal with occupancies 0.49, 0.42 and 0.09, respectively. They differ for the binding modes of the dpm anions on peripheral Fe. They have slightly different D values: $-0.29\,k_B$, $-0.27\,k_B$ and $-0.25\,k_B$, respectively. This differences were disregarded and a single isomer was considered with D given by the weighted mean of the three values. E value is more questioned, since HF-EPR and INS gave fairly different results: $-0.014\,k_B$ and $0.029\,k_B$, respectively. In any case, $|E| \ll |D|$, so that the ground state consists approximately of five doublets, with the doublet $|5, \pm 5\rangle$ having the lowest energy, plus the singlet $|5, 0\rangle$ that has energy $\Delta = 25|D| \approx 7\,k_B$ with respect to the lowest doublet. Calculations based on a ligand field approach [5] showed that the anisotropy for the ground state has both single ion and dipolar contributions with the unique axis quasi-perpendicular to the iron plane.

1.2 Spin fluctuations

Mössbauer spectra from 1.38 to 77 K were collected (Figure2), which display a strong dependence on the temperature. Thus, spin fluctuations due to energy exchanges between spins and thermal vibrations are present. Spectra have then to be evaluated by considering the transitions induced by the atomic vibrations. Phenomenologically, the interaction Hamiltonian was assumed to have the form [8]:

$$H_i = F(t)\, A(S_x, S_y, S_z) \tag{3}$$

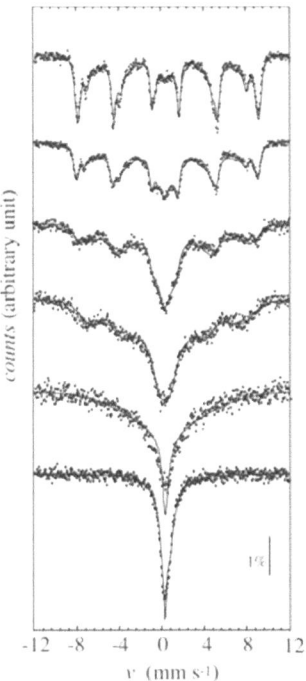

Fig. 2. Mössbauer spectra of the Fe4 molecule at the temperatures of (starting from the top) 1.38, 4.25, 12.5, 25.7, 45, 77 K.

where $F(t)$ is a stochastic function having dimension s^{-1} and of order of the interaction energy (in unit h) and A is a dimensionless function of the spin components of order of unit. For the sake of simplicity, only transitions $M \rightarrow M'$ with $M' = M \pm 1$ were considered. The transition probabilities per unit time are then given by [8]:

$$W_{M,M-1} = |\langle M|A|M-1\rangle|^2 J(\omega_M) \tag{4}$$

$$W_{M-1,M} = W_{M,M-1} \exp\left(\frac{\hbar\omega_M}{k_B T}\right) \tag{5}$$

where ω_M is the energy difference between the states $M-1$ and M, and $J(\omega_M)$ is the Fourier transform of the correlation function of F:

$$J(\omega) = \int_{-\infty}^{\infty} \exp(i\omega t) \langle F(0)F(t)\rangle dt \tag{6}$$

By assuming for $\langle F(0)F(t)\rangle$ an exponential trend: $\langle F(0)F(t)\rangle = F^2 \exp(-|t|/\tau)$, where the decay constant τ is of order of the mean life of the atomic vibration modes that is of the order of 10^{-12}s (Sect. 3, p. 255 of the previous lesson). Since $\omega_M \leq 6\,10^9\,s^{-1} \ll \tau^{-1}$, we have with good approxmation:

$$W_{M,M-1} = \frac{|\langle M|A|M-1\rangle|^2}{|\langle 5|A|4\rangle|^2}W \tag{7}$$

where $W = W_{5,4}$.

As above mentioned, in a fixed cluster state $|S, M\rangle$, the single iron-ion spin $z-$component are not constant in time; only their mean value is definite. Consequently, the Mössbauer nucleus should experience the static hyperfine field corresponding to the mean $z-$component of the electronic spin. Standard techniques [4, 9] provide the projection $\langle m_i \rangle$ of the individual spin s_i on the total spin $\mathbf{S}$. One finds: $\langle m_2 \rangle = \langle m_3 \rangle = \langle m_{3'} \rangle = \dfrac{M - \langle m_1 \rangle}{3}$ with

$$\frac{\langle m_1 \rangle}{M} = -0.4167, \quad \text{and} \quad \frac{M - \langle m_1 \rangle}{3M} = 0.472 \tag{8}$$

When the spin undergoes transitions between its states, the spin mean $z-$component of the iron-ions changes simultaneously between their corresponding values, so that the iron nuclei are subjected to stochastic changes in the hyperfine field. According to this picture, central and peripheral iron nuclei experience fields having different magnitudes, but changing with the same rate. To be precise, the fields for central and peripheral nuclei are proportional to $-0.4167M$ and $0.472M$, respectively.

Since the environment symmetry of the iron ions is lower than octahedral, the electric quadrupole interaction has to be considered. An estimate of the electric field gradient (EFG) at the three non-equivalent (Fe1, Fe2 and Fe3 $\equiv$ Fe3$'$) iron sites showed that one of the EFG principal axes resulted almost perpendicular to the iron plane, i.e. practically parallel to the magnetic anisotropy axis, for all of the sites. It was then convenient to assume this principal axis as z-axis, no matter whether or not V_{zz} was the maximum component of the EFG. With this assumption, $\eta > 1$ values are possible.

The nuclear Hamiltonian of an iron ion can then be written as a sum of the time-independent and stochastically-changing terms V_1 and V_2, respectively (see sect.**??**):

$$V_1 = \frac{eV_{zz}Q_I}{4}\left[I_z^2 - \frac{1}{3}I(I+1) + \frac{\eta}{2}(I_+^2 + I_-^2)\right] \tag{9}$$

$$V_2 = g_I\mu_N I_z B_h(t) \tag{10}$$

where $B_h(t)$ is the hyperfine field which depends on time because of the transitions between spin states. The other symbols have the usual meaning (Sect. 4 of Filoti's lesson).

In order to evaluate the spectrum forms, the formalism of Sect. 4 of previous lesson was used (see also [10]). Since there are 11 spin states and 8 Mössbauer transitions the Blume's matrix has order $11 \times 8 = 88$. Moreover three different sites have to be considered for the iron. The central site (Fe(1)), the apical one (Fe(2)) and the two lateral sites (Fe(3), Fe(3)), the normalized spectrum shapes of which will be denoted by $I_1(\omega)$, $I_2(\omega)$ and $I_3(\omega)$, respectively. For the resultant spectrum we can then write [11]:

$$I(\omega) = \frac{1}{1 + 3\phi} [I_1(\omega) + \phi I_2(\omega) + 2\phi I_3(\omega)] \tag{11}$$

where the parameter ϕ takes into account the fact that the absorption spectra areas may be different for the central and peripheral sites.

Fitting outcomes

The quadrupolar parameters ($\Delta Q = eV_{zz}Q/2$ and η) and the isomer shift (IS) at the three sites and the hyperfine field corresponding to $s_z = 5/2$ were obtained from the fit of the spectra at 1.38 K [12]. At this temperature, relaxation effects on the spectrum shape are negligible, so that the spectrum contains maximum information about the hyperfine parameters. The values of $-0.5(3)$ mms^{-1} (central site), $0.1(1)$ mms^{-1} (apical site) and $-0.3(1)$ mms^{-1} (lateral sites) were obtained for ΔQ and the values of $0.38(5)$ mms^{-1}, $0.37(7)$ mms^{-1} and $0.42(4)$ mms^{-1}, respectively, for IS, which are just characteristic of Fe(III) ions. As far as η is concerning, its value was found greater than 1 at all sites. This means that the maximum EFG components are directed in the iron plane. The value of $W(T)$ obtained by assuming

Table 1. Rate of the transition $M = 5 \rightarrow M = 4$ of Fe4. Values obtained from the fitting of the spectra. χ^2 values are also shown.

T(K)	W(MHz)	χ^2
1.38	2.0(1)	1171
4.25	3.3(1)	1488
12.5	7.5(3)	1128
25.7	37(1)	1016
45.0	267(5)	1428
77.0	1198(40)	1022

the spin-thermal bath Hamiltonian of the form: $H_i = F(t)S_x$ are reported in Table1, together with the χ^2 values. A good fit of the $W(T)$ trend is given by the function

$$W(t) = 9.8(6) \, 10^3 \exp\left(-\frac{161(4)}{T}\right) \tag{12}$$

This is the characteristic trend of the Orbach mechanism involving an excited state with an energy of about 160 k_B [13]. States of this energy are definitely present, since the energy difference between the ground $S = 5$ state and the top $S = 10$ state is about $1200k_B$ ($J \approx 30$K), with 1296 spin states shared in this range. Another possible mechanism could involve vibrational modes of a $160k_B$ centered peak of the vibrational-states density, if present, and would consist in the creation of one and subsequent destruction of another one of these vibrational quanta. However, further studies are necessary in order to clarify this issue.

1.3 The Fe8 molecule

Now, let us consider the single-molecule magnet $[Fe_8O_2(OH)_{12}(tacn)_6]Br_8$, briefly Fe8, Figure 3. The ground state is a $S = 10$ multiplet and the first

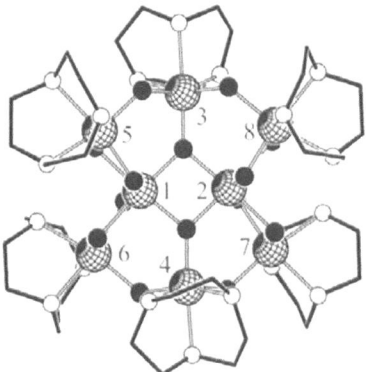

Fig. 3. Schematic structure of the Fe8 molecule. Globes denote Fe(III) ions; black and white circles denote oxygen and nitrogen atoms, respectively; carbon atoms are at the vertices of the black lines. Hydrogen atoms are not indicated.

excited states are three $S = 9$ multiplets at $44\,k_B$, $61.5\,k_B$ and $72\,k_B$ [14]. EPS and INS gave for D the values of $-0.29\,k_B$ and for E the values of $0.056\,k_B$ and $0.044\,k_B$. respectively [15, 16]. By diagonalizing H_S (eq.(1)) in the space of the ground spin states $|10, M\rangle$, the eigenstates displayed in Figure 4 were obtained. The six quasi degenerate lowest doublets do not differ much from the ones corresponding to the axial anisotropy term of H_S, while the nine highest singlets consist of superposition of states having different M, with $\langle M \rangle = 0$.

Three spectra at temperatures lower than 20 K were collected (Figure 5) because at higher temperatures the contribution of the excited spin states are not negligible.

In order to calculate the spectrum shapes, three different sites were considered [12]: one for Fe1 and Fe2, another for Fe3 and Fe4 and the third for

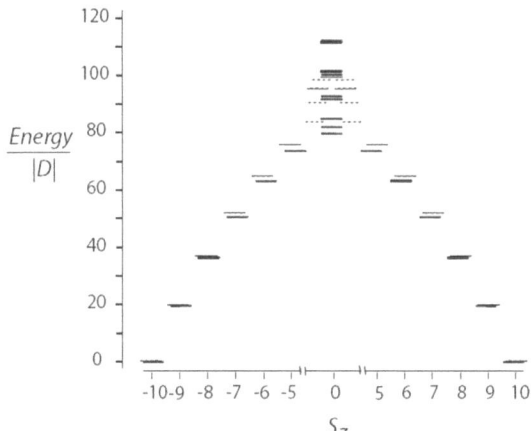

Fig. 4. Level scheme of the ground state of the Fe8 molecule. The six lowest levels (bold lines) are quasi degenerate doublets. The nine highest levels are singlet ($\langle M \rangle = 0$).The dashed lines denote levels corresponding to the only axial term of H_S.

Fe5 to Fe8. However, we met with twofold difficulty. Firstly, since the energy of the lowest excited multiplet is $\approx 44\, k_B$, the calculation of the spectra, disregarding the excited levels is corrected only for $T \ll 44$K. Secondly, as the ground state consists of 21 spin states and there are 8 possible Mössbauer transitions, the Blumes matrix is of order $8 \times 21 = 168$. Therefore, also taking into account the large number of parameters, the fittings of the spectra require a rather long computation time. Fortunately, a simplified theoretical approach, which is based on the particular structure of the spin ground state, can be used. As we have seen above, these states can be ideally divided into two groups: the former consists, with a good approximation, of the six lowest doublets $|10, \pm M\rangle$ with M from 5 to 10 and the latter of the nine singlets. Since in all the singlets M has an average value of zero, their Mössbauer spectrum contributions to the Mössbauer spectrum does not have a magnetic structure and will be the same for all of them, apart from the absorption area. In order to reproduce the spectra, the nine states were considered as a single fictitious state, the occupation of which is the sum of the occupations of the nine states. By using this simplified model, the order of the Blumes matrix is reduced from 168 to 104.

As far as the hyperfine magnetic field is concerned, in the $S = 0$ ground state the iron spins are parallel to each other and, for $M = 10$, $\langle s_z \rangle = \pm 5/2$. The field magnitudes at the three non equivalent iron sites have then the greatest values $B_{5/2^i}$ $i = (1, 2, 3)$. These values were considered as fitting parameters, and the fields for $S_z = M$ were evaluated by scaling in proportion to M: $B_M^i = (M/10)B_{5/2}^i$. Lastly, the presence at each site of an EFG having axial symmetry parallel to the hyperfine field was assumed.

Table 2. Hyperfine parameters obtained from the fitting of the Fe8 lowest temperature spectrum.

Parameter	Central	Apical	Lateral
$\Delta Q \, \mathrm{mm\,s^{-1}}$	0.13(1)	-0.11(2)	0.057(2)
$IS \, \mathrm{mm\,s^{-1}}$	0.25(1)	0.36(1)	0.24(1)
$B_{5/2}$ (T)	47.3(1)	47.9(1)	53.1(1)

According to the procedure followed for Fe4, the hyperfine parameters were obtained from the spectrum at the lowest temperature (in this case, 4.2 K), which displays a well-resolved hyperfine structure. The fitting outlets are reported in Table 2 The transition rate $W_{10,9}$ of the transition $M = 10 \rightarrow M = 9$ was assumed as relaxation parameter. By setting $W = W_{10,9}$ a linear trend was obtained for $W(T)$. To be precise, the best fit gave (in MHz): $W(T) = 2.2(1)T$.

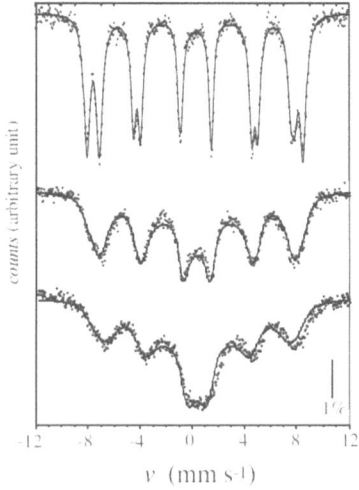

Fig. 5. Mössbauer spectra of the Fe8 molecule at the temperatures of (starting from the top) 4.2, 11 and 18 K.

2 Molecules with $S = 0$ ground spin state

In the previous examples, molecules with high-spin ground states were considered and correlation functions of the spin components were determined in ranges of temperatures where only the ground state is important. In this section, we review two Mössbauer studies of molecules with diamagnetic ground state, one regarding a molecule of six iron(III) ions in coplanar ring-shaped arrangements [3], and the other regarding one of a dimers' family [17].

An antiferromagnetic coupling between iron ions is present so that the spin ground state consists in a singlet ($S = 0$). All the spin multiplets can be fairly well described by means of an Heisemberg Hamiltonian with nearest-neighbours coupling [18]:

$$H = J \sum_{i=1}^{N-1} s_i s_{i+1} + s_N s_1 \tag{13}$$

where s_i is the spin of the i-th magnetic ion, N is the number of magnetic ions, and J the exchange constant.

2.1 hexairon(III) molecule

The six Fe(III) ions are lying on the vertices of a hexagon and a Na^+ ion on the centre, Figure 6.

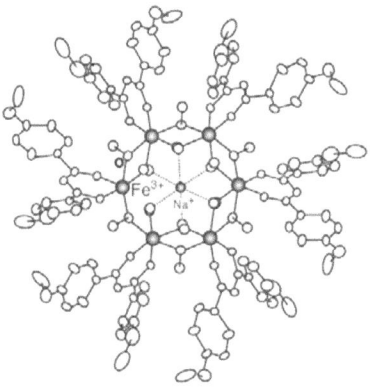

Fig. 6. Structure of the hexairon(III) molecule.

It has S_6 symmetry, with the axis S_6 perpendicular to the iron plane [18]. Susceptibility measurements on a powdered sample gave $J = 28.7 k_B$. Moreover, from single crystal measurements magnetic anisotropy was detected with an easy axis along the S_6 direction [18].

Spin fluctuations in this compound were investigated by using ^{1}H-NMR [19]. $T^{-1}$was found to display a particular trend as a function of the temperature. It presents a sharp maximum, the amplitude of which decreases as the applied magnetic field increases, but its position is field independent. Now, due to energy-level complexity, a quantitative analysis of this trend is very complicate. In order to evade the difficulty of a direct calculation, a different approach in terms of collective q-dependent spin variables was used [20]. The trend of T^{-1} that was found in this way fits fairly well the experimental one. A further in-dept study of ^{1}H-NMR results for a series of molecular rings was performed in ref. [21], were a model that accurately fits the NMR data is developed.

As far as the Mössbauer study is concerned [22], spectra in the range of temperatures from 10 to 200 K consist of quadrupolar doublets, the linewidth of which depends on the temperature as shown in Figure 7: it monotonically increases with the temperature from 0.27 mms^{-1} at 12 K to 0.37 mms^{-1} at 200 K, showing a flat trend from 20 to 60 K. Since the spectra display no magnetic structure, the average of the hyperfine field in a Larmor period has to be zero. That is, we are in the *fast relaxation* conditions [23], in which contributions to the spectra of a single spin multiplet correspond to a null hyperfine field and fluctuations affect only linewidths. Let us now consider transitions between states of different multiples. Until their probabilities per unit time are much smaller than τ_M^{-1}, where τ_M is the mean life of the Mössbauer excited level, spectra consist in the superposition of the single multiplet contributions. Conversely, if the transitions are much faster than τ_M^{-1}, spectra will be a suitable average of the multiplet sub-spectra.

Since the magnetic sites are equivalent, each iron ion has mean spin z-component $\langle s_z \rangle = S_z/N$, where N is the number of magnetic ions in the molecule. The corresponding hyperfine field is then given by:

$$B_z = \frac{1}{N}\frac{S_z}{5/2}B_{hyp}^{5/2} = \frac{2}{5N}S_z B_{hyp}^{5/2} \tag{14}$$

where $B_{hyp}^{5/2}$ is the hyperfine field corresponding to the state $s_z = 5/2$ of the ion Fe(III). According to the theory of the Mössbauer spectra in the presence of fast relaxation [24] the spin fluctuations cause the spectrum lines to be broadened by quantities proportional to

$$J_{rel} = \mu_N^2 \int_0^\infty \langle B_z(0)B_z(t)\rangle dt = \left(\frac{2\mu_N B_{hyp}^{5/2}}{5N}\right)^2 \int_0^\infty \langle S_z(0)S_z(t)\rangle dt \tag{15}$$

2.2 General expression for J_{rel}

Here as follows, a detailed analysis of the correlation function (CF) in terms of physical quantities characterizing the spin fluctuations will be made. Its general expression is:

$$\langle S_z(0)S_z(t)\rangle = \text{Tr}\{\rho S_z(0)S_z(t)\}_{m,s} \tag{16}$$

where $\{\dots\}_{m,s}$ means that the trace is calculated on a complete set of states of the system constituted by the molecule and its surroundings and ρ is the density operator. By making the factorization approximation of the density operator [24], we can carry out the trace on the surrounding states, obtaining:

$$\langle S_z(0)S_z(t)\rangle = \text{Tr}\{\rho_m S_z(0)S_z(t)\}_m \tag{17}$$

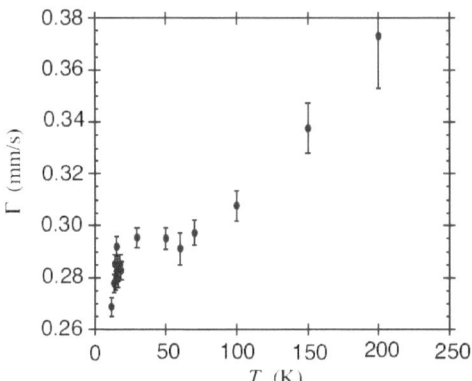

Fig. 7. Linewidth versus temperatures obtained by fitting the spectra with quadrupole doublets.

where ρ_m is the density operator of the spin system.

Let us consider first the transitions within a spin multiplet S of the molecule. In the fast relaxation limit with $\langle B_z \rangle = 0$, we can assume that the spin z—component undergoes simple inversions. Two causes can determine this behaviour: the interaction between spin and thermal bath and the quantum tunnelling [13, 3]. In the first case the rate $w_M^{(S)}$ of the transition between the two states $|S, \pm M\rangle$ increases with the temperature, while in the second case we have coherent transitions with independent temperature frequency $\nu_M^{(S)}$. The contribution of the doublet $|S, \pm M\rangle$ to the CF can be evaluated by means of the procedure described in [12]. We obtain:

$$\langle S_z(0)S_z(t)\rangle_{S,M} = M^2 \exp(-2w_M^{(S)}t) \cos(2\pi\nu_M^{(S)}t) \tag{18}$$

We must consider now the intermultiplet transitions. As above mentioned, if the rate are much smaller than (τ_M^{-1}), spectra will be given by the sum of the multiplet contributions. Conversely, if the transition rates are much faster than τ_M^{-1}, spectra will be given as average of the multiplet sub-spectra.. Since in the fast relaxation limit sub-spectrum shapes differ from each other only for the line broadening, in order to obtain the mean spectrum it is sufficient to carry out the average of the multiplet contributions to the CF (*mean spin approximation*).

In the more general case, spectrum shapes explicitly depend on the transition rates and the calculations become very complicated. As a result, long computation times are expected and great uncertainties regarding the parameter values will result from their correlations. Fortunately, for our cases, preliminary fittings showed the approximation of the mean spin to be fairly good. Therefore, according to the mean spin approximation, we must carry out the average of the single multiplet contributions to the CF:

$$\langle S_z(0)S_z(t)\rangle = 2 \sum_{\text{multiplets}} \frac{W_{E_S}}{2S+1} \sum_{M=1}^{S} M^2 \exp[] - (2w_M^{(S)} + \tau_M^{-1})t] \cos(2\pi\nu_M^{(S)}t)$$

(19)

where W_{E_S} is the occupation probability of the multiplet of energy E_S.

Obviously, the fitting of J_{rel} did not make possible to obtain the very large number of transition rates $w_M^{(S)}$, so that these quantities were replaced by an appropriate mean w value. Frequencies $\nu_M^{(S)}$ can be calculated by means of the procedure of [12], where, by assuming the presence of a transverse anisotropy of Hamiltonian $H_T = E(S_+^2 + S_-^2)/2$, a general expression was obtained for $\nu_M^{(S)}$ as function of E/D. In summary, we write the CF in the form:

$$\langle S_z(0)S_z(t)\rangle = 2\exp(-2w't) \sum_{\text{multiplets}} \frac{W_{E_S}}{2S+1} \sum_{M=1}^{S} M^2 \cos(2\pi\nu_M^{(S)}t) \qquad (20)$$

where $w' = w + \tau_M^{-1}/2$. By replacing the right-hand side of this equation in (14), we obtain:

$$J_{rel} = \left(\mu_N \frac{2}{5N} B_{hyp}^{5/2}\right)^2 \sum_{\text{multiplets}} \frac{W_{E_S}}{2S+1} \sum_{M=1}^{S} M^2 \frac{w'}{w'^2 + (\pi\nu_M^{(S)})^2} \qquad (21)$$

The best fit J_{rel} as function of the temperature by means of (19) should permit to test the possible $w(T)$ trends and also make an estimate of E/D. However, difficulties arose when this procedure was applied to the hexairon molecule. The difficulties were due to the large number of multiplets, the D values of which are not known and change from one multiplet to another, so that the frequencies $\nu_M^{(S)}$ occour to be different for different multiplets. Thus, too many parameters had to be determined and the fitting was problematic. Fortunately, many magnetic properties are qualitatively common also to a series of Fe(III) dimers, where the two Fe(III) ions are antiferromagnetically coupled through diamagnetic groups. In these cases, apart from the ground state $S = 0$, there are only five excited spin multiplets, so that we are able to take into account every contribution. For example, in [25] the spin fluctuations in a dimer were studied in detail through the analysis of the Mssbauer spectra, which will be reviewed here as follows.

2.3 Mössbauer analysis of the dimer [Fe(OMe)(dpm)$_2$]$_2$

The structure of the molecule is shown in Figure 8. The two Fe(III) ions are antiferromagnetically coupled through oxygen bridges, so that the ground state is $S = 0$ and, in order of increasing energy, there are the multiplets

$S = 1, 2, 3, 4, 5$. Moreover, the exchange constant is $J = 27.4\,\mathrm{K}$ and an axial magnetic anisotropy is present, with the constant $D(S)$ depending to S according to [26]:

$$D(S) = -\frac{1}{3.2}\frac{3S(S+1) - 38}{2(2S+3)(2S-1)}D(1) \tag{22}$$

From EPR measurements on a similar dimer the value of about $8\,\mathrm{K}$ was obtained for the multiplet $S = 1$ [27]. More detailed properties of this molecule are reported in ref. [17].

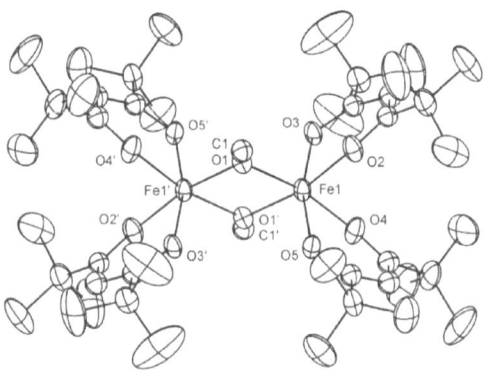

Fig. 8. ORTEP view of the dimer $[\mathrm{Fe(OMe)(dpm)_2}]_2$.

Fifteen spectra in $\pm 6\,\mathrm{mm\,s^{-1}}$ velocity range were collected between 6.5 and 288 K. Some of them are shown in Figure 9 to illustrate the evolution of the spectrum shape with the temperature. They consist of quadrupolar doublets with linewidth increasing with the temperature due to the spin fluctuations. To be more precise, for $T < 50\,\mathrm{K}$ the linewidths are similar and fairly narrow, while, for $T > 100\,\mathrm{K}$, they become wider and wider as the temperature increases. This is expected because for $T < 50\,\mathrm{K}$ the prevalent contribution to the spectra come from the ground singlet, while spin fluctuations become more and more important with the increasing of temperature and, consequently, populations of the spin multiplets. We observe, moreover, that the line broadening is much more evident for the spectrum line corresponding to the more energetic Mössbauer transition.

Since in the spectrum shapes any magnetic structure is absent at all the temperature, we are in *fast relaxation* conditions, at least for $T > 50\,\mathrm{K}$: the detailed theory of the spectrum shape for this dimer in the fast relaxation limit is reported in [25]. From the fitting of the spectra the parameter J_{rel} defined in the previous section was obtained and then fitted by means of (21).

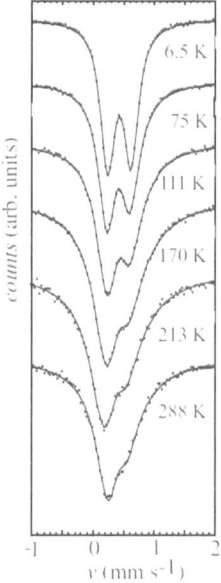

Fig. 9. Set of spectra collected in a range from 6.5 K to room temperature. Solid lines represent the relative fit obtained as described in the text. The velocity range is restricted between -1 and 2 mm s^{-1} to emphasize the line shape evolution.

For this purpose, the presence of a transverse anisotropy of Hamiltonian $H_T = E(S_+^2 + S_-^2)/2$ was assumed, and the tunnel frequencies $\nu_M^{(S)}$ were calculated by means of the Brillouin-Wigner perturbation theory [12], obtaining:

$$\nu_M^{(S)} = F_M^{(S)} \left(\frac{E}{D(S)} \right)^M D(S) \tag{23}$$

where $F_M^{(S)}$ are numerical factors depending on M and S. As far as the trend of ν is concerned, we note that, as a significant linewidth broadening is observed for $T > 100$ K, the microscopic mechanism describing the spin fluctuations has to be at least of the second order. Moreover, there are not other spin states in addition to the spin states considered, so that mechanisms based on Orbach or Raman effects should be excluded. It is therefore reasonable that the basic process consists of a first order Raman effect [13]. We also assumed that only vibrational molecular modes of a narrow band, so that the w trend has to be of the form $w = w_0 \exp(2T_0/k_B T)$, which just describes a first order Raman effect in the presence of vibrational modes of a narrow band around T_0 [28]. The experimental J_{rel} values together with the best fit are shown in Figure 10. We see that the fitting is poor for $T < 50$ K, which is expected, since, being the $S = 0$ ground state prevalently populated in this range, the spectrum linewidts are practically independent of spin fluctuations and the mean spin approximation is not allowed.

Table 3. Values of D and E/D for $S = 1$; moreover, values of τ_M, w_0 and T_0 obtained from the fitting of J_{rel} by eq.(21)

$D(1)$ (K)	$E/D(1)$ (10^{-4})	τ_M^{-1} $(10^7\ s^{-1})$	w_0 $(10^8\ s^{-1})$	T_0 (K)
8.1	8.5	1.3	9.2	732

The outlet values of the parameters are reported in Table 3. We included τ_M^{-1} in the fitting parameters in order to check the foundation of (21). The value of $1.3\ 10^7 \mathrm{s}^{-1}$ is of the same order of magnitude of the value reported in the literature [29]. As far as the values of $D(1)$ and $E/D(1)$ are concerned, these are substantially in agreement with EPR measurements on a similar dimer [27]. The corresponding tunnelling frequencies of the multiplets are of the order of $10^9 \mathrm{s}^{-1}$, i.e. of the same of w_0. Lastly, the value of 732 K for T_0 corresponds to modes of circular frequency $\omega_0 = 8.5\ 10^{13} \mathrm{s}^{-1}$.

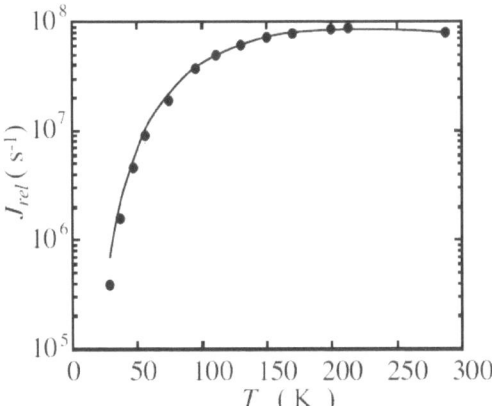

Fig. 10. Full circles denote J_{rel} values obtained from the fitting of the spectra. The solid line is the best fit obtained from eq.(21).

References

1. D. Gatteschi, A. Caneschi, L. Pardi, and R. Sessoli. Science, 265, 1054 (1994).
2. D. D. Awschalom and D. P. Di Vincenzo. Physics Today, 48, 43 (1995).
3. D. Gatteschi, R. Sessoli, and J. Villain in *Molecular Nanomagnet* (Oxford University Press, New York, first edition, 2006).

4. D. Gatteschi and L. Pardi. Gazzetta Chimica Italiana, 123, 231 (1993).

5. A. L. Barra, A. Caneschi, A. Cornia, F. Fabrizi de Biani, D. Gatteschi, C. Sangregorio, R. Sessoli, and L. Sorace. J. Am. Chem. Soc., 121, 5302 (1999).

6. A. Bouwen, A. Caneschi, D. Gatteschi, E. Goovaerts, D. Schoemaker, L. Sorace, and M. Stefan. J. Phys. Chem. B, 105, 2658 (2001).

7. G. Amoretti, S. Carretta, R. Caciuffo, H. Casalta, A. Cornia, M. Affronte, and D. Gatteschi. Phys Rev. B 64, 104403 (2001).

8. A. Abragam in *The Theory of Nuclear Magnetism* (Oxford University Press, first edition, 1961).

9. A. Bencini and D. Gatteschi in *EPR of Exchange Coupled Systems* (Springer, first edition, 1990)

10. M. Blume, Phys. Rev. 174, 351 (1968).

11. A. Caneschi, L. Cianchi, F. Del Giallo, D. Gatteschi, P. Moretti, F. Pieralli, and G. Spina, J. Phys: Condens. Matter 11, 3395 (1999)

12. L. Cianchi, Franco Del Giallo, G. Spina, W. Reiff, and A. Caneschi. Phys. Rev. B 65, 064415 (2002).

13. A. Abragam and B. Bleaney in *Electron Paramagnetic Resonance of Transition Ions* (Dover Publications, New York, first edition, 1986)

14. A. L. Barra, D. Gatteschi, and R. Sessoli, Chem. Eur. J. 6, 1608 (2000).

15. A. L. Barra, P. Debrunner, D. Gatteschi, CH. E. Schulz, and R. Sessoli, Europhys. Lett. 35, 133 (1996).

16. R. Caciuffo, G. Amoretti, A. Murani, R. Sessoli, A. Caneschi, and D. Gatteschi. Phys Rev. Lett. 81, 4744 (1998).

17. F. Le Gall, F. Fabrizzi de Biani, A. Caneschi, P. Cinelli, A. Cornia, A.C. Fabretti, and D. Gatteschi, Inorg. Chim. Acta 262, 123 (1997).

18. A. Caneschi, A. Cornia, A.C. Fabretti, S. Foner, D. Gatteschi, R. Grandi, and L. Schenetti. Chem Eur. J. 2, 1379 (1996).

19. A. Lascialfari, D. Gatteschi, F. Borsa, and A. Cornia, Phys. Rev. B 55, 14341 (1997).

20. A. Lascialfari, D. Gatteschi, F. Borsa, A. Cornia, U. Balucani, M. G. Pini, and A. Rettori, Phys. Rev. B 57, 1115 (1998).

21. S. H. Baek, M. Luban, A. Lascialfari, E. Micotti, Y. Furukawa, F. Borsa, J. van Slageren, and A. Cornia, Phys. Rev. B 70, 134434 (2004).

22. A. Caneschi, M. Capaccioli, L. Cianchi, A. Cornia, Franco Del Giallo, F. Pieralli, and G. Spina, Hyp. Int. 116, 215 (1998).

23. A. M. Afanasev and V. D. Gorobchenko, Sov. Phys. JETP 39, 690 (1974).

24. L. Cianchi, P. Moretti, M. Mancini, and G. Spina, Rep. Prog. Phys. 49, 1243 (1986).

25. L. Cianchi, F. Del Giallo, M. Lantieri, P. Moretti, G. Spina, and A. Caneschi, Phys. Rev. B 69, 014418 (2004).

26. A. Cornia, Private communication.

27. P. ter Heerdt, A. Bouwen, A. Cornia, A. Caneschi, and D. Gatteschi, in ICMM 2002, Valencia.

28. L. Cianchi and M. Mancini, Rivista del Nuovo Cimento 2, 25 (1972).

29. C. Janot in *Leffet Mössbauer et ses applications* (Masson, Paris, first edition, 1972).